T0332601

Theory of Global Random Search

Mathematics and Its Applications (*Soviet Series*)

Volume 65

Theory of Global Random Search

by

Anatoly A. Zhigljavsky
Leningrad University, U.S.S.R.

Edited by

J. Pintér

KLUWER ACADEMIC PUBLISHERS
DORDRECHT / BOSTON / LONDON

Library of Congress Cataloging-in-Publication Data

Zhigli͡avskiĭ, A. A. (Anatoliĭ Aleksandrovich)
[Matematicheskai͡a teorii͡a global'nogo sluchaĭnogo poiska. English]
Theory of global random search / by Anatoly A. Zhigljavsky ; edited by J. Pintér.
p. cm. -- (Mathematics and its applications. Soviet series ; v. 65)
Rev. translation of: Matematicheskai͡a teorii͡a global'nogo sluchaĭnogo poiska.
Includes bibliographical references and index.
ISBN 0-7923-1122-1 (alk. paper)
1. Stochastic processes. 2. Search theory. 3. Mathematical optimization. I. Pintér, J. II. Title. III. Series: Mathematics and its applications (Kluwer Academic Publishers). Soviet series ; v. 65.
QA274.Z4813 1991
519.2--dc20 90-26821

ISBN 0-7923-1122-1

Published by Kluwer Academic Publishers,
P.O. Box 17, 3300 AA Dordrecht, The Netherlands.

Kluwer Academic Publishers incorporates
the publishing programmes of
D. Reidel, Martinus Nijhoff, Dr W. Junk and MTP Press.

Sold and distributed in the U.S.A. and Canada
by Kluwer Academic Publishers,
101 Philip Drive, Norwell, MA 02061, U.S.A.

In all other countries, sold and distributed
by Kluwer Academic Publishers Group,
P.O. Box 322, 3300 AH Dordrecht, The Netherlands.

Printed on acid-free paper

Printed in the Netherlands

SERIES EDITOR'S PREFACE

'Et moi, ..., si j'avait su comment en revenir, je n'y serais point allé.'

Jules Verne

The series is divergent; therefore we may be able to do something with it.

O. Heaviside

One service mathematics has rendered the human race. It has put common sense back where it belongs, on the topmost shelf next to the dusty canister labelled 'discarded non-sense'.

Eric T. Bell

Mathematics is a tool for thought. A highly necessary tool in a world where both feedback and non-linearities abound. Similarly, all kinds of parts of mathematics serve as tools for other parts and for other sciences.

Applying a simple rewriting rule to the quote on the right above one finds such statements as: 'One service topology has rendered mathematical physics ...'; 'One service logic has rendered computer science ...'; 'One service category theory has rendered mathematics ...'. All arguably true. And all statements obtainable this way form part of the raison d'être of this series.

This series, *Mathematics and Its Applications*, started in 1977. Now that over one hundred volumes have appeared it seems opportune to reexamine its scope. At the time I wrote

> "Growing specialization and diversification have brought a host of monographs and textbooks on increasingly specialized topics. However, the 'tree' of knowledge of mathematics and related fields does not grow only by putting forth new branches. It also happens, quite often in fact, that branches which were thought to be completely disparate are suddenly seen to be related. Further, the kind and level of sophistication of mathematics applied in various sciences has changed drastically in recent years: measure theory is used (non-trivially) in regional and theoretical economics; algebraic geometry interacts with physics; the Minkowsky lemma, coding theory and the structure of water meet one another in packing and covering theory; quantum fields, crystal defects and mathematical programming profit from homotopy theory; Lie algebras are relevant to filtering; and prediction and electrical engineering can use Stein spaces. And in addition to this there are such new emerging subdisciplines as 'experimental mathematics', 'CFD', 'completely integrable systems', 'chaos, synergetics and large-scale order', which are almost impossible to fit into the existing classification schemes. They draw upon widely different sections of mathematics."

By and large, all this still applies today. It is still true that at first sight mathematics seems rather fragmented and that to find, see, and exploit the deeper underlying interrelations more effort is needed and so are books that can help mathematicians and scientists do so. Accordingly MIA will continue to try to make such books available.

If anything, the description I gave in 1977 is now an understatement. To the examples of interaction areas one should add string theory where Riemann surfaces, algebraic geometry, modular functions, knots, quantum field theory, Kac-Moody algebras, monstrous moonshine (and more) all come together. And to the examples of things which can be usefully applied let me add the topic 'finite geometry'; a combination of words which sounds like it might not even exist, let alone be applicable. And yet it is being applied: to statistics via designs, to radar/sonar detection arrays (via finite projective planes), and to bus connections of VLSI chips (via difference sets). There seems to be no part of (so-called pure) mathematics that is not in immediate danger of being applied. And, accordingly, the applied mathematician needs to be aware of much more. Besides analysis and numerics, the traditional workhorses, he may need all kinds of combinatorics, algebra, probability, and so on.

In addition, the applied scientist needs to cope increasingly with the nonlinear world and the extra mathematical sophistication that this requires. For that is where the rewards are. Linear models are honest and a bit sad and depressing: proportional efforts and results. It is in the nonlinear world that infinitesimal inputs may result in macroscopic outputs (or vice versa). To appreciate what I am hinting at: if electronics were linear we would have no fun with transistors and computers; we would have no TV; in fact you would not be reading these lines.

There is also no safety in ignoring such outlandish things as nonstandard analysis, superspace and anticommuting integration, p-adic and ultrametric space. All three have applications in both electrical engineering and physics. Once, complex numbers were equally outlandish, but they frequently proved the shortest path between 'real' results. Similarly, the first two topics named have already provided a number of 'wormhole' paths. There is no telling where all this is leading - fortunately.

Thus the original scope of the series, which for various (sound) reasons now comprises five subseries: white (Japan), yellow (China), red (USSR), blue (Eastern Europe), and green (everything else), still applies. It has been enlarged a bit to include books treating of the tools from one subdiscipline which are used in others. Thus the series still aims at books dealing with:

- a central concept which plays an important role in several different mathematical and/or scientific specialization areas;
- new applications of the results and ideas from one area of scientific endeavour into another;
- influences which the results, problems and concepts of one field of enquiry have, and have had, on the development of another.

A very large part of mathematics has to do with optimization in one form or another. There are good theoretical reasons for that because to understand a phenomenon it is usually a good idea to start with extremal cases, but in addition - if not predominantly - all kinds of optimization problems come directly from practical situations: How to pack a maximal number of spare parts in a box? How to operate a hydroelectric plant optimally? How to travel most economically from one place to another? How to minimize fuel consumption of an aeroplane? How to assign departure gates in an aeroport optimally? etc., etc. This is perhaps also the area of mathematics which is most visibly applicable. And it is, in fact, astonishing how much can be earned (or saved) by a mathematical analysis in many cases.

In complicated situations - and many practical ones are very complicated - there tend to be many local extrema. Finding such a one is a basically well understood affair. It is a totally different matter to find a global extremum. The first part of this book surveys and analyses the known methods for doing this. The second and main part is concerned with the powerful technique of random search methods for global extrema. This phrase describes a group of methods that have many advantages - whence their popularity - such as simple implementation (also on parallel processor machines) and stability (both with respect to perturbations and uncertainties) and some disadvantages: principally relatively low speed and not nearly enough theoretical background results. In this last direction the author has made fundamental and wide ranging contributions. Many of these appear here for the first time in a larger integrated context.

The book addresses itself to both practitioners who want to use and implement random search methods (and it explains when it may be wise to consider these methods), and to specialists in the field who need an up-to-date authoritative survey of the field.

The shortest path between two truths in the real domain passes through the complex domain.

J. Hadamard

La physique ne nous donne pas seulement l'occasion de résoudre des problèmes ... elle nous fait pressentir la solution.

H. Poincaré

Never lend books, for no one ever returns them; the only books I have in my library are books that other folk have lent me.

Anatole France

The function of an expert is not to be more right than other people, but to be wrong for more sophisticated reasons.

David Butler

Bussum, 18 February 1991

Michiel Hazewinkel

CONTENTS

CHAPTER 4. STATISTICAL INFERENCE IN GLOBAL RANDOM SEARCH. 114

FOREWORD

The investigation of any computational method in such a form as it has been realized on a certain computer is extremely complicated. As a rule, a simplified model of the method (neglecting, in particular, the effects of rounding and errors of some approximations) is being studied. Frequently, stochastic approaches based on a probabilistic analysis of computational processes are of great efficiency. They are natural, for instance, for investigating high-dimensional problems where the deterministic solution techniques are often inefficient. Among others, the global optimization problem can be cited as an example of problems, where the probabilistic approach appears to be very fruitful.

The English version of the book by Professor Anatoly A. Zhigljavsky proposed to the Reader is devoted primarily to the development and study of probabilistic global optimization algorithms in which the evolution of probabilistic distributions corresponding to the computational process is studied. It seems to be the first time in the literature that rigorous results grounding and optimizing a wide range of global random search algorithms are treated in a unified manner. A thorough survey and analysis of the results of other authors together with the clearness of the presentation are great merits of the book.

A. Zhigljavsky is one of the representatives of the Leningradian theoretical probabilistic school studying Monte Carlo methods and their applications. In spite of his young age, he also participates in writing well-known monographs on experimental design theory and, besides, on detection of abrupt changes of random processes.

Certainly, the book is going to be interesting and useful for a great number of mathematicians dealing with optimization theory as well as users employing optimization methods for solving various applied problems.

Professor Sergei M. Ermakov
Leningrad University
USSR

PREFACE

Optimization is one of the most important fields in contemporary computational and applied mathematics: significant efforts of theoreticians as well as practicioners are devoted to its investigation.

If discrete and multicriteria optimization problems are neglected, then optimization theory can be divided into local and global optimization. The former may be regarded as being almost *completed.* On the contrary, the latter is now at the height of its development that is confirmed by observing the dynamics and contents of the related publications.

Part 1 of the present book attempts to elucidate the current state-of-art in global optimization theory. The author admits that it is impossible to describe all known solution approaches, but supposes that the present review is rather complete and may be of interest, since most recent books (viz., Dixon and Szegö, eds. (1975, 1978), Strongin (1978), Evtushenko (1985), Fedorov ed. (1985), Zhigljavsky (1985), Zilinskas (1986), Pardalos and Rosen (1987), Törn and Zilinskas (1989)) as well as surveys (e.g. Sukharev (1981), Archetti and Schoen (1984), Rinnooy Kan and Timmer (1985)) are devoted mainly to particular directions in global optimization.

The variety of global optimization problem statements is rather great: it was heuristically, numerically, and theoretically justified that different methods may be suitable for different classes of problems. Global random search methods occupy a peculiar place among them, as sometimes they offer the unique way of solving complicated problems.

A random choice of solutions is traditional under uncertainty: recall vital situations when a choice was conditioned by coin-tossing. Application of random decisions in global optimization algorithm construction leads to global random search algorithms which are popular among users as well as theoreticians. Part 2 of the book thoroughly describes and investigates such algorithms.

Apparently, the main popularity reasons of global random search methods among users are due to the following attractive features which many of them possess: the structure of these methods is rather simple and they can be easily realized as subroutines (in multiprocessor computers as well), they are rather insensitive to irregularity of the objective function behavior and the feasible region structure, as well as to the presence of noise in the objective function evaluations and also to the growth of dimensionality. Besides, it is easy to construct methods that guarantee global convergence or to realize various adaptation ideas. As for theoreticians, the global random search methods represent a rich and not very complicated subject matter for investigation and possible inclusion of statistical concepts and procedures.

Of course, the global random search methods have also their drawbacks. First, the standard convergence rate of these methods is slow and can be generally increased only if the probability of failure in reaching the extremum is allowed to increase. Second, practically efficient methods usually involve heuristic elements and thus have corresponding disadvantages. Third, their efficiency can be often increased by means of increasing complexity and decreasing their randomness. Thus, the usage of global random search can be beneficial primarily in cases, when the optimization problem is complicated enough but the objective function evaluation is not very time-consuming.

The global random search theory is connected (occasionally, rather intensively) with several other branches of mathematical statistics and computational mathematics: some of them are considered in Part 3 of the book.

The author's own research results compose nearly half of the mathematical contents of the book: they are accompanied by proofs unlike the results of other authors whose proofs can be found in the corresponding references.

Many sections of the book can be read almost independently. A large part of its exposition is intended not only for theoreticians, but also for users of the optimization algorithms.

The Russian variant of the book published in 1985 was considerably revised. Thus, Chapters 1-3 and Section 4.4 were completely rewritten; various recent results of other authors were reconsidered. On the other hand, the exposition of a number of original results of minor importance for global random search theory was reduced.

I am indebted to Professor S.M. Ermakov for his valuable influence on my scientific interests. I am grateful to Professor A. Zilinskas for his help in reviewing the current state of global optimization theory and to Dr. Pintér for careful reading of the manuscript that had led to its substantial improving. I also wish to thank Professor G.A. Mikhailov and Professor F. Pukelsheim for many helpful discussions on the subject of Section 8.2 and E. P. Andreeva for the help in translation and for the careful typing of the manuscript.

Anatoly A. Zhigljavsky
Leningrad University
USSR

LIST OF BASIC NOTATIONS

$\mathcal{R}^n$	is the n-dimensional real Euclidean space; $\mathcal{R}=\mathcal{R}^1$
$\mathcal{X}$	is the feasible region, usually a compact subset of $\mathcal{R}^n$ of nonzero Lebesgue measure, having a sufficiently simple structure ($\mathcal{X}$ is considered as a measurable space)
$\mathcal{B}$	is the σ–algebra of Borel subsets of $\mathcal{X}$
ρ	is a metric on $\mathcal{X}$
μ_n	is the Lebesgue measure on $\mathcal{R}^n$ or on $\mathcal{X}\subset\mathcal{R}^n$
f	is the objective function given on $\mathcal{X}$
$\mathcal{F}$	is the set of possible objective functions f
$Lip(\mathcal{X},L)$	is the set of functions f on $\mathcal{X}$ satisfying the Lipschitz condition:

$$|f(x)-f(z)| \le L\|x-z\| \qquad \text{for all } x,z \in \mathcal{X}$$

where $\|\cdot\|$ is the Euclidean norm

$Lip(\mathcal{X},L,\rho)$ is the set of functions f satisfying

$$|f(x)-f(z)| \le L\rho(x,z) \qquad \text{for all } x,z \in \mathcal{X}$$

$\min f(x)$ is a short-hand notation for

$$\min_{x\in\mathcal{X}} f(x)$$

the same concerns the operations inf, max, sup

$\int g(x)dx$ is a short-hand notation for

$$\int_{\mathcal{X}} g(x)dx$$

$$z^* = \arg\min_{x\in Z} g(z)$$

is arbitrary global minimizer of a function g on the set Z, i.e. a point $z^*\in Z$ such that

$$g(z^*) = \min_{z\in Z} g(z)$$

the same concerns the operation max

$x^*=\arg\min f(x)$ or $x^*=\arg\max f(x)$ (depending on the context)

$B(x,\varepsilon,\rho) = \{z\in\mathcal{X} : \rho(x,z) \le \varepsilon\}$

$B(x,\varepsilon) = \{z\in\mathcal{X} : \|x,z\| \le \varepsilon\}$

$B(\varepsilon) = B\{x^*,\varepsilon\}$

$A(\varepsilon) = \{x\in\mathcal{X} : |f(x^*)-f(x)| \le \varepsilon\}$

$\mathbf{1}_A$ is the indicator of a set A, $\mathbf{1}_{\{x\in A\}}$ is the indicator of an event $\{x\in A\}$, i.e.

$$1_A(x) = 1_{\{x\in A\}} = \begin{cases} 1 & \text{if} \quad x\in A \\ 0 & \text{if} \quad x\notin A \end{cases}$$

$C_i^j = i!/(j!(i-j)!)$ for $i \geq j$

$\Gamma(u)$ is the gamma – function: $\Gamma(u) = \int_0^\infty t^{u-1}e^{-t}dt$

$\lceil a \rceil$ is the smallest integer not less than a

vrai sup η is the essential supremum of a random variable η, i.e.
vrai sup $\eta = \inf\{a: \Pr\{\eta \leq a\}=1\}$,

the notation $a_N \sim b_N$ $(N\to\infty)$ expresses the fact that the limit

$$\lim_{N\to\infty} a_N/b_N$$

exists and equals 1 (the relation $\sim$ is called asymptotic equivalence)

the notation $a_N \asymp b_N$ $(N\to\infty)$ expresses the fact that

$$0 < \liminf_{N\to\infty} a_N/b_N \leq \limsup_{N\to\infty} a_N/b_N \leq \infty$$

(the relation $\asymp$ is called weak equivalence)

$\emptyset$ is the empty set

I_m is the unit matrix of order $m\times m$

independent (or repeated) sample is a sample consisting of independent realizations of a random variable (vector)

random variables (vectors) and their realizations are denoted by the same symbol.

PART 1. GLOBAL OPTIMIZATION: AN OVERVIEW

CHAPTER 1. GLOBAL OPTIMIZATION THEORY: GENERAL CONCEPTS

This chapter is of introductory character: it considers various statements of the global optimization problem, the most commonly used types of prior information concerning the objective function and the feasible region, the main solution approaches, several classes of practical problems, and an algorithm comparison dilemma.

1.1 Statements of the global optimization problem

The most common form of the global optimization problem is the following. Let $\mathcal{X}$ be a set called the feasible region and $f: \mathcal{X} \to \mathbb{R}^1$ be a given function called the objective function, then it is required to approximate the value

$$f^* = \inf_{x \in \mathcal{X}} f(x) \tag{1.1.1}$$

and usually also a point in $\mathcal{X}$ at which f takes a value close to f^*.

This problem is called the global minimization problem. The maximization problem, in which the value

$$M = \sup_{x \in \mathcal{X}} f(x)$$

is to be approximated, can be treated analogously and can obviously be derived from the minimization problem by substituting $-f$ for f.

A procedure of constructing a sequence $\{x_k\}$ of points in $\mathcal{X}$ converging to a point in which the objective function value equals or approximates the value f^* is called a global minimization method (or algorithm). The types of convergence may be different, e.g. convergence with respect to values of f or convergence with some probability. As in practice the amount of computational effort is always restricted, hence only a finite subsequence of $\{x_k\}$ can be generated. For constructing these, one usually tries to reach a desirable (or optimal) accuracy, spending the smallest possible (or a bounded) computational effort.

Depending on the statement of the optimization problem (1.1.1), prior information f and $\mathcal{X}$, as values of f (and perhaps, its derivatives) at previous points of the sequence (and occasionally also at some auxiliary points) may be used for constructing a search point sequence $\{x_k\}$. Sometimes it is supposed that f is some regression function: this way, the evaluations of f may be subjected to random noise.

As a rule, the global optimization problem is stated in such a way that a point $x^* \in \mathcal{X}$ exists in which the minimal value f^* is attained. We shall call an arbitrary point $x^* \in \mathcal{X}$ with the property $f(x^*)=f^*$ a global minimizer and denote it by

$$x^* = \arg\min_{x \in \mathcal{X}} f(x) \tag{1.1.2}$$

or, more simply, by $x^*=\arg\min f$. In general, this point is not necessarily unique.

As a rule, the initial minimization problem is stated as the problem of approximating a point x^* and its value f^*. Sometimes the value f^* is required only, but a point x^* is not: naturally such problems are somewhat simpler.

Approximation of a point $x^*=\arg\min f$ and the value f^* is usually interpreted as finding a point in the set

$$A(\delta) = \{x \in \mathcal{X}: | f(x)-f(x^*)| \leq \delta\} \tag{1.1.3}$$

or in the set

$$B(\varepsilon) = B(x^*, \varepsilon, \rho) = \{x \in \mathcal{X}: \rho(x,x^*) \leq \varepsilon\} \tag{1.1.4}$$

where ρ is a given metric on $\mathcal{X}$, δ and ε determine the accuracy with respect to function (or argument) values.

The complexity of the optimization problem is determined mainly by properties of the feasible region and the objective function. There is a duality concerning the properties of $\mathcal{X}$ and f. If explicit forms of f and $\mathcal{X}$ are known and f is complicated, then the optimization problem can be reformulated in such a way that the objective function transforms into a simple one (for instance, linear) but the feasible region becomes complicated. The opposite type of reformulation is possible, too. Usually global optimization problems with relatively simple-structured sets $\mathcal{X}$ are considered (as they are, in general, easier to solve even if the objective function is complicated).

Unlike the local optimization problems, a global one can not be solved in general, if $\mathcal{X}$ is not bounded. The boundaries of $\mathcal{X}$ correspond to the prior information concerning the location of x^*. The wider boundaries $\mathcal{X}$ has, the larger is the uncertainty of the location of x^* (i.e. the more complicated is the problem). Supposing the closedness of $\mathcal{X}$ and the continuity of f in a neighbourhood of a global minimizer we ensure that $x^* \in \mathcal{X}$, i.e. the global minimum of f is attained in $\mathcal{X}$.

Typically, $\mathcal{X}$ can be endowed with a metric ρ, but usually there are many different metrics and there is no natural way of choosing a representative amongst them. The ambiguity of the metric choice is connected, for instance, with the scaling of variables. The properties of the selected metric influence the features of many optimization algorithms: therefore its selection must be performed carefully. In the case $\mathcal{X} \subset \mathbb{R}^n$ it is supposed that ρ is the Euclidean metric unless otherwise stated.

If $\mathcal{X}\subset\mathbb{R}^n$, then the dimensionality n of $\mathcal{X}$ determines the complexity of an extremal problem to large extent. One-dimensional problems (the case $\mathcal{X}\subset\mathbb{R}^1$) are thoroughly investigated (see Section 2.3.1); multidimensional problems ($\mathcal{X}\subset\mathbb{R}^n$, n>1) are however, of the main interest. Optimization problems in infinite dimensional (functional) spaces are usually reduced to finite dimensional problems (see Section 6.2.1). Discrete problems in which $\mathcal{X}$ is a discrete set are rather specific: to solve some classes of such problems special methods are needed. Nevertheless, general discrete optimization methods are constructed in the same way as the commonly-used finite dimensional ones (see Section 6.2.3). For omitting superfluous technical details, we shall suppose in Chapters 1-4 that $\mathcal{X}$ is finite-dimensional and compact.

As it was mentioned above, the structure of $\mathcal{X}$ is assumed to be relatively simple. Different algorithms require various structural simplicity features of $\mathcal{X}$. In a considerable number of algorithms (e.g. see Section 2.2) $\mathcal{X}$ is supposed to be the unit cube

$$\mathcal{X} = [0, 1]^n . \tag{1.1.5}$$

Some of these algorithms may be reformulated for a wider class of sets. But even in case of a hyperrectangle

$$\mathcal{X} = [a_1, b_1] \times ... \times [a_n, b_n] \tag{1.1.6}$$

such a reformulation may be ambiguous and may require a lot of care (as the metric on the cube differs from the metric induced by the corresponding metric on a hyperrectangle after transforming it into the cube). There are classes of algorithms (in particular, random search algorithms) which require only weak structural restrictions concerning the set $\mathcal{X}$. Below either the type of $\mathcal{X}$ is explicitly indicated or its structure is supposed to be simple enough. In many cases it suffices that $\mathcal{X}$ is compact, connected, and it is the closure of its interior (the last property guarantees that $\mu_n(\mathcal{X})>0$ and $\mu_n(B(x,\varepsilon))>0$ for all $x\in\mathcal{X}$, $\varepsilon>0$).

Sometimes the set $\mathcal{X}$ is defined by constraints and has a complicated structure. In such cases the initial optimization problem is usually reduced to the problem on a set of simple structure by means of standard techniques worked out in local optimization theory (namely, penalty functions, projections, convex approximation and conditional directions techniques). Special algorithms for such problems can also be created, see Sections 6.1 and 2.3.4.

The degree of complexity of the main optimization problem is partially determined by properties of the objective function. Typically, one should select a functional class $\mathcal{F}$ which f belongs to, before selecting or constructing the optimization algorithm. In practice, $\mathcal{F}$ is determined by prior information concerning f. In theory, the setting of $\mathcal{F}$ corresponds to the choice of a model of f. The wider $\mathcal{F}$ is, the wider the class of allowable practical problems is and the less efficient the corresponding algorithms are.

The widest reasonable functional class $\mathcal{F}$ is the class of all measurable functions. It is too wide, however, and thus unsuitable in modelling global optimization problems. In a sense, the same is true for the class $\mathcal{F} = C(\mathcal{X})$ of all continuous functions (and for classes of continuously differentiable functions $C^1(\mathcal{X})$, $C^2(\mathcal{X})$, etc. as well): this results from the existence of such two continuous (or differentiable) functions whose values coincide at any fixed collection of points but their minima may differ by any magnitude. On the other hand, the class of uniextremal functions is too narrow because the corresponding extremal problems can be successfully treated in the frames of local optimization theory.

Unlike in local optimization, the efficiency of many global optimization algorithms is not much influenced by the possibility and computational demand of evaluation/estimation of the gradient ∇f or the Hessian $\nabla^2 f$ of f, since the main aim of a global optimization strategy is to find out global features of f (while the smoothness characterizes the local features only).

Naturally, the computational demand of evaluating the derivative of f influences the efficiency of a global optimization strategy only as much as a local descent routine is a part of the strategy, see Section 2.1.

The computational demand of evaluating f is of great significance in constructing global optimization algorithms. If this demand is high, then it is worth to apply various sophisticated procedures at each iteration for receiving and using information about f. If this demand is small or moderate, then the simplicity of programming and the absence of complicated auxiliary computations are characteristics of great importance.

Global optimization problems in presence of random noise concerning evaluations of f are even more complicated. Only a small portion of the known global optimization algorithms may be applied to this case. Such algorithms will be pointed out individually; hence, it will usually be supposed that f is evaluated without noise.

The selection of a global optimization algorithm must be also based on a desired accuracy of a solution. If the accuracy required is high, then a local descent routine has to be used in the last stage of a global optimization algorithm, see Section 2.1.1. Some methods exist (see Section 2.2) in which the accuracy is an algorithm parameter. It should be indicated in this connection that under a fixed structure of an algorithm, the selection of its parameters must be influenced by a desired accuracy. For example, Section 3.3 describes an algorithm whose order of convergence rate will be twice improved, if the parameters of the algorithm are chosen depending on the accuracy.

Concluding this section, let us formulate a general statement of the optimization problem which will be taken automatically if opposite or additional details are not inserted. Let the compact set $\mathcal{X} \subset \mathbb{R}^n$ ($n \geq 1$) have a sufficiently simple structure; furthermore, let a bounded (from below) function $f: \mathcal{X} \to \mathbb{R}^1$ belonging to a certain functional class $\mathcal{F}$ be given. It is then required to approximate (perhaps, with a given accuracy) a point x^* and the value f^* by using a finite number of evaluations of f (without noise).

1.2 *Types of prior information about the objective function and a classification of methods*

This section describes various types of prior information concerning the objective function used in the construction and investigation of global optimization methods as well as a classification of the principal methods and approaches in global optimization, represented in a tabular form.

1.2.1 Types of prior information

The role of prior information - concerning the objective function f and the feasible region $X \subset \mathbb{R}^n$ - in choosing a global optimization algorithm is hard to overestimate. Having more information, more efficient algorithms can be constructed. Section 1.1 described the information that is considered as always available. But it does not suffice to construct an algorithm solving an extremal problem with a given accuracy over a finite time. Therefore it is worth of taking into account the more specific properties of the objective function while constructing global optimization algorithms.

There exist various types of prior information about f determining functional class $\mathcal{F}$. Usually $\mathcal{F}$ is determined by some conditions on f : a number of typical cases are listed below.

a) $\mathcal{F} \subset C(X)$.

a') $\mathcal{F} \subset C^1(X)$.

a'') $\mathcal{F} \subset C^1(X)$ and the gradient ∇f of $f \in \mathcal{F}$ can be evaluated.

a''') $\mathcal{F} \subset C^2(X)$.

a^{iv}) $\mathcal{F} \subset C^2(X)$ and Hessian $\nabla^2 f$ of $f \in \mathcal{F}$ can be evaluated.

a^{v}) $\mathcal{F} \subset C^k(X)$.

b) $\mathcal{F} \subset Lip(X, L, \rho)$ where L is a constant and ρ is a metric on X.

b') $\mathcal{F} \subset Lip(X, L)$ where L is a constant.

c) $\mathcal{F} \subset \{f \in C^1(X):\ \nabla f \in Lip(X, L)\}$ for some constant L.

c') $\mathcal{F} \subset \{f \in C^2(X):\ \|\nabla^2 f\| \leq M\}$ for some constant M.

d) $\mathcal{F} \subset \{f(x, \theta),\ \theta \in \Theta\}$ where Θ is a finite dimensional set: that is, f is given in a parametric form.

d') f is a polynomial of degree not higher than p.

d'') f is (not positive definite) quadratic.

e) f is a rational function.

e') f is given in an algebraic form.

f) $\mathcal{F}$ is a set of functions which can be approximated by functions from a certain class $\mathcal{F}_0$.

g) f is concave.

g') $f = f_1 - f_2$ where f_1 and f_2 are concave.

h) f has exactly ℓ local minimizers.

h') f has not more than ℓ local minimizers.

i) $f = f_1 + f_2$ where f_1 is a uniextremal function and $\|f_2\|$ is much less than $\|f_1\|$.

i') $f = f_1 + f_2$ where f_1 depends on no more than $s<n$ variables and $\|f_2\|$ is much less than $\|f_1\|$.

i'') $f = f_1 + f_2$ where f_1 is linear and $\|f_2\|$ is much less than $\|f_1\|$.

j) f is a separable function, i.e.

$$f(x_1,\ldots,x_n) = \sum_{i=1}^{n} f_i(x_i). \tag{1.2.1}$$

j') For appropriately defined two-dimensional functions $f_2,\ldots,f_n$ the representation

$$f(x_1,\ldots,x_n) = \sum_{i=2}^{n} f_i(x_{i-1}, x_i) \tag{1.2.2}$$

is valid.

k) $\mathcal{F} = \{f(x,\omega), \omega \in \Omega\}$ where (Ω, Ξ, P) is an underlying probability space, i.e. $\mathcal{F}$ is the set of realizations of a random process or field.

k') The prior distribution of a global minimizer x^* is given.

l) $\mathcal{F}$ is axiomatically given, i.e. it is supposed that the objective function satisfies some axioms.

m) The measure of domains of attraction of all local minimizers is not less than a fixed value $\gamma>0$.

m') The measure of the domain of attraction of a global minimizer is not less than a fixed value $\gamma>0$.

m'') $\mu_n\{A(\delta)\}/\mu_n\{\mathcal{X}\}\geq\gamma$ where δ and γ are given positive values, μ_n is the Lebesgue measure, and $A(\delta)$ is defined by (1.1.3).

n) The global minimizer x^* of f is unique.

n') There exists such $\delta_o>0$ that for any $\delta\in(0,\delta_o)$ the set $A(\delta)$ is connected and
$\mu_n\{A(\delta)\}>0$.

o) The smoothness conditions of the type a) - c) are fulfilled in a vicinity of the global minimizer.

p) There exist positive numbers ε, β, c_1, c_2 such that the inequalities

$$c_1\|x-x^*\|^\beta \leq f(x)-f(x^*)\leq c_2\|x-x^*\|^\beta$$

are valid for all $x\in B(\varepsilon)$.

p') There is a homogeneous function $H(z)$ of degree $\beta>0$ such that

$$f(x)-f(x^*)=H(x-x^*)+o(\|x-x^*\|^\beta)$$

for $\|x-x^*\|\to 0$.

q) f is evaluated with a random noise.

Let us shortly comment on these conditions.

The most familiar condition is b). There are known optimal and nearly optimal algorithms (in a well-defined sense) for the case when b) holds. An unavoidable drawback of the algorithms corresponding to this case is that the choice of the metric ρ may be quite arbitrary; further, the exact (smallest) Lipschitz constant $\mathcal{L}$ is typically unknown, while the numerical characteristics of the algorithms are greatly influenced by the choice of ρ and $\mathcal{L}$, see Section 2.2.

Assumptions a) - j) are often supposed to hold, when constructing and investigating deterministic global optimization algorithms which do not contain random elements. These algorithms are shortly described in Sections 2.1, 2.2, 2.3.

No fewer number of suggestions are encountered for constructing and investigating probabilistic algorithms, based on diverse statistical models of the objective function or of the search procedure. Probabilistic models of $f \in \mathcal{F}$ are the basis of many Bayesian, information-statistical and axiomatically constructed algorithms described mainly in Section 2.4 and using the conditions b), f), k) - l) and some others. Probabilistic models of the search process correspond to global random search algorithms: their investigation may be closely related to conditions m) - q), k'), d), f), h), h') and some others.

Information concerning these suppositions and cases of using them is contained in Table 1 of the next section and in the following chapters of the book.

1.2.2 Classification of principal approaches and methods of the global optimization

This section contains a table classifying the principal approaches and methods of global optimization as well as some additional explanation.

Certainly, it is impossible to enumerate all global optimization methods in an unambigous fashion: Table 1 provides a possible classification. Not all notations used are common: alternative terms for these and related methods may be found in the corresponding sections and references.

The first two columns of Table 1 give the names of approaches and methods (or groups of methods). The first five approaches mostly include deterministic methods. A family of probabilistic methods includes the methods of the last two approaches, together with methods based on smoothing the objective function and screening of variables, random direction methods, many commonly used versions of multistart and the candidate points method, as well as some versions of a number of other methods. Let us note again that the classification of approaches and methods in Table 1 is somewhat arbitrary, since some methods (in particular, random covering, random multistart, polygonal line methods and the method based on smoothing the objective function) may be in various groups.

The third column of Table 1 gives typical conditions (from among those described in Section 1.2.1) imposed on f for construction and investigation of methods. It should be noted that the condition collections for the majority of methods are not full or accurate since different versions of some methods require slightly different suggestions, and not for all methods are known the precise conditions ensuring their convergence.

According to Section 1.1, the feasible region $\mathcal{X}$ is mostly assumed to be a compact subset of $\mathbb{R}^n$, $n \geq 1$, having a nonzero Lebesgue measure and a relatively simple structure, unless additional details are given. The fourth column of Table 1 contains such details required for the realization of the corresponding method.

The fifth column provides the number of the section describing or studying the method. There are many corresponding references in these sections. Of course, it is impossible to mention all works devoted to global optimization and, as intended, those works are referred that contain much information related to the corresponding subject.

The sixth and seventh columns reflect the state of theoretical and numerical foundation of the methods. Under theoretical basis of a method we mean the existence of theoretical results on convergence, rate of convergence, optimality, and decision accuracy depending on the dimension of $\mathcal{X}$ as well as recommendations on the choice of method parameters. Needless to say, there is an element of subjectivity in our notes (as e.g. since the time of finalizing this book, new results became known).

1.2.3 General properties of multiextremal functions

Not all the works connected with global optimization theory deal with construction, investigation or application of optimization methods. For example, some recent investigations are devoted to asynchronous or parallel global optimization algorithms, but the author does not know impressive results in this field. Another group of works studies general properties of multiextremal functions: these works investigate the integral representations (see Section 2.3.3) and derive conditions on f ensuring that its every local minimum is also global. Outline the latest topic.

Let y be a real number and

$$\mathcal{L}_f(y) = \{x \in \mathcal{X}: f(x) \le y\}$$

be a level set of f . Consider the set

$$G = \left\{y \in \mathcal{R}^1: \mathcal{L}_f(y) \neq 0\right\}.$$

Zang and Avriel (1975) state that every local minimum of f is a global minimum of f if and only if the point-to-set mapping $\mathcal{L}_f: \mathcal{R}^1 \to \mathcal{B}$ is lower semicontinuous at any point $y \in G$. The lower semicontinuity of $\mathcal{L}_f$ at a point $y \in G$ means that if $x \in \mathcal{L}_f(y)$, $\{y_i\} \subset G$, $y_i \to y$ $(i \to \infty)$ then there exist a natural number K and a sequence $\{x_i\}$ such that $x_i \in \mathcal{L}_f(y_i)$, $i \ge K$, and $x_i \to x (i \to \infty)$.

The next result of Gabrielsen (1986) is closely connected with the preceding one.

Proposition 1.2.1. Let $f: \mathcal{R}^n \to \mathcal{R}^1$ be a function and $\mathcal{X} = \{x \in \mathcal{R}^n: f(x) < c\}$ where c is a real number. Suppose that $\mathcal{X}$ is open, connected, and not empty; f is continuously differentiable in $\mathcal{X}$;for every $\varepsilon > 0$ there exists a compact subset $\mathcal{D}_\varepsilon$ of $\mathcal{X}$ such that $f(x) \ge c - \varepsilon$ for all $x \notin \mathcal{D}_\varepsilon$; every stationary point of f is a strict local minimizer of f. Then there exists only one point $x^* \in \mathcal{X}$ such that $\nabla f(x^*) = 0$ and f attains the global minimum at this point.

The main condition in Proposition 1.2.1 is that every stationary point of f would be a local minimizer. If f is twice continuously differentiable in $\mathcal{X}$ then this condition takes the next form: if $x \in \mathcal{X}$ and $\nabla f(x) = 0$ then the Hessian $\nabla^2 f(x)$ is a positive definite matrix.

Proposition 1.2.1 presents a sufficient criterion of uniextremality of a function f. Gabrielsen (1986) used it to investigate unimodality of the likelihood function which plays a prominent role in mathematical statistics. Demidenko (1988) used almost the same

Table 1. Principal Approaches and Methods of Global Optimization

Approach	Method	Conditions on f	Conditions on X	Section	Theoretical Ground	Amount of Numerical Results	N o t e s
1	2	3	4	5	6	7	8
Based on the use of local search techniques	multistart	a) -a'^{V}), h) -h'), m) -m'')	–	2.1.3	elementary	different for various versions	in the pure form is used very seldom
	candidate points (using a cluster analysis technique)	a) -a'^{V}), h) -h'), m) -m'')	–	2.1.3	elementary	sufficient	numerical results indicate high efficiency
	tunneling	a') -a'^{V}), h'	–	2.1.4	different for various versions but not sufficient on the whole	sufficient	the method is prospective but hard to judge its efficiency
	Branin	a'^{V})	–	2.1.5	not sufficient	sufficient	generally is not sufficiently reliable and efficient
	heavy ball	a'')	–	2.1.5	poor	sufficient	has poor efficiency and reliability
	based on solution of differential equations	a'') a''')	–	2.1.5	different for various versions	different for various versions	efficient versions exist for some classes of problems
	based on smoothing the objective function	o), i), probably q)	–	2.1.6	poor	few	no results showing on high efficiency

Table 1. cont.

1	2	3	4	5	6	7	8
Covering methods	grid search	b), b'), c') a^{V})	mainly $X = [0, 1]^n$	2.2.1	high	sufficient	optimal with respect to some criteria, but practically inefficient
	polygonal line	b), b')	–	2.2.2	high	sufficient	efficient for the case of small dimensions, but requires a considerable amount of auxiliary computations
	Evtushenko	b), b'), c')	$X = [0, 1]^n$	2.2.2	relatively high	sufficient	efficient enough for the case n=1 and sometimes also for the case of small dimensions
	Strongin	b), b')	$X \subset [0, 1]^n$	2.3.1	relatively high	sufficient	efficient for small n
	sequentially best (Sukharev)	b), b')	–	2.2.3	high	no results	has a theoretical significance only
Branch and bound methods	based on use of convex minorants and concave majorants	e')	hyperrectangle	2.3.4	relatively high	small	complicated for realization, for each objective function requires a separate construction
	interval	e'), probably q)	hyperrectangle	2.3.4	relatively high	different for various versions	complicated realization in the multidimensional case
	concave minimization on a convex set	d"), g), g'), i")	convex, polyhedron	2.3.4	relatively high	different for various versions	the feasible region may have a high dimension

Table 1. cont.

1	2	3	4	5	6	7	8
Based on dimension reduction	coordinate-wise minimization	a), b)	–	2.3.2	poor	small	analogous to the corresponding local minimization algorithm
	random directions	b), d')	–	2.3.2	poor	sufficient	inefficient except the case d') with $p \le 4$
	multistep dimension reduction	b), j), j')	–	2.3.2	relatively high	different for various versions	efficient for some particular cases (for instance, case j))
	based on use of Peano curve type mappings	b)	$X \subset [0,1]^n$	2.3.2	relatively high	different for various versions	no results showing high efficiency in the case $n \ge 3$
	based on screening of variables	i'), a)	hyperrectangle	2.3.2	poor	small	there are examples of efficient solution of complicated problems
Based on approximation and integral representations	based on approximation of objective function	f), a^v)	–	2.3.3	relatively high	sufficient	some versions are optimal, but numerical results indicate poor efficiency
	based on integral representation	m''), n') o), p)	–	2.3.3	different for various versions	small	no results showing high efficiency
Based on statistical and axiomatic models	Bayesian	k)	mainly $X \subset R^1$	2.4.2 2.4.3	relatively high	sufficient for the case n=1	efficient, but can become cumbersome (especially in multidimensional cases)
	information -statistical	b), k')	mainly $X \subset R^1$	2.4.5	relatively high	no results	has no practical significance
	axiomatic	l)	–	2.4.4	relatively high	sufficient	there are many examples of efficient solution for some classes of optimization problem

Table 1. cont.

1	2	3	4	5	6	7	8
Global random search	random sampling	m'), o)	–	3.1.1 3.1.2	high	sufficient	poor efficiency, but the method serves as a component and a basic guideline for some other global random search methods
	random covering	b), m')	–	3.1.4	relatively high	few	more efficient than the preceding method
	simulated annealing	n), o) possibly q)	–	3.3.2	relatively high	not much	mostly of theoretical importance
	based on solution of stochastic differential equations	a'), a'''), possibly q)	–	3.3.3	relatively high	few	mostly of theoretical importance, perspective of practical use is not clear enough yet
	Markovian	o), m''), possibly q)	–	3.3	relatively high	not much	mostly of theoretical importance
	random multistart	a) -a' V), h) -h'), m) -m'')	–	2.1.3 4.5	relatively high	sufficient	the main version of multistart
	branch and probability bound	o), p), p'), sometimes d), k), k'), n), m''), q)	–	4.3	relatively high	sufficient	there are a number of variants, some of them proved to be highly efficient
	generation methods	n'), o), q) and others	–	Ch.5	different for various versions	sufficient	inefficient for simple problems, but promising for complicated multidimensional problems

criterion for verifying the uniextremality of the sum of deviation squares (that determine the least squares estimators) and other functions arising in identification of nonlinear stochastic models. The last mentioned work also generalized the above criterion to the case where the set $\mathcal{X}=\{x\in\mathbb{R}^n: f(x)<c\}$ is not connected and investigated the property of local uniextremality, i.e. the uniextremality of f on the connected parts of $\mathcal{X}$.

1.3 Comparison and practical use of global optimization algorithms

Let us suppose that a user looks for a computer program or a package of computer programs that would solve optimization problems on a particular computer with an acceptable accuracy in a suitable time. If the problems are complicated then he looks for efficient programs or algorithms and needs the results of algorithm comparison.

This section discusses the problem of comparing global optimization algorithms and partly answers the question above: i.e. how the user should choose a suitable algorithm or program. Numerical comparison is considered firstly.

1.3.1 Numerical comparison

Numerical comparison of global optimization algorithms consists of accomplishing numerical experiments on a computer via solving some (test or practical) multiextremal problems. Numerical comparison is not a particular feature of global optimization, but an important part of investigation in various fields of applied mathematics. In local optimization, classes of test functions, efficiency criteria, and conditions of numerical experiments are standardized, see Crowder et al. (1978), Hock and Schittlowski (1981) or Jackson and Mulvey (1978). Although some recommendations of the above works are suitable also for global optimization, an analogous standardization does not exist at present.

Unfortunately (but naturally), it is very hard to draw accurate conclusions on the efficiency of (global) optimization algorithms from results of numerical experiments. The reasons of this are diverse: (i) the results are unavoidable subjective, (ii) the classes of optimization problems on which a given algorithm (or program) is efficient are scarcely formalized, and (iii) there are many different efficiency criteria. Let us briefly comment on these reasons.

An algorithm and the corresponding computer program are of different nature. Two programs realizing the same algorithm but made by different prigrammers may be of different efficiency: this is the first ground for subjectivity in numerical comparison of results.

Most of the global optimization algorithms depend on parameters whose choice is ambiguous and hardly formalizable. This is the second basis for subjectivity and the explanation of the fact that the author of an algorithm may usually exploit it much more efficiently than an ordinary user.

Poor reporting accuracy and deficiency of numerical results represent the third ground for subjectivity. These cause difficulties in estimating the influence on algorithms efficiency of optimization problem features such as dimension, number of local minimizers, size of domain of attraction of the global minimizer, etc.

In order to overcome another difficulty, connected with the diversity of computers on which the numerical experiments are performed, it is universally adopted to express the computing time in standard time units which correspond to 1000 evaluations of the so-called Shekel function

$$f(x) = \sum_{i=1}^{m} 1/\left(\|x - a_i\|^2 + c_i \right), \qquad x \in \mathcal{R}^n \tag{1.3.1}$$

for n=4 and m=5 where a_i and c_i (i=1,...,m) are fixed vectors and numbers.

It is very hard to determine a class of optimization problems on which a given algorithm is numerically efficient. To do this, one experiments on various test (or practical) multiextremal functions that hypothetically reflect several specific function features. It is a difficult problem to choose a collection of multivariate test functions reflecting required function features. Standard collections of test functions (see e.g. Dixon and Szegö, eds. (1978), Walster et al. (1985) and some other works) include functions with a rather simple structure and thus are hardly able to reflect delicate features. Let us mention only the test function of Csendes (1985)

$$f(x_1,\dots,x_n)=\sum_{i=1}^{n}x_i^6\left(2+\sin\frac{1}{x_i}\right),\qquad x_i\in[-1,1]\backslash\{0\},\qquad (1.3.2)$$

which has a countable infinity of local minimizers. The work on standardizing of global optimization test function classes is being investigated by other researchers, thus we shall stop here further discussion of the topic.

The third reason for doubtfulness of numerical comparison is due to the existence of various efficiency criteria used in global optimization. The most common efficiency criterion is the time (expressed in standard units) required to reach a given solution accuracy. Sometimes the number of objective function evaluations is used instead of time. Reliability is the second important criterion that is very difficult to estimate in complicated setups. The simplicity of programming and the computer memory required are some other quality criteria of global optimization algorithms.

It follows from the above that numerical comparison studies *per se* are hardly able to measure the efficiency of global optimization algorithms appropriately and are not able to satisfy all requirements of the users. In many cases, a rigorous theoretical study gives more information on the efficiency. Let us indicate below several efficiency criteria for the theoretical comparison of global optimization algorithms.

1.3.2 Theoretical comparison criteria

(i) Domain of algorithm convergence. It is the set of optimization problems for which a point sequence generated by the algorithm converges to a global minimizer. The wider the domain of convergence is, the more universal is the algorithm. As it will be clear later, the convergence domain can be determined for a great number of algorithms. The difficulty of using the domain of convergence as a quality criterion arises due to the variety of convergence types, especially so for probabilistic algorithms.

(ii) Speed of convergence. It is studied only for some groups of algorithms and serves mainly for comparison of algorithms whithin these groups since the variety of types of convergence speed and methods of their estimation is great (much more than that of the convergence domains).

(iii) Inaccuracy. Let d^N be a deterministic rule (i.e. algorithm) of a sequential generation of N points $x_1,\dots,x_N$ from $\mathcal{X}$ and

$$f_N^*=\min_{1\le i\le N} f(x_i).$$

The inaccuracy (error) of an algorithm d^N for a function f is usually defined as

$$\varepsilon(f,d^N) = f^*_N - f^* \tag{1.3.3}$$

and the same for a functional class $\mathcal{F}$ can be defined as

$$\varepsilon(d^N) = \sup_{f \in \mathcal{F}} \varepsilon(f,d^N). \tag{1.3.4}$$

If it is possible to determine a measure $\lambda(df)$ on the functional class $\mathcal{F}$, then one can use the inaccuracy

$$\varepsilon(d^N) = \int_{\mathcal{F}} \varepsilon(f,d^N)\lambda(df) \tag{1.3.5}$$

instead of (1.3.4). For probabilistic algorithms of Section 2.4, the mean value $E\{\varepsilon(d^N)\}$ is usual for measuring in accuracy, replacing $\varepsilon(d^N)$ defined by (1.3.4) or (1.3.5).

Inaccuracies of type (1.3.4) or (1.3.5) are important efficiency characteristics of global optimization algorithms, but their utility range is very restricted since they can be evaluated only for a small part of methods d^N and functional classes $\mathcal{F}$.

(iv) Optimality. According to the ordinary definition of optimality (in the minimax sense), an algorithm is called optimal, if it minimizes the inaccuracy (1.3.4) for a fixed N or the number N for a fixed value of $\varepsilon(d^N)$. For some functional classes $\mathcal{F}$ (of the type $\mathcal{F}=\mathcal{L}ip(\mathcal{X},\mathcal{L},\rho)$) optimal methods exist and can be constructed. Various numerical results and theoretical studies show, however, that the above optimality property gives almost nothing from the practical point of view. This is connected with the fact that the minimax-optimal method gives good result for the *worst* function from a given class $\mathcal{F}$, but for an ordinary one it may be much worse than some other algorithm.

The use of the inaccuracy measure (1.3.5) leads to the Bayesian concept of optimality. Some other concepts of optimality (stepwise, Bayesian stepwise, asymptotical, asymptotical by order, etc.) also exist, but the observation that theoretical optimality of a particular method does not guarantee its practical efficiency, is valid again.

Many works in global optimization theory are devoted to the problem of optimal choice of algorithm parameters or their components: in essense, these works consider the optimality of parameters over narrow classes of algorithms.

All the abovesaid leads to the conclusion that even if a method may be theoretically investigated, the comparison of its efficiency to the efficiency of a method belonging to another approach is complicated and hardly formalizable because of the great variety of algorithm characteristics and optimization circumstances. Further problems arise when heuristic methods are investigated, see Zanakis and Evans (1981).

In conclusion, while studying the efficiency of a global optimization method, it is worthwhile to use a composite approach including the determination of a class of optimization problems that is being solved by the method, the investigation of theoretical properties of the method, and its numerical study. We shall mainly be concerned with the theoretical study in this book.

1.3.3 Practical optimization problems

It is hard to find branches of science or engineering that do not induce global optimization problems. Many such problems arise e.g. in the following fields: optimal design, construction, identification, location, pattern recognition, control, experimental design, etc. Instead of detailed description of the corresponding classes of optimization problems or particular ways of their solution we refer to a number of papers in the Journal of Optimization Theory and Applications, and also to Dixon and Szegö, eds. (1975, 1978), Batishev (1975), Mockus (1967), Zilinskas (1986), Zhigljavsky (1987) and make a few additional comments.

Experimental design theory (see Ermakov et al. (1983), Ermakov and Zhigljavsky (1987)) leads to a wide class of complicated multiextremal problems: some of them occasionally regarded as test ones. For instance, Hartley and Rund (1969), Ermakov and Mitioglova (1977), and Zhigljavsky (1985) used the function

$$f(x) = \det \left\| 1, x_i, x_{6+i}, x_i^2, x_i x_{6+i}, x_{6+i}^2 \right\|_{i=1}^{6} \tag{1.3.6}$$

that is to be maximized on the set

$$\mathcal{X} = \{x = x_1, \ldots, x_{12}) : 1 \le x_j \le 6 \ \ (j = 1, \ldots, 12)\}. \tag{1.3.7}$$

A maximizer of (1.3.6) is the determinant of a 6×6 matrix that depends on 12 variables: we remark that corresponds to the optimal saturated $\mathbf{D}$-optimal design for the second-degree polynomial regression

$$\theta_1 + \theta_2 x + \theta_3 y + \theta_4 x^2 + \theta_5 xy + \theta_6 y^2$$

on the square $(x,y) \in [1,6]\times[1,6]$. The function (1.3.6) is complicated to optimize, since it has several thousands of local maximizers, it has a great number of global maximizers, and its relatively high dimension eliminates the possibility of using most of the deterministic methods. Note that one can use the feasible region

$$\mathcal{X} = \{x = (x_1, \ldots, x_{12}) : 1 \le x_1 \le \ldots \le x_6 \le 6,\ 1 \le x_j \le 6 \ \ (j = 7, \ldots, 12)\}$$

instead of (1.3.7), furthermore, that test problems similar to (1.3.6) - (1.3.7) were treated in Bates (1983), Bohachevsky et al. (1986), Haines (1987).

The more capacity contemporary computers have, the greater is the actuality of optimization in simulation models. These are optimization problems of functions subjected to a random noise that can be controlled. The optimization problems of mathematical models with an objective function solving a differential/integral equation are closely related to the above-mentioned ones. Their main peculiarity is connected with the nonrandomness of the noise that can be controlled as well.

Formally, a great number of optimization problems arising in economics, optimal design, and optimal construction are not represented in the standard form described above since (i) the feasible region can be either fully or partly discrete, and (ii) the problem can be of multicriterial character. However, a number methods described in the book may be applied to such problems. Indeed, it was pointed out that the discreteness of $\mathbf{X}$ does not exclude the possibility of applying the majority of global optimization approaches. Moreover, multicriterion optimization problems are usually reduced to a number of global optimization ones for objective functions which are transformations of the initial criteria. Since these functions are defined on the same feasible region, the grid methods of Section 2.2.1 are rather efficient for solving these problems. For further discussion see Section 6.2.

CHAPTER 2. GLOBAL OPTIMIZATION METHODS

According to the classification of Section 1.2.2 this chapter describes some principal approaches and methods of global optimization except the global random search methods which Part 2 is devoted to. We shall consider the minimization of the objective function f (belonging to a given functional set $\mathcal{F}$ and evaluated without a noise) on the feasible region $\mathcal{X}$ (a compact subset of $\mathbb{R}^n$, having a sufficiently simple structure).

Section 2.1 deals with global optimization algorithms based on the use of local methods. They are widespread in practice, but many of them are not theoretically investigated to a considerable extent.

Section 2.2 is devoted to the covering methods including passive grid searches. The situation here is opposite to the above: the covering methods are thoroughly studied theoretically, but their practical importance is not great.

Section 2.3 considers one-dimensional optimization algorithms, reduction and partition techniques. The most attractive of them are the branch and bound methods. Under various (generally, substantial) prior information about f they combine the practical efficiency with theoretical elegance.

Section 2.4 treats the approach based on stochastic and axiomatic models of the objective function. Here the balance of the theoretical and practical significance of the algorithms does not hold. Most of them are cumbersome and time-consuming when being realized.

2.1 *Global optimization algorithms based on the use of local search techniques*

The bulk of the theory and methodology of local optimization have actually been developed around the turn of 1960's. By that time, global optimization theory as such did not exist and global optimization methodology almost had not been discussed. Thus it is not surprising that most global optimization algorithms of the sixties and seventies, having a practical importance, were based on the use of local optimization methodology. Over the past decade the situation changed, but global algorithms based on local methods remained popular both in theory and practice.

Section 2.1.1 presents some local algorithms. Section 2.1.2 describes general ways of using local algorithms to construct global optimization methods. One of the most popular global optimization algorithm is called multistart. Its basic form and its modifications are discussed in Section 2.1.3. Section 2.1.4 is devoted to tunneling algorithms consisting of sequential local minimization of functions constructed by iteratively transforming the objective function. Section 2.1.5 is concerned with algorithms of transition from some local minima to others. The construction of such algorithms is closely connected with solving some kinds of differential equations or simultaneous differential equations. Section 2.1.6 considers algorithm consisting of local minimization of a function obtained by means of smoothing the objective function.

2.1.1 *Local optimization algorithms*

First various setups of the local optimization problem are considered.

Let $\mathcal{X}$ be a measurable subset of $\mathbb{R}^n$, $n \geq 1$, and $f: \mathcal{X} \to \mathbb{R}^1$ be a lower bounded measurable function (objective function).A point x_* is called a local minimizer of f if there exists an $\varepsilon>0$ such that $f(x) \leq f(x_*)$ for all $x \in \mathbf{B}(x_*, \varepsilon)$. If f has a unique local minimizer, then it coincides with the global minimizer of f, i.e. $x^*=\arg\min f$ and function f is then referred to as uniextremal. The local optimization problem is the one of an approximate determination of a local minimizer of f for which values (and sometimes derivatives as well) at any point $x \in \mathcal{X}$ can be evaluated.

If $\mathcal{X} \in \mathbb{R}^n$, then the problem is called the unconstrained local optimization problem; in the opposite case it is constarined and is also called the mathematical programming problem. In the particular case, when f and $\mathcal{X}$ are convex, it is called the convex programming problem.

In local optimization algorithms a sequence of points $x_1, x_2, \ldots$ from $\mathcal{X}$ converging to x_* under certain assumptions is constructed. The way of constructing this sequence depends on properties of f and $\mathcal{X}$, information of f and $\mathcal{X}$ used at each iteration, and the degree of simplicity desired in the construction.

The point sequence $\{x_j\}$ in unconstrained local optimization algorithms is constructed by a recursive relation

$$x_{k+1} = x_k + \gamma_k s_k, \qquad k=1,2,\ldots$$

where $x_1 \in \mathcal{X}$ is an initial point, s_k is the search direction, and $\gamma_k \geq 0$ is the step-length.

Local minimization algorithms differ in the way of constructing $\{\gamma_k\}$ and $\{s_k\}$ that usually results in descent (or relaxation) algorithms for which the inequalities $f(x_{k+1}) \leq f(x_k)$ hold for all $k=1,2,\ldots$. To this end, it is necessary to find $\{s_k\}$ such that $s_k' \nabla f(x_k) < 0$ for each $k=1,2,\ldots$ and to choose γ_k in a suitable way. Two such ways of selecting γ_k are the most well-known. The first one is to choose

$$\gamma_k = \arg\min_{\gamma \geq 0} f_k(\gamma) \tag{2.1.1}$$

where $f_k(\gamma) = f(x_k + \gamma s_k)$ is the one-dimensional function determined by f, x_k and s_k. This way is primarily of theoretical significance.

The second way is as follows. Let $\gamma_1 > 0$ and $\beta > 1$ be some real values, γ_{k-1} be the preceding step-length, s_k be the search direction at x_k. If the inequality

$$f(x_k + \gamma_{k-1} s_k) < f(x_k) \tag{2.1.2}$$

holds then set

$$\gamma_k = \gamma_{k-1}\beta^i$$

where

$$i=\max\{j=1,2,... \mid f(x_k+\gamma_{k-1}\beta^j s_k)< f(x_k)\}$$

or

$$i=\max\{j=1,2,... \mid f(x_k+\gamma_{k-1}\beta^j s_k)< f(x_k+\gamma_{k-1}\beta^{j-1}s_k)\}.$$

If (2.1.2) does not hold then set

$$\gamma_k = \gamma_{k-1}\beta^{-i},$$

$$i=\min\{j=1,2,... \mid f(x_k+\gamma_{k-1}\beta^{-j}s_k)\le f(x_k)\}.$$

In the case $\beta=2$ this way is referred to as the bisection method.

Depending on the information used for constructing s_k, local optimization algorithms can be divided into three groups, viz., direct search algorithms that make use of values f only; first-order ones, using also first derivatives of f; and second-order ones, using also second derivatives of f. (Of course, f is supposed to be smooth enough while considering the latter two groups of methods.)

First and second-order algorithms are mostly special cases of the generalized gradient algorithm in which $s_k=-A_k\nabla f(x_k)$ and A_k $(k=1,2,...)$ are some positive definite matrices. For the gradient method $A_k=I_n$ for all $k\geq 1$ where I_n is the unit $(n\times n)$-matrix. If, in addition, γ_k are chosen as in (2.1.1), then the algorithm is called the steepest descent method. In the Newton method $A_k=[\nabla^2 f(x_k)]^{-1}$. In variable-metric methods A_k are recursive approximations of $[\nabla^2 f(x_k)]^{-1}$. One of the most well-known variable-metric method is due to Davidon, Fletcher and Powell, for this method (2.1.1) holds and

$$A_{k+1} = A_k(A_k a_k a'_k A'_k)/(a'_k A_k a_k) + \eta_k \eta'_k/(\eta'_k a_k)$$

where

$$a_k = \nabla f(x_{k-1}) - \nabla f(x_k), \qquad \eta_k = x_{k+1} - x_k, \qquad A_1 = A_{n+1} = A_{2n+1} =...= I_n.$$

A general class of first-order local minimization algorithms is composed by the method of conjugate directions. This class contains e.g. a version of the Davidon-Fletcher-Powell algorithm and the Fletcher-Reeves method, for the latter (2.1.1) holds and

$$s_{k+1} = -\nabla f(x_{k+1}) + \left[\|\nabla f(x_{k+1})\| / \|\nabla f(x_k)\|\right]^2 s_k,\ A_1 = A_{n+1} = A_{2n+1} = .. = I_n.$$

The direct search algorithms usually apply the bisection method for determination of step-lengths. At each k-th iteration of many search algorithms, that of the two directions +e or -e is taken as s_k for which f decreases. In the coordinate-wise search algorithms e belongs to the collection $\{e_1,...,e_n\}$ of the unit coordinate vectors. The cyclic coordinate-wise descent selects e_i sequentially, but the random coordinate-wise search does it at random. The random search algorithm with uniform trial draws e as a realization of a random vector uniformly distributed on the unit sphere $S=\{x\in\mathbb{R}^n:\ \|x-x_k\|=1\}$. In random searches with learning, e is a realization of a random vector whose distribution depends on the previous evaluations of f. The random m-gradient method ($1<m<n$) selects

$$s_k = -\sum_{i=1}^{m} q_i\left[f(x_k + \alpha q_i) - f(x_k)\right] / \alpha$$

where α is a small positive value and $q_1,...,q_m$ are orthonormalized vectors constructed by means of the orthogonalization procedure from m independent realizations of random vectors uniformly distributed on S.

A thorough description and investigation of local optimization technique can be found in many textbooks, see, for instance, Avriel (1976), Dennis and Schnabel (1983) or Fletcher (1980). Among others Fedorov (1979), Demyanov and Vasil'ev (1985), Mikhalevitch et al. (1987) treated the nondifferentiable case. It should be noted that the local optimization routines available in contemporary software packages are suitable for most practical needs.

2.1.2 *Use of local algorithms in constructing global optimization strategies*

If the objective function f is multiextremal, then the application of a local optimization algorithm leads to a local extremum which is generally not a global one. Thus, in order to find a global extremum, one has to use another technique ensuring global optimization.

The local optimization techniques, however, takes an important place in the global optimization methodology. This is mainly due to the usual construction of global optimization strategies consisting of two stages: global and local. At the global stage, the objective function is evaluated at points located, so to speek, to *cover* $\mathcal{X}$: its aim is to reach a small neighbourhood of a global optimizer.

The local stage of global optimization algorithms may be expressed in explicit or implicit form and used one or more times. An implicit form of the local stage is present in a global algorithm when several points are chosen in neighbourhoods of some points obtained previously. An explicit form of the local stage is present in the following cases: if a crude approximation of a global optimizer is obtained and it has to be refined; or if one needs to simply obtain values of the objective function as small as possible (in some particular algorithms considered below).

Ordinarily local algorithms are aimed at locating a global optimizer x* making more precise the approximation of $f^* = f(x^*)$. To do this, one should apply a local descent routine starting from an initial point $x_{(0)}$ that is an approximation of x* as obtained by a global optimization method.Roughly speaking, in that case the aim of the global stage is the localization of x* and the aim of the local one is finding the precise location of x*. Applying the above approach, it is necessary to assume that the point $x_{(0)}$ belongs to the neighbourhood of x* in which f is sufficiently smooth and has only one local minimum.

The efficiency of a considerable number of global minimization methods depends on closeness of the values f_k^* and f^* for all $k \geq 1$. In these methods it is natural to make some iterations of a local descent right away after obtaining a new record point.

Occasionally, if iterations of a local descent are easily computable, then it is adventegous to start them from several points obtained at the global stage. This problem for random optimization technique is discussed later in Section 3.2.

Note that if a local optimization algorithm is part of a global optimization strategy, then it should be supposed that the objective function is subjected to some local conditions besides the global ones. Moreover, it can be concluded from the cited properties of local optimization algorithms (see Section 2.1.1) that it is especially advantageous to include them into a global optimization strategy, if the derivatives of the objective function may be evaluated without much effort.

2.1.3 Multistart

The technique under consideration is concerned with the method recently called multistart which has been historically the first and for a long time the only widely used method of global optimization. This method consists of multiple (successive or simultaneous) searches of local extrema starting at different initial points. The initial points are frequently chosen as elements of a uniform grid (such grids are described later in Section 2.2.1)

To use multistart, one should impose suggestions on the number of local optimizers of the objective function and also on its smoothness.

The main difficulty for practical use of multistart is the following: to obtain a global optimizer with high reliability one should choose the number of initial points much greater than the number of local optimizers (which is usually unknown). So the main part of computer efforts would be taken for attaining local extrema repeatedly. If it is supposed that the local optimizers of the objective function are rather far from each other (a heuristic suggestion), then the basic version of multistart is often modified by one of the two ways described below.

The first version is to surround every evaluated local minimizer by a neighbourhood and attribute the next property to it: if a point of search attains the neighbourhood then it moves into the local minimizer. The proper use of the corresponding algorithm is possible only if one is able to choose neighbourhoods which either are subsets of the corresponding domain of attraction of the local minimizers or surely do not contain a global optimizer. In the last case the global minimization algorithm is a covering method and is studied in Section 2.2.

The second way is called the candidate points method and consists in simultaneous local descents from many initial points and joining neighbouring points. The joining action is equivalent to substituting such points by one of them (whose objective function value is the smallest). The problem of joining belongs to cluster analysis; this way any clustering method may be used for this purpose. One of the most simple and popular methods of joining is the so-called nearest neighbour method based on the heuristic supposition that the distance $\rho_{ij}=\rho(z_i,z_j)$ (in metric ρ) between points z_i and z_j belonging to a neighbourhood of a local minimizer is less than the distance between the points belonging to the neighbourhoods of different minimizers.

Algorithm 2.1.1. (nearest neighbour method).

1. Let K points $z_1,...,z_K$ be given. Assume that all of them belong to various clusters, i.e. the number of clusters equals K (here the word *cluster* is associated with *neighbourhood of a local minimizer*).
2. Find *the nearest* pair of points (z_i,z_j), i.e. such a pair for which

$$\rho_{ij} = \min_{k,\ell=1,...,K,k\neq\ell} \rho_{k\ell}$$

3. If the distance ρ_{ij} between the *nearest neighbours* z_i and z_j does not exceed a certain small $\delta>0$, then the points z_i and z_j are incorporated and the corresponding clusters are unified; hence the number of clusters K decreases by 1 (i.e. K-1 is substituted for K).
4. If the distance between the *nearest neighbours* exceeds δ or K=1 (only one cluster remains) then the algorithm stops. Otherwise go to Step 2.

In Algorithm 2.1.1 the metric ρ is usually chosen to be Euclidean. The goodness of Algorithm 2.1.1 depends on the choice of δ as well. This number must be small enough, at any rate, less than any distance between a pair of local minimizers.

The typical candidate points method is as follows.

Algorithm 2.1.2 (candidate points method).

1. Obtain N points $x_1,...x_N$ generating them uniformly on $\mathcal{X}$ (i.e. $x_1,...x_N$ are the elements of a uniform random sample.
2. By performing several iterations of a local descent routine from initial points $x_1,...x_N$ obtain points $z_1,...,z_N$ (the number of the local descent iterations from the points x_i may depend on values of the objective function $f(x_i)$, i=1,...,N).
3. Apply a cluster analysis method (for instance, Algorithm 2.1.1) to the points $z_1,...,z_N$. Let m be the number of clusters obtained.
4. Select representatives $x_1,...x_m$ from the clusters (a natural selection criterion is the objective function value). If m=1, or the number of the local descent iterations exceeds a fixed number then go to Step 5. Otherwise put N=m and go to Step 2.
5. Suppose we are in the neighbourhood of a global minimizer. In case of necessity, apply a local optimization routine which has high speed of convergence in the neighbourhood of an extremal point.

Algorithm 2.1.2 is heuristic, its efficiency depends on parameters and auxiliary methods choice. The number N must be significantly greater than the expected number of local minima of the objective function. The choice of Algorithm 2.1.1 as a cluster anlysis method and a local optimization routine available in a computer software is usual choice. Note that quasirandom grids have almost the same prevalences as random ones for selection of the initial points in candidate points methods.

Algorithm 2.1.2 and some similar methods (see Törn (1978), Spircu (1978), Batishev and Lubomirov (1985)) are appealing in case of some complicated multiextremal problems. The well-known method of Boender et al. (1982) is also based on the above principles and proved to be efficient for solving a number of multivariate global optimization problems. Its recent version goes under the name of multi-level single linkage method and is summarized as follows.

Algorithm 2.1.3 (Multi-level single linkage method)

1. Set $k=1$, $\Xi=\emptyset$.
2. Generate N uniformly distributed on $\mathcal{X}$ random points, evaluate f at each one, and add them to the point collection Ξ.
3. Select points from Ξ as initial ones for local descent.
4. Perform local minimization from all initial points selected.
5. Check a stopping rule. If the algorithm is not terminated, then return to Step 2 substituting k+1 for k.

At each k-th iteration the next selection procedure is applied at Step 3 of Algorithm 2.1.3. First, cut off the sample Ξ and choose γkN point of Ξ with the lowest function values where γ is a fixed number in (0,1]. Each chosen point x is selected as an initial point for a local descent if it has not been used as an initial point at a previous iteration, and if there is no neighbour sample point $z\in \Xi\cap B(x,\eta_k)$ with lower function value where the critical distance η_k is given by

$$\eta_k = \pi^{-1/2}[\,\mu_n(\mathcal{X})\,\Gamma(n/2+1)\delta(\log kN)/kN\,]^{1/n},$$

where δ is positive constant.

The works of Rinnooy Kan (1987), Rinnooy Kan and Timmer (1985,1987a,b) contain much more details on the theme of using cluster analysis techniques in global optimization algorithms. Note that statistical inferences in the multistart in which the initial points are the elements of the random grid will be studied in Section 4.5 under the term *random multistart*.

2.1.4 *Tunneling algorithms*

Over the past decade the so-called tunneling algorithms gained popularity. Their essence consists in the successive search for a new point $x_{(0)}$ in the set

$$\mathcal{U}(x_*) = \{x\in\mathcal{X}: f(x)\le f(x_*)\}\setminus\{x_*\} \tag{2.1.3}$$

after obtaining a record point x_*, and then descending from the point $x_{(o)}$.

A tunneling algorithm consists of two stages: a minimization stage and a tunneling one. These stages are used sequentially. In the minimization stage a local descent routine is used for finding a local minimizer x_* of the objective function. In the tunneling stage, the so-called tunneling function (it was called penalty or filled function as well) T(x) is determined. This function attains a maximum (may be, a local maximum) at x_*, has continuous first derivatives at all points (probably, except x_*) and depends on f, x_*, and a number of parameters that are automatically chosen by the algorithm. After constructing the tunneling function, a point $x_{(o)}$ from the set (2.1.3) is sought for at the tunneling stage. Then one proceeds to the minimization stage and finds a local minimizer of f starting at $x_{(o)}$. The obtained local minimizer is a new record point and the above iteration may be repeated. The search stops if it does not succeed in finding a point $x_{(o)} \in \mathcal{U}(x_*)$ in a given number of algorithmic steps.

Vilkov et al (1975), Levy and Montalvo (1985) determined the tunneling function by

$$T(x) = T(x, x_*, \alpha) = (f(x) - f(x_*))/\|x - x_*\|^{\alpha} \qquad (2.1.4)$$

where $\alpha>0$ is a fixed parameter. If there are some record points $x_*^{(1)}, \ldots x_*^{(\ell)}$ with the same objective function value then Levy and Montalvo (1985) proposed to construct the tunneling function by

$$T(x) = (f(x) - f(x_*))/\left[\prod_{i=1}^{\ell} \left\|x - x_*^{(i)}\right\|^{\alpha_i}\right].$$

Ge (1983) determined the tunneling function (calling it a *filled function*) via

$$T(x) = T(x, x_*, \eta, \beta) = \frac{1}{\eta + f(x_*)} \exp\left\{- \|x - x_*\| / \beta^2\right\} \qquad (2.1.5)$$

where η and β are parameters, $\beta>0$,

$$\eta + f(x_*) > 0. \qquad (2.1.6)$$

Ge and Qin (1987) modified (2.1.5) in different ways and presented five new tunneling functions, viz.,

$$\frac{1}{\eta + f(x)} \exp\left\{-\|x - x_*\| / \beta^2\right\},$$

$$-\left\{\beta^2 \log[\eta + f(x)] + \|x - x_*\|^p\right\}, \quad p = 1, 2$$

$$-(f(x) - f(x_*)) \exp\left\{\beta^2 \|x - x_*\|^p\right\}, \quad p = 1, 2.$$

To find a point $x_{(o)} \in \mathcal{U}(x_*)$ in the tunneling stage needs performing a local minimization of the tunneling function starting at an initial point x_o belonging to a rather small neighbourhood of the record point x_*. If the local minimizer of T does not belong to the set (2.1.3) or the search trajectory goes out of $\mathcal{X}$ not attaining $\mathcal{U}(x_*)$ then it is advisable to change values of parameters (parameter α in case of (2.1.4) and parameters η, β in case of (2.1.5)) and to descend into a local minimizer of a modified function T. If after several changes of values of parameters a point from the set $\mathcal{U}(x_*)$ is not found then it is advisable to perform the same actions using another initial point x_o. For example, it is natural that such points are sequentially selected from the collection of points

$$x_0 = x_* \pm \gamma e_i, \quad i = 1, \ldots, n, \tag{2.1.7}$$

where γ is a small number, e_i are the coordinate vectors.

Vilkov et al. (1975) showed that if $\mathcal{X}$ is an interval, function f is continuously differentiable and has a finite number of local minimizers, the tunneling function has the form (2.1.4) and its minimization under all parameter values $\alpha>0$ is performed, then the above defined tunneling method converges to a global minimizer of the objective function f. For the multidimensional case, a similar statement on the convergence of the tunneling method using function (2.1.4) is absent.

Let us more thoroughly consider the form (2.1.5) for the tunneling function. Suppose that f is continuously differentiable, $\mathcal{L}$ is a bound for the Euclidean norm of its gradient, i.e. such a constant that the inequality $\|\nabla f(x)\| \leq \mathcal{L}$ for all $x \in \mathcal{X}$ holds, x^* is a global minimizer of f, x_* is a local minimizer, $\mathcal{S}(x_*)$ is the subset of the domain of attraction of x_* containing points x such that

$$(x - x_*)' \nabla f(x) \geq 0,$$

$\mathcal{D}(x_*)$ is the radius of the set $\mathcal{S}(x_*)$, i.e. the shortest distance from x_* to the boundary of $\mathcal{S}(x_*)$, $\mathcal{D}$ is the smallest radius of all subsets $\mathcal{S}(x_*)$, i.e.

$$\mathcal{D} = \min_{x_*} \mathcal{D}(x_*)$$

where the minimum is taken over all local minimizers of f.

The next two statements are due to Ge (1983).

Proposition 2.1.1. If the inequalities (2.1.6) and

$$0 < \beta/(\eta + f(x_*)) \leq 2\mathcal{D}(x_*)/\mathcal{L} \tag{2.1.8}$$

hold, then the function (2.1.5) can not have any stationary point in the set

$$\{x \in \mathcal{X}: f(x) > f(x_*)\} \tag{2.1.9}$$

Proposition 2.1.2. If (2.1.6) holds and the ratio $\beta^2/(\eta+f(x_*))$ is small enough to assure the inequality

$$\beta^2/(\eta + f(x^*)) \leq 2\mathcal{D}/\mathcal{L} \tag{2.1.10}$$

then the function (2.1.5) has no local minimizers in the interior of $\mathcal{X}$.

Moreover it is evident that if (2.1.6) holds then x_* is a local maximizer of the function (2.1.5).

Proposition 2.1.1 shows that if the ratio $\beta^2/(\eta+f(x_*))$ is small enough (satisfies the inequality (2.1.8)) then the function (2.1.5) has no stationary points in the set (2.1.9). Hence, any local descent algorithm applied to the function (2.1.5) and starting from a neighbourhood of x_* should either arrive to the boundary of $\mathcal{X}$ or reach the set $\mathcal{U}(x_*)$.

On the other hand, Proposition 2.1.2 yields that if this ratio is too small (satisfies the inequality (2.1.10)), then any local descent algorithm applied to the tunneling function (2.1.5) should arrive to the boundary of $\mathcal{X}$. Hence, in this case the aim of the tunneling stage can not be reached and further search of a global minimizer of f would be impossible.

The main difficulty of constructing a suitable version of the tunneling algorithm is connected with the problem of choosing its parameters η,β (or, respectively, α and α_i). Under their favourable choice the convergence of the algorithm may be ensured. Let us now formulate the basic version of the Ge algorithm.

Algorithm 2.1.4. (Ge (1983)).

1. Find a local minimizer x_* of the function f starting at an arbitrary point x_1; set k=1.

2. Form the tunneling function T by (2.1.5) (for the modified tunneling functions presented above the algorithm is analogous, see Ge and Qin (1987)).

3. Sequentially starting at initial points (2.1.7) descend to local minimizers of the tunneling function T until either a local minimizer of T or a point from $\mathcal{U}(x_*)$ is found. If none of the local minimizers of T or a point from $\mathcal{U}(x_*)$ is found then go to Step 8.

4. Use a local descent routine for function f, starting at the point obtained at Step 3. Let z_* be the obtained local minimizer of f.

5. If $f(z_*) \le f(x_*)$ then substitute z_* for x_* and go to Step 2.
6. If $f(z_*) > f(x_*)$ and $k<K$ (where K is a given number) then increase values of η and β, so that the ratio $\beta^2/(\eta+f(x_*))$ decreases. Go to Step 3 but use the initial point

$$x_0 = x_* + \gamma(z_* - x_*)/\| z_* - x_* \|$$

instead of (2.1.7) and z_* instead of x_* in formula (2.1.5).
7. If $f(z_*) > f(x_*)$ and $k \ge K$ then the algorithm stops and gives the point x_* as a global minimizer of f.

8. Decrease η to increase the ratio $\beta^2/(\eta+f(x_*))$, replace k by k+1 and go to Step 2.

In Algorithm 2.1.4 the choice of a number K (defining the stopping criterion) and change rules for the values of parameters η, β are heuristic, but they influence considerably the efficiency of the algorithm.

Some more information about the tunneling method realization may be found in Levy and Montalvo (1985) and Ge and Qin (1987). The analysis of numerical results yields that the tunneling method can not be referred to as a very efficient one (in particular, it does not always succeed in finding the global minimizer). The main difficulties of realizing the tunneling method are the following: the tunneling function under certain values of parameters (α in (2.1.4) and η, β in (2.1.5)) are *flat* and close to zero in a considerable part of $\mathcal{X}$; local minimization of a tunneling function needs to be carefully performed in order not to pass over a minimizer; trajectories of a local optimization of T often arrive the boundary of $\mathcal{X}$; the termination of the search is problematic (in essence, the stopping problem is equivalent to the main optimization problem). Nevertheless, the tunneling methods have only been recently created and the progress in the field of increasing their numerical efficiency is quite probable.

2.1.5 *Methods of transition from one local minimizer into another*

As already noted, local optimization theory is better developed than its global counterpart. Hence the attempts of some authors to modify local methods in such a way that a search trajectory might pass over from a local minimizer to another, in order to find a global minimizer, seem to be natural. The idea has served as a base for creating some heuristic algorithms (including the algorithms of this and the preceeding section). In this section we shall suppose that the objective function f is sufficiently smooth (as a rule, twice continuously differentiable).

A simple transition algorithm may be constructed by means of alternating descents to local minimizers with ascents to local maximizers. As an initial direction of an ascent (descent) the latter direction of the preceding descent (respectively, ascent) is natural to use in this algorithm. Its disadvantage consists in the possibility of cycling at a collection of some local minimizers and failing to reach a global minimizer. Another disadvantage is that a lot of evaluations of the objective function and its derivatives are wasted on investigating nonprospective subsets (in particular, on ascents to maximizers). The former disadvantage may be removed by means of surrounding minimizers by ellipsoids in order to prevent waste descents (Treccani et al. (1972)). But already in the bivariate case the above algorithm is rather complicated for realization (see Corles (1975)) and in case of greater dimensions the possibility of its efficient realization is indeed problematic.

Many global minimization methods of this section originate from investigating properties of solutions of various differential equations. One of the first such attempts is the *heavy ball* algorithm (Pshenichnij and Marchenko (1967)) according to which the search trajectory coincides with the trajectory of a ball motion on the surface generated by the objective function. Globality of the *heavy ball* algorithm associated with that of a moving ball by inertia may pass over *flat hollows* (but may stop in one of them).

The search trajectories for a general class of algorithms including the *heavy ball* algorithm and algorithms from Zhidkov and Schedrin (1968), Incerti et al. (1979), Griewank (1981) are discrete approximations of solutions of second order ordinary differential equations having the form

$$\mu(t)x''(t) + \nu(t)x'(t) = -\nabla f(x(t)), \qquad t \geq t_0 \tag{2.1.11}$$

subject to the initial conditions

$$x(t_0) = x_0, \qquad x'(t_0) = z_0 \in R^n, \qquad \|z_0\| = 1$$

where $\mu(t)$ and $\nu(t)$ are functions of time t. According to classical mechanics the equation (2.1.11) represents Newton's law for a particle of mass $\mu(t)$ in a potential f subject to a dissipative force $-\nu(t)x'(t)$. At the initial time $t=t_0$, a particle is at a point x_0 and has a motion direction z_0. Given suitable assumptions on functions f, μ, ν, any trajectory converges to a local minimizer of f tending to pass over *flat* minima.

Let c be a certain upper bound for $f^*=\min f$ (i.e. $c \geq f^*$). Griewank (1981) showed that the search trajectory obtained from (2.1.11) under

$$\mu(t) = [f(x(t)) - c]/e, \qquad \nu(t) = -[\nabla f(x(t))]'$$

can not converge to a local minimizer with objective function value greater than c; further, that a point $x_{(0)} \in \mathcal{X}$ will be reached with the value $f(x_{(0)}) \leq c$. The latter algorithm seems to be the most promising of the mentioned group.

Let us consider another class of global optimization algorithms based on solving differential equations. The main representative of it is the method developed by Branin (1972) and Branin and Ho (1972).

Let f be twice continuously differentiable and assume that not only the values $f(x)$ can be evaluated but also the values of the gradient $g(x)=\nabla f(x)$ and the Hessian $H(x)=\nabla^2 f(x)$ for all $x \in \mathcal{X}$. Consider a system of simultaneous differential equations

$$\frac{d}{dt} g(x(t)) = s\, g(x(t)) \tag{2.1.12}$$

subject to the initial condition $g(x(0))=g_0$ where s is a constant taking on either value +1 or -1. The solution of the system (2.1.12) has the form $g(x(t))= g_0 e^{st}$ and if $s=1$, $t\to -\infty$ or $s=-1$, $t\to\infty$ it converges to $g=0$. It means that the trajectory corresponding to a solution tends to a stationary point of f. Branin method consists of sequentially solving (2.1.12) (in order to attain a stationary point of f) and alternating the sign of s (in order to pass over from one stationary point to another). To solve the system (2.1.12) rewrite it as follows

$$x'(t) = sH^{-1}(x(t))\, g(x(t)) \tag{2.1.13}$$

Let $A(x)$ be the adjoint matrix of $H(x)$ which determines the inverse of the Hessian by the formula $H^{-1}(x)=A(x)/\det H(x)$. Then (2.1.13) may be replaced by the system

$$x'(t) = sA(x(t))\, g(x(t)). \tag{2.1.14}$$

The latter is determined for all x. It is obtained from (2.1.13) by changing the time in case $\det H(x)\neq 0$ and by inverting the time while passing over to the points of neglecting det $H(x)$. If (2.1.14) is used then the sign of the constant s should be alternated when attaining a stationary point as well as at a point of neglecting det $H(x)$.

An ordinary method of numerically solving these first order differential equations is based on the use of discrete approximation

$$x_{k+1} = x_k + \gamma_k \dot{x}_k, \qquad k = 0,1,\ldots \tag{2.1.15}$$

where

$$\dot{x} = \frac{dx}{dt}, \qquad x_0 = x(0)$$

is an arbitrary point from $\mathcal{X}$, $\gamma_0, \gamma_1, \ldots$ is a certain sequence of nonnegative numbers. Applying (2.1.15) to solve (2.1.13) we obtain

$$x_{k+1} = x_k + s\gamma_k H^{-1}(x_k) g(x_k), \qquad k = 0, 1, \ldots$$

that is the Newton method with a variable step length γ_k. The system (2.1.14) may be solved analogously.

A comprehensive description of the Branin method and its modifications as well as numerical results are contained in Hardy (1975), Gomulka (1978). Treccani (1978) studied a function f whose contours are topologically equivalent to spheres and the Branin method for minimizing it may not converge. Anderson and Walsh (1986) proposed a simply realized version of the Branin method for minimization of two-dimensional functions of special kind.

Branin's idea of changing a system of simultaneous differential equations after reaching a stationary point is used in Yamashita (1976) to solve a minimization problem on a set

$$\mathcal{X} = \{x \in \mathbb{R}^n : h(x) = 0\} \text{ where } h = (h_1,\dots,h_m): \mathbb{R}^n \to \mathbb{R}^m,\ m < n,$$

the functions f and h_j are three times continuously differentiable and the matrix $B(x)$ with elements $B_{ij}(x)=\partial h_j/\partial x_i$ is of full rank. His work shows that the local optimizers (maximizers as well as minimizers) of f under restrictions $h(x)=0$ are stable states of trajectories corresponding to systems of differential equations

$$\dot{x} = B(x)\lambda(x) - sg(x), \qquad \frac{dh(x)}{dt} = -h(x)$$

where $g(x)=\nabla f(x)$, $s\in\{-1,1\}$, $\lambda:\mathbb{R}^n\to\mathbb{R}^m$ is an unknown vector function determined by the system (λ is an analog to the Lagrange multipliers). Eliminating λ from the system, we are led to

$$\dot{x} = -sP(x)g(x) - Q(x)h(x)$$

where

$$P = I_n - B(B'B)^{-1}B', \qquad Q = B(B'B)^{-1}.$$

The discretization (2.1.15) may be used to solve this system. Sequentially alternating the constant s sign after reaching a stable state allows to find several local minimizers of f under restrictions $h(x)=0$. Certainly, the above approach does not guarantee that a global minimizer is reached.

All algorithms based on solving differential equations have the following disadvantages. First, there are no general results on their convergence to a global optimizer. Hence, one may guarantee that a global minimizer is found only if it is assured that all local minimizers are found. Second, the above algorithms are relatively complicated for realization and investigation. In particular, it is necessary to evaluate the objective function derivatives (for some algorithms also the Hessian) and the possibility of using finite difference approximations instead of the derivatives is not obvious. It is difficult to draw a general conclusion on the efficiency of the above algorithms. It should be noted, however, that there are numerical examples which demonstrate that for finding a global optimizer the Branin method requires fewer evaluations of the objective function than the tunneling algorithms. (Note however that Branin method requires also evaluations of the first and second derivatives).

2.1.6 Algorithms based on smoothing the objective function

Let a function f satisfy the Lipschitz condition and be presented as $f=f_1+f_2$, where f_1 is a uniextremal function and the supremum-norm of f_2 is much less than the norm of f_1, i.e. $\sup|f_2| \ll \sup|f_1|$. According to Katkovnik (1976), the smoothed function

$$\hat{f}(x,\beta) = \int f(z)p(x-\beta z)dz \qquad (2.1.16)$$

is uniextremal for sufficiently large $\beta>0$, while its global minimizer is close to x^* for sufficiently small $\beta>0$. Here $p(z)$ is the kernel of the smoothing operator, being a continuously differentiable, unimodal, symmetric density of a probability distribution on $\mathfrak{R}^n$. The case of the nondifferentiable kernel

$$p(z) = 1_A(z), \qquad (2.1.17)$$

where $A=[-a,a]^n$, is analogous and considered in Mikhalevitch et al (1987).

In view of the above qualitative results the following global minimization concept is sensible: find the minimizer of the uniextremal function (2.1.16) for some $\beta>0$ and starting from this point try to descend into a minimizer of f. From the point of view of local optimization theory, the extended algorithm below is sensible, too: the values of β are to be sequentially decreased (slowly enough), while the number of steps increases (i.e. a minimizer is more closely approached). In case of (2.1.17), the latter algorithm was investigated. But even for this rather simple kernel, the problem of clear determination of the functional classes for which the algorithm converges is not clarified.

The impossibility of the exact evaluation of the smoothed function (2.1.16) is the main computational difficulty encountered when realizing the above conceptual algorithms. Instead of the exact values, Monte Carlo estimators of the form

$$\hat{f}(x,\beta) \approx \frac{1}{N}\sum_{j=1}^{N}\left[f(\xi_j)p(x-\beta\xi_j)\right]/\varphi(\xi_j) \qquad (2.1.18)$$

are usually applied, where $\xi_1,\ldots,\xi_N$ are independent realizations of a random vector with a probability density $\varphi(x)$ which is positive on $\mathcal{X}$. The use of (2.1.18) implies that the evaluations of (2.1.16) are subject to random errors. Thus, to minimize (2.1.16) it is necessary to use a stochastic approximation type algorithm. The solution of the problem would be facilitated, if analogously to (2.1.18), one could use the Monte-Carlo estimators

$$\nabla_x \hat{f}(x,\beta) \approx \frac{1}{N}\sum_{j=1}^{N}\left[f(\xi_j)\,\nabla p(x-\beta\xi_j)\right]/\varphi(\xi_j)$$

of the gradient values of (2.1.16): here $\nabla p(z)$ is the gradient of the density $p(z)$.

Poor numerical and not sufficiently well-founded theoretical results concerning the above algorithms do not permit to draw a clear conclusion on their efficiency.

2.2 Set covering methods

In practice, the variation rate of an objective function is usually bounded. The set $\mathcal{F}=Lip(\mathcal{X},L,\rho)$ of Lipschitz functions represents a well-known example of a class of such functions. This set contains those functions which satisfy the inequality

$$|f(x)-f(z)|\leq L\,\rho(x,z)$$

for all $x, z\in\mathcal{X}$ (again, ρ is a fixed metric defined on $\mathcal{X}$).

Let the variation rate of the objective function f be known. The knowing its value $f(x_i)$ at a point $x_i\in\mathcal{X}$ a set

$$\mathcal{X}_i\subset\{x\in\mathcal{X}: f(x)\geq f(x_i)-\delta\} \tag{2.2.1}$$

can be determined where δ is a fixed positive number. If the points $x_1,...,x_N$ are chosen in $\mathcal{X}$ in such a way that the subsets $\mathcal{X}_1,..., \mathcal{X}_N$ form a covering of the set $\mathcal{X}$ (that is $\mathcal{X}\subset\bigcup_{i=1}^{N}\mathcal{X}_i$),
then the global minimization problem (1.1.1) is solved with accuracy δ, with respect to the function value. In this case, the inequality

$$f_N^*-\delta\leq f^* \tag{2.2.2}$$

hold, where

$$f_N^*=\min_{1\leq i\leq N} f(x_i), \qquad f^*=\min f. \tag{2.2.3}$$

Methods of selecting points $x_1,...,x_N$, having the above property are called covering methods and are the main subject of this section.

A covering method may be described in terms of either subsets $\mathcal{X}_1,..., \mathcal{X}_N$ or points $x_1,...,x_N$. In the former case one should try to construct the sets $\mathcal{X}_i$ of a maximal volume and to reach a simple structure of the sets

$$\mathcal{X}\setminus\bigcup_{i=1}^{k}\mathcal{X}_i$$

for all $k=1,2,...$. In the latter case more formal methods to analyse the quality of point collections $\Xi_N=\{x_1,...,x_N\}$ or sequences $\{x_1,x_2,...\}$ are generally used.

If the way of choosing points $x_i\in\Xi_N$ or sets $\mathcal{X}_i$ $(i=1,...,N)$ does not depend on the values which the objective function takes at the points $x_j\in\Xi_N$ $(j\neq i)$ then the point set $\Xi_N=\{x_1,...,x_N\}$ is called a grid and the corresponding minimization algorithm is called a grid (or passive covering) algorithm: these will be considered in Section 2.2.1.

Section 2.2.2 describes sequential (active) covering algorithms in which all the previously chosen points $x_1,...,x_{k-1}$ and function values $f(x_1),..., f(x_{k-1})$ may be used when choosing the next point x_k $(k=2,3,...)$.

Section 2.2.3 deals with the problem of the global optimization algorithms optimality and of the practical usefulness of optimal algorithms.

2.2.1 *Grid algorithms (Passive coverings)*

A grid is a point set $\Xi_N = \{x_1,...,x_N\}$ constructed independently of the function f values $f(x_i)$, $i=1,...,N$. A (passive) grid algorithm is a global minimization method consisting of constructing a grid Ξ_N, computing the values $f(x_i)$ for all $x_i \in \Xi_N$ and choosing the point

$$x_N = \arg\min_{x_i \in \Xi_N} f(x_i) \qquad (2.2.4)$$

as an approximation of a global minimizer $x^*=\arg\min f$ (and the value

$$f^*_N = f(x^*_N)$$

as an approximation of the minimal value $f^*=\min f$). It is also possible to add a local descent routine with the initial point x_N^* or to construct an approximation of f with the use of the global minimizer of the approximation instead of (2.2.4).

The grid algorithms are well-known in the global optimization theory: a number of theoretical studies have been devoted to them. This is the consequence of the following attributes of the grid algorithms:

a) simplicity of construction,
b) optimality in some well-defined sense,
c) simplicity of investigation,
d) simplicity of realization on a multiprocessor computer.

The attribute a) is relative: this section shows that only a few grids for simply structured set $\mathcal{X}$ (usually for cubes or hyperrectangles only) are constructed simply. But even if a grid is not simply constructed, the corresponding grid algorithm still remains simple and only one of its stages is difficult for realization. Let us note that once this stage is done, it can be used repeatedly, if one solves a multicriterion optimization problem or several global optimization problems for the same set $\mathcal{X}$.

The attribute b) is studied in Section 2.2.3 in detail. It turns out that for some types of functional classes $\mathcal{F}$ there exist grid algorithms that are optimal, typically in the minimax (worst case) sense. But from the practical point of view, this optimality property gives almost nothing: the reason being that the objective functions appearing in practice are usually not like the functions at which optimal algorithms perform best. Nonoptimal algorithms hence may be much better, than optimal ones, for most other functions from a given functional class $\mathcal{F}$.

The attributes c) and d) follow by the fact that the location of the grid points is independent of the function f values. The attribute d) is evident and c) is discussed in this section and Section 2.2.3.

Grid algorithms may be characterized in many different ways. One of the most important is the formation of a cover of $\mathcal{X}$ by the balls

$$B(x_i,\varepsilon,\rho)=\{x\in\mathcal{X}: \rho(x,x_i)\le\varepsilon\},\quad i=1,\dots,N, \tag{2.2.5}$$

where $x_i\in\Xi_N$, ρ is a metric on $\mathcal{X}$, ε is a radius of the balls. If the sets (2.2.5) form a cover, i.e.

$$\mathcal{X}\subset\bigcup_{i=1}^{N}B(x_i,\varepsilon,\rho) \tag{2.2.6}$$

then any point from $\mathcal{X}$ (and the global minimizer x^* too) belongs to at least one ball from the collection (2.2.5). If $f\in Lip(\mathcal{X},L,\rho)$ then, under fulfilling (2.2.6), the minimization problem is solved with the accuracy $\delta=L\varepsilon$ with respect to values of f, since

$$\left|f_N^*-f^*\right|=\left|f(x_N^*)-f(x^*)\right|\le\left|f(x_j)-f(x^*)\right|\le\sup_{x\in B(x_j,\varepsilon,\rho)}\left|f(x_j)-f(x)\right|=L\varepsilon$$

where x_j is the centre of the ball from the collection (2.2.5) which the global minimizer x^* belongs to. Similar results hold also for some other functional classes $\mathcal{F}$ (see Evtushenko (1985), p. 473).

Thus, the grid algorithms have some theoretically attractive properties; on the other hand, they have a basic disadvantage, too. Namely, they completely neglect the information on the objective function that is obtained during the search process. This disadvantage makes the grid algorithms utterly inefficient, especially for large-dimensional and complicated optimization problems. In practice, grid algorithms are efficiently used only for searching the global extremum for several objective functions given on the same set $\mathcal{X}$ (in particular, in multicriteria optimization, see Sobol and Statnikov (1981)); or as parts of algorithms using random grids or deterministic ones constructed similarly to the global random search methods of Chapter 5, see Galperin (1988), Niederreiter and Peart (1986), Galperin and Zheng (1987). Nevertheless the theoretical significance of the grid algorithms is non-negligible: this is caused by their minimax optimality, simplicity and use as a pattern in comparative studies of global optimization algorithms.

A degree of uniform distribution in a certain metric is a quality criterion of a grid. Let us introduce the appropriate notions.

Let μ be a finite measure on the feasible region $\mathcal{X}$ ($\mathcal{X}$ is considered as a measurable space). A point sequence

$$\{x_k\}_{k=1}^{\infty}$$

is called uniformly distributed in measure μ in the set $\mathcal{X}$, if $x_k\in\mathcal{X}$ for all $k=1,2,\dots$ and the asymptotic relationship

$$\lim_{N\to\infty} S_N(A)/N = \mu(A)/\mu(\mathcal{X}) \qquad (2.2.7)$$

is valid for any measurable subset $A \subset \mathcal{X}$, where $S_N(A)$ is the number of points x_k $(1 \le k \le N)$ belonging to A. If $\mu = \mu_n$ is the Lebesgue measure and (2.2.7) is valid, then the sequence $\{x_k\}$ is called uniformly distributed in $\mathcal{X}$. A grid Ξ_N is called uniform (in measure μ), if it contains the first N points of a uniformly distributed sequence (in measure μ).

Uniformity is an asymptotic property: degrees of grid uniformity are usually expressed in terms of dispersion and discrepancy defined as follows.

Let ρ be a metric on $\mathcal{X}$. The value

$$d_\rho(\Xi_N) = \sup_{x \in \mathcal{X}} \min_{x_i \in \Xi_N} \rho(x, x_i) \qquad (2.2.8)$$

is called the ρ-dispersion of a grid $\Xi_N = \{x_1, \ldots, x_N\}$. If ρ is the Euclidean metric then (2.2.8) is called dispersion and is denoted by $d(\Xi_N)$.

The dispersion is one of most generally used characteristics of grid uniformity. If the unit ball in metric ρ

$$\{x \in \mathbb{R}^n : \rho(0,x) \le 1\} \qquad (2.2.9)$$

is symmetrical with respect to change of variable order, then the ρ-dispersion is a characteristic of the uniformity of grids.

The above mentioned property of the unit ball symmetricity is valid e.g. for the Euclidean metric

$$\rho(x,z) = \left[\sum_{i=1}^{n} (x(i) - z(i))^2\right]^{1/2} \qquad (2.2.10)$$

and, more generally, for all ℓ_p-metrics

$$\rho(x,z) = \sum_{i=1}^{n} |x(i) - z(i)|, \qquad (p = 1) \qquad (2.2.11)$$

$$\rho(x,z) = \left[\sum_{i=1}^{n} |x(i) - z(i)|^p\right]^{1/p}, \qquad 1 < p < \infty \qquad (2.2.12)$$

$$\rho(x,z) = \max_{1 \le i \le n} |x(i) - z(i)| \qquad (p = \infty). \qquad (2.2.13)$$

(The last metric is also called cubic.) However, the above property is not valid e.g. for the metric

$$\rho(x,z) = \sum_{i=1}^{n} a_i |x(i) - z(i)| \tag{2.2.14}$$

where the numbers $a_1,...,a_n$ are nonnegative and not all equal to a fixed number. In formulae (2.2.10) through (2.2.14) the values x(i) and z(i) for i=1,...,n are the coordinates of points $x,z\in\mathfrak{R}^n$.

The importance of ρ-dispersion as a characteristic of a grid global optimization algorithm is explained by the inequality

$$\left|f_N^* - f^*\right| \leq \mathcal{L} d_\rho(\Xi_N) \tag{2.2.15}$$

which is valid for any function $f\in \mathcal{L}ip(\mathcal{X},\mathcal{L},\rho)$. Generally, the inaccuracy of a grid algorithm may be estimated with the help of ρ-dispersion and the modulus of continuity

$$\omega_\rho(t) = \sup_{\substack{x,y\in \mathcal{X} \\ \rho(x,y)<t}} |f(x) - f(y)| \tag{2.2.16}$$

of function f in metric ρ. Indeed, simple calculations of Niederreiter (1986) lead to the inequality

$$\left|f_N^* - f^*\right| \leq \omega_\rho(d_\rho(\Xi_N)). \tag{2.2.17}$$

Inequality (2.2.15) is a special case of (2.2.17), since $\omega_\rho(t)\leq\mathcal{L}t$ for any $f\in \mathcal{L}ip(\mathcal{X},\mathcal{L},\rho)$. In this connection it should also be noted that if prior information about the location of the optimal points is not available then one should prefer uniform grids to non-uniform ones.

If $\mathcal{X}$ is the unit cube, that is

$$\mathcal{X} = [0,1]^n , \tag{2.2.18}$$

then the value

$$D_N(\Xi_N) = \sup_B \left|S_N(B)/N - \mu_n(B)\right| \tag{2.2.19}$$

where the supremum is taken over the set of all hyperrectangles

$$B=[0,b_1]\times...\times[0,b_n] , \quad 0<b_j\leq 1 \quad (j=1,...,n)$$

is also a uniformity characteristic of a grid Ξ_N: this value is called discrepancy.

In case of (2.2.18), dispersion and discrepancy are closely related. In general, small discrepancy values correspond to small dispersion. Niederreiter (1983) proved that for any grid Ξ_N the inequalities

$$\frac{1}{2}\sqrt{n/e}\, N^{-1/n} \le d(\Xi_N) \le \sqrt{n}\left[D_N(\Xi_N)\right]^{1/n} \qquad (2.2.20)$$

are valid. This way, if the evaluation of the dispersion is practically impossible, then the inequalities (2.2.20) are used for estimating the rate of dispersion.

The next property of grids is important from the practical point of view. A grid is said to be composite if it keeps its features, when the number of points N is changed into N+1 (for each N). Many known grids do not posses this property.

Let us define the most popular uniform grids (their majority can be defined for the case (2.2.18) only). If the feasible set $\mathcal{X}$ is not a cube, then a uniform grid can be defined as follows: find a cube Y containing the set $\mathcal{X}$, construct a grid Ξ in Y and form a grid $\Xi\cap\mathcal{X}$ in $\mathcal{X}$.

A cubic grid Ξ_N^1 in $\mathcal{X}=[0,1]^n$ contains $N=p^n$ points

$$\left(\frac{i_1+1/2}{p}, \frac{i_2+1/2}{p}, \ldots, \frac{i_n+1/2}{p}\right), \qquad i_1, i_2, \ldots, i_n \in \{0,1,\ldots,p-1\},$$

where p is a fixed natural number. To construct such a grid, one may divide the cube $\mathcal{X}$ into p^n equal subcubes by dividing every side of $\mathcal{X}$ into p equal parts and choose the grid points as the centres of the subcubes.

A rectangular grid Ξ_N^2 for the case of the hyperrectangle

$$\mathcal{X} = [a_1,b_1]\times\ldots\times[a_n,b_n]$$

is constructed by dividing every side $[a_i,b_i]$ of $\mathcal{X}$ into p_i parts of lengths $\ell_i=(b_i-a_i)/p_i$ and choosing the points

$$\left(a_1+\ell_1(i_1+1/2), a_2+\ell_2(i_2+1/2), \ldots, a_n+\ell_n(i_n+1/2)\right) \in \mathcal{X}$$

where $i_k \in \{0,1,\ldots,p_k-1\}$ for each $k=1,\ldots,n$, $N=p_1,\ldots,p_n$. Of course, all cubic grids are also rectangular.

A rectangular grid Ξ_N^2 is uniform in the case $p_k=\text{const}(b_k-a_k)$, $k=1,\ldots,n$, only. In this context it should be mentioned that the uniformity property is not invariant to scale transformations. For example, if a cube is transformed into a (non-cube) hyperrectangle, then a uniform grid in the cube will induce a nonuniform grid in the hyperrectangle.

Cubic and rectangular grids are the simplest. They have some optimality properties (as it will be shown below), but they are not composite.

We call a grid random, if it consists of N independent realizations of a random vector uniformly distributed on $\mathcal{X}$. Random grids Ξ_N^3 are simple to construct (not only for the case (2.2.18) but also for many other types of sets), being also uniform and composite; on the other hand their uniformity characteristics (ρ-dispersion and discrepancy) are far from optimal.

Random grids are widespread in global optimization theory and in practice: there are two reasons for this. First, if $\mathcal{X}$ is neither a cube nor a hyperrectangle, then tremendous difficulties may be faced during the construction of grids having good uniformity characteristics. Second, if the values of f in random points are known, then one may use mathematical statistics procedures to obtain information on the function f and the location of its extremal points.

Grids Ξ_N^i (i=4,5,6) described below are defined on the unit cube $\mathcal{X}=[0,1]^n$ and are called quasirandom grids; their elements being called quasirandom points. This nomenclature refers to the application of these points in a lot of algorithms similar to Monte-Carlo methods that use random points. The uniformity of quasirandom grids is better, than that of random ones, but slightly worse than optimal. From the practical point of view, quasirandom grids are preferable to the optimal ones (including cubic grids): the reasons for this preference will be given in Section 2.2.3.

The Hammersley-Holton grids Ξ_N^4 form an important class of quasirandom grids: they consist of the N first terms of the Holton sequence that is defined as follows. For integers $\eta \geq 2$ and $k \geq 1$ let

$$k = \sum_{i=0}^{\ell} a_i \eta^i$$

be the η-adic expansion of k, the function φ_η be defined by

$$\varphi_\eta(k) = \sum_{i=0}^{\ell} a_i \eta^{-i-1} \quad ,$$

and $p_1,\dots,p_n$ be n distinct prime numbers. Then the i-th term x_i of a Holton sequence is

$$x_i = (\varphi_{p_1}(i),\dots,\varphi_{p_n}(i)), \quad i = 1,2,\dots$$

Lattice grids Ξ_N^5 consist of the points

$$x_i = (\{a_1 i/N\},\{a_2 i/N\},\dots,\{a_n i/N\}), \quad i = 1,\dots,N$$

where $\{.\}$ denotes the fraction part of a number, and $a_1,\dots,a_n$ are to be suitably chosen from tables (see Korobov (1963), Hua and Wang (1981)). Lattice grids are rather popular in applied mathematics, note that they are not composite.

The computation of binary Π_τ-grids and η-adic Π_o-grids Ξ_N^6 is more complicated, see Sobol (1969, 1985), Faure (1982). The η-adic Π_τ-grids Ξ_N^6 are composite and have the additional uniformity property. If the number N of the grid points is fixed then the grids Ξ_N^4 and Ξ_N^6 are sometimes modified so as i/N is substituted for the last n's coordinate of the grid points x_i. This gives the grids that are not composite but possess a little bit better uniformity characteristics, see Niederreiter (1978), Faure (1982).

An intermediate place between the random and Π_τ-grids is occupied by stratified sample grids $\Xi_N{}^7$. These grids contain the elements of a stratified sample in $\mathcal{X}$. To obtain a grid $\Xi_N{}^7$ for $N=m\ell$ (where m and ℓ are natural numbers), any measurable set $\mathcal{X}$, and any finite measure μ on $\mathcal{X}$, one must divide $\mathcal{X}$ into m parts $\mathcal{X}_i$ $(i=1,...,m)$ of equal measure

$$\mu(\mathcal{X}_i)=\mu(\mathcal{X})/m,\ \bigcup_{i=1}^{m}\mathcal{X}_i=\mathcal{X},\quad \mu(\mathcal{X}_i\cap\mathcal{X}_j)=0,\ i\neq j,$$

and sample ℓ times each probability distribution P_i $(i=1,...,m)$ determined by

$$P_i(A)=m\mu(A\cap\mathcal{X}_i)/\mu(\mathcal{X})$$

for measurable subsets $A\subset\mathcal{X}$.

Stratified sample grids are not composite but may be constructed for a wide class of sets $\mathcal{X}$ and surpass the random grids by some characteristics that are important for global optimization, see Section 4.4.

An important place in global optimization theory is occupied by the grids $\Xi_N{}^8$ consisting of the centres of those balls $B(x_i,\varepsilon,\rho)$ that form a minimal cover of the set $\mathcal{X}$. Here $\varepsilon>0$ is a fixed number, ρ is a metric and N is equal to the minimal number of balls forming a cover of $\mathcal{X}$. The grids $\Xi_N{}^8$ are not composite and it is usually very hard to construct them. The exception is the case of $N=p^n$, and the fulfilment of relations (2.2.13) and (2.2.18) for which the grids $\Xi_N{}^1$ and $\Xi_N{}^8$ coincide. The theoretical importance of grids $\Xi_N{}^8$ is connected with the optimality (in case $\mathcal{F}=\mathcal{L}ip(\mathcal{X},\mathcal{L},\rho)$) of the grid algorithms using them, see Section 2.2.3.

Some other optimal and nearly optimal (for other functional classes $\mathcal{F}$) grid optimization algorithms are known, see Ganshin (1979), Ivanov et al. (1985), Nefedov (1987), Zaliznjak and Ligun (1978). These grids are usually rather difficult to construct thus being not very suitable for practical purposes.

Let us cite here statements on the dispersion and discrepancy of the cubic, random and quasirandom grids defined on the unit cube $\mathcal{X}=[0,1]^n$: Proofs of these statements can be found in Niederreiter (1978) and Sobol (1969).

For cubic grids $\Xi_N{}^1$ one has

$$\mathbf{D}_N(\Xi_N^1)=\tfrac{1}{2}N^{-1/n},\qquad d(\Xi_N^1)=\frac{\sqrt{n}}{2}N^{-1/n}. \tag{2.2.21}$$

For random grids $\Xi_N{}^3$ with any probability (less than one)

$$\mathbf{D}_N(\Xi_N^3)=O(N^{-1/2}),\qquad d(\Xi_N^3)=O(N^{-1/2n}),\ N\to\infty. \tag{2.2.22}$$

For quasirandom grids Ξ_N^i, i=4,5,6

$$\mathcal{D}_N(\Xi_N^i) = O(N^{-1}\log^n N), \quad d(\Xi_N^i) = O(N^{-1/n}\log N), \quad N \to \infty. \qquad (2.2.23)$$

Besides, for η-adic Π_τ-grids Ξ_N^6 with $N_j = \eta^j$

$$\mathcal{D}_{N_j}(\Xi_{N_j}^6) = O(N_j^{-1}\log^{n-1} N_j), \quad j \to \infty.$$

The general term in the asymptotic expression (2.2.23) has the form $B(n)N^{-1}\log^n N$ where

$$\log B(n) = O(n \log n), \quad n \to \infty$$

for grids Ξ_N^4;

$$\log B(n) = O(n \log\log n), \quad n \to \infty$$

for binary Π_τ-grids and optimal values of $\tau = \tau(n)$; finally,

$$B(n) \to 0, \quad n \to \infty$$

for η-adic Π_0-grids with minimal values of $\eta = \eta(n)$.

The above mentioned expressions lead to the following conclusions. Under the criterion $\mathcal{D}_N$ the best grid from the above mentioned is the η-adic Π_0-grid, while the other quasirandom grids are also good. The cubic grid Ξ_N^1 is optimal for n=1, but is poor for large dimensions; it is worse than the random grid for $n \geq 3$. At the same time, for any dimension n the order of convergence of $\{d(\Xi_N)\}$ to zero for $n \to \infty$ is optimal for cubic grids and equals $N^{-1/n}$. For quasirandom grids this order is $N^{-1/n}\log N$, $N \to \infty$, i.e. it is almost optimal.

The dispersion criterion d is judged as more significant than discrepancy in global optimization theory. Numerical and theoretical studies have shown, however, that the goodness of cubic grids for global optimization algorithms is questionable. For example, Section 2.2.3 demonstrates that the dispersion criterion must not be the unique optimality criterion applied to grids.

2.2.2 *Sequential covering methods*

The grid algorithms described above are the simplest covering methods. Their covering strategy is independent of the computed values of f, i.e. these algorithms are passive. On the contrary, sequential (active) covering methods use the information obtained during the search: these algorithms are of great theoretical and practical interest.

Let us consider first the main idea of covering methods supposing that they designed for optimizing an objective function with a fixed accuracy $\delta>0$, with respect to its values.

Let the function f be evaluated at points $x_1,...,x_k$. We shall call the value

$$f_k^* = \min\{f(x_1),\dots,f(x_k)\}$$

a record and a point x_k^* with the function value $f(x_k^*) = f_k^*$ a record point. Let us define the set

$$Z_k = \left\{x \in X: f_k^* - \delta \leq f(x)\right\}. \tag{2.2.24}$$

Obviously

$$f_k^* - \delta \leq \inf_{x \in Z_k} f(x).$$

Points of Z_k are not of interest for further search since the record f_k^* in the set Z_k can be improved not more than by δ. Consequently, the search may be continued on the set $X \setminus Z_k$ only. In particular, if

$$X \subset Z_k \tag{2.2.25}$$

then the initial problem is approximately solved, since $f_k^* - f^* \leq \delta$, i.e. the record point x_k^* can be taken as an approximation for x^*.

Thus, the construction of a covering method is reduced to the construction of a point sequence $\{x_k\}$ and a corresponding set sequence $\{Z_k\}$ until the condition (2.2.25) is fulfilled for some k.

Let us note that the record sequence $\{f_k^*\}$ is decreasing and the set sequence $\{Z_k\}$ is increasing, i.e.

$$f_{k+1}^* \leq f_k^* \quad \text{and} \quad Z_k \subseteq Z_{k+1} \quad \text{for all } k = 1,2,\dots \tag{2.2.26}$$

Let us note also that the efficiency of covering methods depends significantly on the closeness of records f_k^* and the optimal value f^*, since the size of the sets Z_k crucially depends on this values. Consequently, to improve the efficiency of a covering method one should use a local minimization technique just after obtaining a new record point. It gives us an opportunity to decrease the incumbent records and, hence, to increase the size of the sets Z_k.

Let us consider the covering method for the case $\mathcal{F}=\mathcal{L}ip(\mathcal{X},\mathcal{L},\rho)$; methods for functions which have a gradient satisfying the Lipschitz condition are constructed analogously, see Evtushenko (1985), p. 472.

In the case $\mathcal{F}=\mathcal{L}ip(\mathcal{X},\mathcal{L},\rho)$ for every $x,y\in\mathcal{X}$ one has

$$f(y)-\mathcal{L}\rho(x,y)\le f(x). \tag{2.2.27}$$

Hence, the inequality

$$f_k^*-\delta\le f(x) \tag{2.2.28}$$

is valid for any $x\in\mathcal{X}$ satisfying

$$f_k^*-\delta\le f(y)-\mathcal{L}\rho(x,y)$$

for some $y\in\mathcal{X}$. Hence, it follows that the inequality (2.2.28) holds for any $x\in B(x_j,\eta_{jk},\rho)$ where $1\le j\le k$, $\eta_{jk}=(f(x_j)-f_k^*+\delta)/\mathcal{L}$. Thus, we may take

$$\mathcal{Z}_k=\bigcup_{j=1}^{k}B(x_j,\eta_{jk},\rho)=\left\{x\in\mathcal{X}:\max_{1\le i\le k}\left[f(x_i)-\mathcal{L}\rho(x,x_i)\right]\ge f_k^*-\delta\right\}. \tag{2.2.29}$$

For all $k\ge1$, a new point x_{k+1} can be obtained in different ways. For instance, to choose new points one may use any grid point. Devroye (1978) proposed random grids for the above aims, see later Algorithm 3.1.4.

In the class of covering methods based on (2.2.29) the method proposed by Piyavskii (1967, 1972) and, independently, by Shubert (1972) is the most popular. This method selects the point

$$x_{k+1}=\arg\min_x\left[\max_{1\le i\le k}\left[f(x_i)-\mathcal{L}\rho(x,x_i)\right]\right] \tag{2.2.30}$$

that is a minimizer of the Lipschitzian minorant

$$f^{(k)}(x)=\max_{1\le i\le k}\left[f(x_i)-\mathcal{L}\rho(x,x_i)\right] \tag{2.2.31}$$

of the objective function f at each k-th iteration. Minimization in (2.2.30) is done either on $\mathcal{X}$ or on $\mathcal{X}\setminus\mathcal{Z}_k$.

Following the Russian literature, we shall call the method (2.2.30) a polygonal line method. This nomenclature originates from the fact that (2.2.31) is a polygonal line, in the case when $n=1$ and ρ is the Euclidean metric. The deficiency of the polygonal line method lies in the complexity of the auxiliary extremal problems (2.2.30). Wood (1985) described a constructive (but cumbersome) multidimensional variant of the method for the Euclidean metric case. Lbov (1972) proposed to solve the auxiliary extremal problems by the simplest global random search algorithm. Lbov's variant of the polygonal line method is itself a global search algorithm and similar to the Devroye (1978) algorithm mentioned above. A scheme generalizing the polygonal line method was developed also by Meewella and Mayne (1988); related but different approaches will be treated in Section 2.3.

To simplify the basic covering method, one may use some subsets $\mathcal{X}_k$ of $\mathcal{Z}_k$ instead of $\mathcal{Z}_k$. An algorithm using hyperrectangles as subsets was proposed and studied by Evtushenko (1971, 1985). The one-dimensional variant of the Evtushenko algorithm has the form

$$x_1 = a, \qquad x_{k+1} = x_k + (f(x_k) - f_k^* + 2\delta)/\mathcal{L}. \tag{2.2.32}$$

If $x_k > b$ then the iteration (2.2.32) is terminated. Here $\mathcal{X}=[a,b]$ and the subsets $\mathcal{X}\backslash\mathcal{Z}_k$ are intervals. The most unfavourable case for algorithm (2.2.32) is the case of a decreasing function f in which points (2.2.32) are at equal distances.

Brent (1973) proposed a similar algorithm for the one-dimensional functional class $\mathcal{F}=\{f: |f''(x)| \leq M\}$ where M is to be known. There are other similar one-dimensional algorithms in Beresovsky and Ludvichenko (1984), Vasil'ev (1988).

Evtushenko (1985) constructed a multidimensional variant of his algorithm (2.2.32). The sets $\mathcal{X}\backslash\mathcal{Z}_k$ in it are hyperrectangles and the sets $\mathcal{X}_k$ are unions of cubes with centres at points being obtained by a one-dimensional global optimization algorithm. Evtushenko's algorithm is cumbersome but it does not require great computer memory or complicated auxiliary computations. The numerical study of the algorithm indicates that for dimensions $n>3$ it requires an exhaustive number of evaluations of f, and thus is inefficient.

There are strong objections against covering methods to be applied directly in the multidimensional case. First, the methods are cumbersome and complicated to realize. Second, their efficiency considerably depends on the prior information about f, i.e. on the choice of the functional class $\mathcal{F}$. In the most important case $\mathcal{F}=Lip(\mathcal{X},\mathcal{L},\rho)$, the functional class $\mathcal{F}$ is determined by a Lipschitz constant $\mathcal{L}$ and the metric ρ. The inclusion $f \in Lip(\mathcal{X},\mathcal{L},\rho)$ for some constant $\mathcal{L}$ and some metric ρ is usually a plausible conjecture, but the explicit (minimal) value of the constant $\mathcal{L}$ is generally unknown, depending also on the choice of $\mathcal{X}$, the metric ρ and variable scales. Section 2.2.3 will demonstrates that the unfortunate choice of variable scales leads to inefficiency of optimal algorithms (even for known Lipschitz constant $\mathcal{L}$). The same criticism is valid for any covering method.

Third, the number of objective function evaluations is excessive in multidimensional case. Let us analyse this number for the case $\mathcal{F}=Lip(\mathcal{X},\mathcal{L})$. Let $\mathcal{X}$ be a ball with radius η. The volume of $\mathcal{X}$ is

$$\mu_n(\mathcal{X}) = \eta^n \pi^{n/2} / \Gamma(n/2+1)$$

where Γ denotes the gamma-function. Let $x_1,...x_N$ be the points of function evaluations and $M=\max f$. One may only guarantee that $f(x)>f^*$ for $x\in X$, if f has been evaluated at a point x_i from the ball $B(x,(f(x_i)-f^*)/\mathcal{L})$. Hence the balls

$$B(x_i,(f(x_i)-f^*)/\mathcal{L}),\quad i=1,...,N,$$

must cover X to assure that the global minimizer has not been overlooked. The joint volume of these N balls is smaller than

$$N\left(\frac{M-f^*}{\mathcal{L}}\right)^n \pi^{n/2}/\Gamma(n/2+1).$$

Thus, for these N balls to cover X we require the fulfillment of

$$N > (\eta/(M-f^*))^n\mathcal{L}^n.$$

If the derivatives of f in the direction of the global minimizer do not equal $(-\mathcal{L})$ everywhere, then $\mathcal{L}>(M-f^*)/\eta$: this way the computational effort required increases exponentially with n.

Covering methods are theoretically built for the case $\mathcal{F}=Lip(X,\mathcal{L},\rho)$ with known value of the Lipschitz constant $\mathcal{L}$. In practical optimization problems, however, this constant is usually unknown. This deficiency is not crucial, since the Lipschitz constant may be estimated during the search. While estimating the Lipschitz constant, one has to bear in mind that if it is underestimated then the method will not be reliable and if it is overestimated, then the amount of computational work needed to achieve a fixed accuracy will exponentially increase.

Let $f\in Lip(X,\mathcal{L},\rho)$ be evaluated at k points $x_1,...x_k$. Then

$$\mathcal{L}_k=\eta_k \max_{1\le i<j\le k}\left\{\left|f(x_i)-f(x_j)\right|/\rho(x_i,x_j)\right\} \qquad (2.2.33)$$

is a usual estimator of the Lipschitz constant $\mathcal{L}$. Here $\{\eta_k\}$ is a nonincreasing sequence of numbers which exceeds 1 and are heuristically chosen. In case of $\eta_k=1$, the values (2.2.33) generally underestimate the constant $\mathcal{L}$.

In the one-dimensional case the Lipschitz constant estimator is built more easily. To provide it, one has to renumber the points $x_1,...x_k$ in their increasing order and then compute

$$\mathcal{L}_k=\eta_k \max_{2\le j\le k}\left\{\left|f(x_j)-f(x_{j-1})\right|/\rho(x_j,x_{j-1})\right\}. \qquad (2.2.34)$$

2.2.3 *Optimality of global minimization algorithms*

If the dimension n of the set $\mathcal{X}$ or the required solution accuracy increases, then the number of objective function evaluations will rapidly increase for every grid algorithm. For example, for the cubic grid algorithm we have

$$N \geq (\sqrt{n}/2\varepsilon)^n$$

where ε is the accuracy in the argument values, in Euclidean metric. Numerical results for the covering methods described above prove that for n>3 all of them have poor efficiency. In this connection the question of best possible covering methods is of (mainly theoretical) interest.

Let the number of steps (i.e. evaluations of the objective function f) be *a priori* bounded by a number N. Every deterministic global minimization algorithm $d^N=(d_1,...,d_{N+1})$ is determined by mappings

$$d_k: (\mathcal{X} \times \mathcal{R})^{k-1} \to \mathcal{X}, \qquad k = 1,\ldots,N+1.$$

That is, first a point x_1 is chosen then the points x_k (k=2,...,N) can depend on all preceeding arguments and corresponding function values: $x_k=d_k(\Sigma_k)$, where

$$\Sigma_k = \{x_1, f(x_1), \ldots, x_{k-1}, f(x_{k-1})\}.$$

The estimator of an optimizer x^* is the point $x_{N+1}=d_{N+1}(\Sigma_{N+1})$. It is usually determined by $x_{N+1}=x_N^*$ and its inaccuracy defined by

$$\varepsilon(f,d^N) = f(x_{N+1}) - f^*$$

is taken as a measure of inaccuracy for method d^N. The inaccuracy of a method d^N on a certain class $\mathcal{F}$ is usually defined by

$$\varepsilon(d^N) = \sup_{f \in \mathcal{F}} \varepsilon(f,d^N) \qquad (2.2.35)$$

corresponding to the minimax approach for measuring the efficiency of algorithms. (Let us remark right here that other approaches exist, too: e.g. the Bayesian efficiency measure may lead to a class of stochastic algorithms, as will be described in Section 2.4.

Let $\mathcal{D}(N)$ be the set of all global minimization algorithms in which the number of steps does not exceed N and $\Delta(N)$ be the set of all N-point grid algorithms in which $x_{N+1}=x_N^*$. It is clear that $\Delta(N) \subset \mathcal{D}(N)$. An algorithm d_*^N is said to be optimal (in minimax sense), if

$$d_*^N = \arg \min_{d^N \in \mathfrak{D}(N)} \varepsilon(d^N). \tag{2.2.36}$$

A grid algorithm $d_{(*)}^N$ is said to be the optimal grid algorithm, if

$$d_{(*)}^N = \arg \min_{d^N \in \Delta(N)} \varepsilon(d^N). \tag{2.2.37}$$

An algorithm $d^N \in \mathfrak{D}(N)$ is said to be asymptotically optimal if

$$\varepsilon(d^N) \sim \varepsilon(d_*^N) \quad \text{for } N \to \infty, \text{ i.e.}$$

$$\lim_{N\to\infty} \varepsilon(d^N)/\varepsilon(d_*^N) = 1.$$

An algorithm d^N is said to be optimal in order, if

$$\varepsilon(d^N) \asymp \varepsilon(d_*^N) \qquad \text{for } N \to \infty, \text{ i.e. } \varepsilon(d^N)/\varepsilon(d_*^N)$$

does not exceed a constant for all N.

There exists another way of optimality definition of global optimization algorithms. Its essence is in fixing a bound ε_o for the inaccuracy (2.2.35) and then minimizing the number of steps N in a set of algorithms d^N satisfying $\varepsilon(d^N) \leq \varepsilon_o$. This approach makes possible to introduce definitions similar to those considered above.

Let us assume now that $\mathfrak{F} = Lip(\mathfrak{X}, \mathfrak{L}, \rho)$. Then Sukharev (1971, 1975) proved that an optimal passive algorithm (see (2.2.37)) is the grid algorithm built by the grid $\Xi_N{}^8$ consisting of the centres of balls $B(x_i, \varepsilon, \rho)$ that form a minimal cover of $\mathfrak{X}$. For this algorithm, the guaranteed accuracy

$$\min_{d^N \in \Delta(N)} \varepsilon(d^N)$$

is equal to the radius of the balls (forming the optimal cover of $\mathfrak{X}$) multiplied by $\mathfrak{L}$.

In the mentioned works it is also proved that the optimal passive algorithms (2.2.37) are also optimal - in minimax sense - among the sequentiel ones (2.2.36), that is

$$\min_{d^N \in \Delta(N)} \max_{f \in \mathfrak{F}} \varepsilon(f, d^N) = \min_{d^N \in \mathfrak{D}(N)} \max_{f \in \mathfrak{F}} \varepsilon(f, d^N) \tag{2.2.38}$$

(The maximum on $\mathfrak{F}$ is attained at a *saw-tooth* function whose values at all points $x_1,...,x_N$ are equal, details can be found in the papers cited.)

In terms of game theory, the above result may be explained as follows. A researcher chooses a minimization algorithm which is known to his *enemy* (say, to *nature* or to an *oracle*). The latter selects then the most unsuitable function for the algorithm in $\mathfrak{F}$. This is the *saw-tooth* function mentioned, having a maximal rate of variation and equal values at points generated by the algorithm. This way the *enemy* eliminates the possibility of collecting valuable information concerning the objective function. So, passive algorithms are not worse (in the worst case sense), than sequential ones.

The supposition of choosing the objective function by an *enemy* who knows the optimization algorithm is, of course, doubtful in practical optimization problems. The conclusion obtained (concerning passive algorithm optimality) causes additional doubts concerning the adequacy of the minimax approach. Note in this context that this approach is of great interest for uniextremal or convex functions, see Kiefer (1953), Chernousko (1960); for multiextremal problems however, this approach calls for additional problem specifications.

Sukharev (1975) also studied a different but related concept of best algorithms. To construct these algorithms, he supposed that *nature* may deviate from its optimal strategy: this concept leads to a certain sequential algorithm. It is constructed as follows: let the number N of objective function evaluations be fixed and k evaluations have been already done at points $x_1,...,x_N$. Then the point x_{k+1} is chosen as the centre of a ball coming into the optimal cover of the set $\mathfrak{X}\backslash\mathfrak{X}_k$, where

$$\mathfrak{X}_k = \bigcup_{i=1}^{k} B(x_i, \eta + (f(x_i) - f_k^*)/\mathfrak{L}, \rho)$$

and η is the fixed lower bound of the radii of balls which form the optimal N-balls cover. For the worst function the points $x_1,...,x_N$ generated by this type of algorithm coincide with the optimal grid algorithm points. For the other functions, the accuracy of the sequentially best algorithm may be better. But the main problem of Sukharev's algorithm lies in its very complicated construction, since optimal covers ought to be built on every step of the algorithm.

If $N=p^n$, where p is a natural number, and the metric $\rho=\rho_0$ is cubic (i.e. (2.2.13) is fulfilled), then the cubic grid algorithm is the minimax optimal global minimization algorithm for the Lipschitz functional class $\mathfrak{F}=\mathfrak{F}_0=\mathfrak{L}ip(\mathfrak{X},\mathfrak{L},\rho_0)$. The same algorithm is also optimal for the more general functional class $\mathfrak{F}_s$, $s\geq 0$, consisting of functions whose derivatives up to order s are from $\mathfrak{F}_0$, see Ivanov (1971, 1972). In the latter case, the algorithm has a guaranteed accuracy of order $O(\mathfrak{L}/N^{(s+1)/n})$ for $N\to\infty$. Ivanov et al. (1985) described another approach for the construction of an optimal in order algorithm in $\mathfrak{F}_s$. It consists of minimizing a spline that is built using the values of f at uniformly chosen grid points.

As stated above, optimal (in the above sense) deterministic algorithms can be built under the supposition of knowing the minimization algorithm by an *enemy* . The *researcher* may make an attempt to increase his *gain* (that is, the guaranteed accuracy) by using a randomized strategy. In this case it is supposed that *nature* does not know the *researcher's* strategy, but knows its statistical characteristics. Sukharev (1971) considered passive randomized algorithms σ^N for the case $\mathcal{F}=\mathcal{F}_0$. They are determined by the joint probability distribution of N random vectors in $\mathcal{X}$. We denote these distributions by $\sigma^N(d\Xi_N)$. The mean accuracy of a randomized algorithm σ^N for a function f is the value

$$\varepsilon(f,\sigma^N) = \int_{\mathcal{X}^N} \varepsilon(f,\Xi_N)\sigma^N(d\Xi_N).$$

We call a randomized algorithm minimax-optimal, if it satisfies the relation

$$\sigma_*^N = \arg\min_{\sigma^N}\left[\max_{f\in\mathcal{F}_0} \varepsilon(f,\sigma^N)\right]$$

Sukharev (1971) proved that in the one-dimensional case, for $\mathcal{X}=[a,b]$, the optimal randomized algorithm σ_*^N is determined by a random choice of the grids

$$\Xi_{N,1} = \left\{0, \frac{2}{2N-1}, \ldots, \frac{2N-2}{2N-1}\right\} \text{ and } \Xi_{N,2} = \left\{\frac{1}{2N-1}, \frac{3}{2N-1}, \ldots, 1\right\}$$

with equal probabilities 0.5. This algorithm has the guaranteed mean accuracy

$$\max_{f\in\mathcal{F}} \varepsilon(f,\sigma^N) = \mathcal{L}/[2(2N-1)].$$

Note that this value is almost twice less, than the guaranteed accuracy $\mathcal{L}/2N$ of the optimal deterministic algorithm: the latter is the cubic grid algorithm for the grid

$$\Xi_N^1 = \left\{\frac{1}{2N}, \frac{3}{2N}, \ldots, \frac{2N-1}{2N}\right\}.$$

Sukharev (1971) also showed that for all $n\geq 1$, the inequality

$$\min_{\sigma^N} \max_{f\in\mathcal{F}_0} \varepsilon(f,\sigma^N) \geq \frac{n}{n+1} \min_{d^N\in\Delta(N)} \max_{f\in\mathcal{F}_0} \varepsilon(f,d^N)$$

is fulfilled: by this inequality it follows that if the dimension n of $\mathcal{X}$ increases, the improvement of the guaranteed accuracy will decrease.

The construction of optimal sequential randomized algorithms is interesting, but rather complicated. The one-step optimal randomized algorithm is known only in the one-dimensional case, see Sukharev (1981). The construction problem for this algorithm is reduced to that of the optimal strategy in a matrix game on the unit square.

More information concerning minimax-optimal algorithms can be found e.g. in Fedorov, ed. (1985), Ivanov (1972), Strongin (1978), Sukharev (1971, 1975, 1981) or Schoen (1982).

Let us note here that in spite of Shubert's (1972) assertion the polygonal line algorithm (2.2.30) is not one-step optimal (as easily seen already for N=2); following Sukharev (1981), a slight modification of the algorithm is one-step optimal. The sequential algorithms will be studied later.

The aim of further study in this section is to show that minimax-optimal global minimization algorithms may have very poor efficiency in realistic optimization problems. The results below question again the practical significance of the above and similar (minimax-type) optimality criteria and the optimal algorithms derived.

Let $\mathcal{X}=[0,1]^n$, $\mathcal{F}=\mathcal{L}ip(\mathcal{X},\mathcal{L})$ and let the objective function f depend on s variables ($s<n$) with coordinate indices $i_1,...,i_s$ ($1\le i_1< ... <i_s\le n$) but not on all n variables. Let us denote by $K=K(i_1,...,i_s)$ the corresponding s-dimensional cube, that is the s-face of the cube $\mathcal{X}$ in which the variables of indices $i_1,...,i_s$ vary from zero to one and all others are equal to zero.

A grid Ξ_N on $\mathcal{X}$ induces the grid $\Xi_N(s)$ on K consisting of projections of the grid Ξ_N points onto $\mathcal{X}$.

For the dispersion of a grid $\Xi_N(s)$ we have

$$\tfrac{1}{2}\sqrt{s/e}\,N^{-1/s}\le d(\Xi_N(s))\le\sqrt{s}\left[\mathfrak{D}_N(\Xi_N(s))\right]^{1/s} \qquad (2.2.39)$$

(instead of (2.2.20)) and the inequality

$$f_N^*-f^*\le\mathcal{L}\,d(\Xi_N(s)) \qquad (2.2.40)$$

replacing (2.2.15).

In fact the case of a function defined on an n-dimensional set, but depending only on s variables is a hardly probable case in practice. However, it is usual that an objective function depends on all n variables, but the degree of dependence is different: there is a group of essential variables that influences the function behaviour more intensively than the others. In other words, in complicated cases one may expect that the objective function f has the form $f=h+g$ where $h\gg g$ and h depends only on s variables ($s<n$). But in this case the accuracy of a grid-type minimization algorithm depends on values $d(\Xi_N(s))$ for different $s<n$, $i_1,...,i_s$ and not on the value $d(\Xi_N)$ only. Thus, the collection $\{d(\Xi_N(s))\}$ for different $s\le n$, $i_1,...,i_s$ is a natural vector criterion of a grid algorithm. Let us use this indicator to compare the cubic, random, and quasirandom grids.

If $N=p^n$ and $\Xi_N=\Xi_N^1$ is a cubic grid, the grid $\Xi_N^1(s)$ contains p^s distinct points only. This gives

$$d(\Xi_N^1(s))=\tfrac{1}{2}\sqrt{s}\,N^{-1/n}, \qquad (2.2.41)$$

see Sobol (1982). The rate of decrease of the values $d(\Xi_N(s))$ for $N\to\infty$ is not influenced by s. For small values s, this rate is much worse than the optimal rate $N^{-1/s}$.

At the same time, for every N the projections $\Xi_N(s)$ of random Ξ_N^3 and quasirandom Ξ_N^i (i=4,5,6) grids contain N distinct points. So, for random grids with any probability less than one there holds

$$\mathfrak{D}_N(\Xi_N^3(s)) = O(N^{-1/2}), \qquad N \to \infty,$$

and for quasirandom grids

$$\mathfrak{D}_N(\Xi_N^i(s)) = O(N^{-1}\log^s N), \qquad N \to \infty, \qquad i = 4,5,6.$$

These relations together with the right side of the inequality (2.2.39) imply that for random grids (with any probability less than one) we have

$$d(\Xi_N^3(s)) = O(N^{-1/2s}), \qquad N \to \infty, \tag{2.2.42}$$

and for quasirandom grids Ξ_N^i, i=4,5,6

$$d(\Xi_N^i(s)) = O(N^{-1/s}\log N), \qquad N \to \infty. \tag{2.2.43}$$

By (2.2.41) - (2.2.43) we arrive at the conclusion that the quasirandom grids qualitatively surpass the random and cubic ones by the criterion $d(\Xi_N(s))$ for all $s<n$ and the random grids surpass the cubic ones for $s<n/2$. The rate of decrease of $d(\Xi_N(s))$ for $N\to\infty$ is nearly optimal for quasirandom grids for all $s\leq n$.

Thus, for a functional subset $\mathfrak{F}\subset Lip(\mathfrak{X},\mathfrak{L})$ under consideration the cubic grids are worse (in the above sense) than quasirandom ones and even may be worse than random grids. Recall that the cubic grid minimization algorithm is optimal for the case $\mathfrak{F}=Lip(\mathfrak{X},\mathfrak{L},\rho_0)$ where ρ_0 is the cubic metric and it is optimal in order for $\mathfrak{F}=Lip(\mathfrak{X},\mathfrak{L})$.

We shall demonstrate now that a similar situation takes place when it is supposed that $f\in Lip(\mathfrak{X},\mathfrak{L},\rho)$, but in fact $f\in Lip(\mathfrak{X},\mathfrak{L},\rho_1)$ where ρ and ρ_1 are metrics for which $Lip(\mathfrak{X},\mathfrak{L},\rho_1)\subset Lip(\mathfrak{X},\mathfrak{L},\rho)$.

Let $\mathfrak{X}=[0,1]^n$ and $f\in\mathfrak{F}_1=Lip(\mathfrak{X},\mathfrak{L},\rho_1)$ where the metric $\rho=\rho_1$ is defined by (2.2.14); furthermore let

$$\max\{a_1,...,a_n\}\leq 1, \quad \min\{a_1,...,a_n\}<1.$$

The relation $f\in\mathfrak{F}_1$ follows by $f\in\mathfrak{F}_2= Lip(\mathfrak{X},\mathfrak{L},\rho_2)$ where the metric $\rho=\rho_2$ is defined by (2.2.11). The condition $f\in\mathfrak{F}_2$ is a typical condition of f in theory as well as in practice (because the precise information about f is always absent and while formulating the Lipschitz condition on f one usually chooses a metric from the collection (2.2.10) -

(2.2.13) which contains metrics having the unit ρ-ball symmetry property discussed earlier).

The relation $f \in \mathcal{F}_2$ means that

$$|f(x) - f(z)| \le \mathcal{L} \sum_{i=1}^{n} |x(i) - z(i)|$$

for all $x, z \in \mathcal{X}$; the relation $f \in \mathcal{F}_1$ is equivalent to

$$|f(x) - f(z)| \le \sum_{i=1}^{n} \mathcal{L}_i |x(i) - z(i)| \tag{2.2.44}$$

where $\mathcal{L}_i = a_i \mathcal{L} \le \mathcal{L}$ for all $i=1,\ldots,n$. For any function $f \in \mathcal{F}_2$ the true (minimal) constants $\mathcal{L}_i$ exist. They are usually unknown but it is precisely them (not $\mathcal{L}$) that determine the true accuracy of a global optimization algorithm.

We shall suppose that $a_i > 0$ for each $i=1,\ldots,n$ (the opposite case was considered earlier) and introduce a^o and a^1 as the arithmetical and geometric mean values

$$a^o = (a_1 + \ldots + a_n)/n, \qquad a^1 = (a_1 \times \ldots \times a_n)^{1/n}.$$

According to (2.2.15) the value $d_{\rho_1}(\Xi_N)$ is a natural quality characteristic of a grid Ξ_N for the case $\mathcal{F} = Lip(\mathcal{X}, \mathcal{L}, \rho_1)$.

If $a_i > 0$ for each $i=1,\ldots,n$ and the natural numbers n, N are arbitrary, then (see Sobol (1987))

$$d_{\rho_1}(\Xi_N) \ge \frac{1}{2} (n!)^{1/n} a^1 N^{-1/n} \tag{2.2.45}$$

for any grid Ξ_N,

$$d_{\rho_1}(\Xi^1_N) = \frac{1}{2} n\, a^o N^{-1/n} \tag{2.2.46}$$

for a cubic grid $\Xi_N{}^1$, and

$$d_{\rho_1}(\Xi^6_N) \le c\, n\, a^1 N^{-1/n} \tag{2.2.47}$$

for an η-adic Π_τ-grid $\Xi_N{}^6$ where the constant c does not depend on n, N.

The formulas (2.2.45), (2.2.46) lead to the following conclusions. If all the values $\mathcal{L}_i$ in (2.2.44) are positive, then cubic grids are optimal in order, but the ratio of the right sides of (2.2.45) and (2.2.46) can be arbitrarily small (for sufficiently large n). If some values $\mathcal{L}_i$ equal zero, then the cubic grids are not optimal in order (this corresponds to the above case). If $\mathcal{L}_1=...=\mathcal{L}_n$ then the ratio of the right sides of (2.2.45) and (2.2.46) exceeds e^{-1} for all n and N: this is again an evidence of high (theoretical) efficiency of cubic grids for this case.

Comparing now the right sides of (2.2.45) and (2.2.47), we find that quasirandom Π_τ-grids are optimal in order, uniformly with respect to values $\mathcal{L}_1=...=\mathcal{L}_n$. Since these values are usually unknown and may be rather different from each other, uniformly optimal (in order) grids are to be preferred to optimal grids for the case $\mathcal{L}_1=...=\mathcal{L}_n$. In particular, according to the above criterion, Π_τ-grids are better than cubic ones (though in the classical minimax sense the opposite judgement holds).

2.3 One-dimensional optimization, reduction and partition techniques

As it is clear by now, the multidimensional multiextremal optimization problem is a fairly complicated issue in computational mathematics. It is natural that some authors have made attempts to reduce the problem to simpler ones. In Section 2.1 we considered methods based on using local optimization techniques. Mathematical statistics procedures are widely used in the global random search methods: these will be investigated in Chapter 4. This section considers methods of global multidimensional search based on partition techniques and on the use of one-dimensional methods, and algorithms deducing the optimization problem to other problems of computational mathematics (approximation, interpolation, evaluation of integrals, etc.). First we shall review some well-known methods of one-dimensional global optimization.

2.3.1 One-dimensional global optimization

We shall suppose below that $\mathcal{X}$ is an interval $\mathcal{X}=[a,b]$, $-\infty<a<b<\infty$.

Much attention in the literature has been paid to the one-dimensional global optimization problem: this is associated with the relative simplicity of the problem and also with the existence of a number of multidimensional algorithms based on multiple use of one-dimensional optimization methods.

Many different one-dimensional global optimization algorithms exist: they are known to be practically efficient under various *a priori* conditions concerning the objective function.

For the most frequently investigated case $\mathcal{F}=\mathcal{L}ip(\mathcal{X},\mathcal{L},\rho)$, where the Lipschitz constant $\mathcal{L}$ is known, a number of methods have been described in Section 2.2.

When $\mathcal{F}=\mathcal{L}ip(\mathcal{X},\mathcal{L},\rho)$ and the Lipschitz constant is not known, it may be estimated e.g. by formula (2.2.34): this way, proper modifications of the above methods are applicable. For some classes of functions, being more smooth than Lipschitzian, algorithms similar to the above mentioned one-dimensional methods have been built by Brent (1973), Ganshin (1977), Berezovsky and Ludvichenko (1984).

In principle one may construct more or less efficient one-dimensional algorithms, using any general global optimization technique. Some methods obtained in this way (e.g. global random search methods) are relatively inefficient in one-dimensional case.Others seem to be the most efficient in the one-dimensional case (including the majority of covering, interval and Bayesian methods).

There exist many global optimization algorithms which are applicable only in the one-dimensional case. For example, Batishev (1975) and Pevnyi (1982) considered algorithms based on a spline approximation of the objective function; Jacobsen and Torabi (1978) reduced the problem of one-dimensional global optimization of a function represented as a sum of convex and concave functions to a sequence of convex functions minimization problems.

The following general scheme comprises a considerable number of (one-dimensional) adaptive partition algorithms in global optimization.

Algorithm 2.3.1. (Pintér (1983))

1. Set $x_1=a$, $x_2=b$, $k=2$ and evaluate $f(x_1)$, $f(x_2)$.

2. Renumber in the increasing order k points $x_1,...,x_k$ with objective function values $f(x_i)$, i=1,...,k. Then we have $a=x_1 < x_2 < ... < x_k=b$.

3. For all subintervals $\Delta_i=[x_i,x_{i+1}]$, i=1,..,k-1, of the interval $\mathcal{X}=[a,b]$ calculate the interval characteristic function value

$$\mathcal{R}(i) = \mathcal{R}\left(x_i, x_{i+1}, f_k^*, f(x_i), f(x_{i+1})\right)$$

which depends on the vertices of subinterval Δ_i, the best function value f_k^* found so for, and function values at the vertices. Choose the index (one of the indices)

$$j = \arg \max_{1\le i\le k-1} \mathcal{R}(i)$$

4. If the length of subinterval Δ_j is less than or equal to a fixed small positive number ε (that is $x_{j+1}-x_j\le\varepsilon$) then the algorithm will stop (other numerical stopping criteria may also be used). Otherwise take $x_{k+1}=S(j)$, where

$$S(j) = S\left(x_j, x_{j+1}, f_k^*, f(x_j), f(x_{j+1})\right)$$

is a function independent of the objective function values not belonging to the subinterval Δ_j except the value f_k^* (the same property has the function $\mathcal{R}$).

5. Evaluate $f(x_{k+1})$, substitute k+1 for k and return to Step 2.

The convergence of the above general partition scheme was investigated in Pintér (1983) and was generalized for the multidimensional case in Pintér (1986a,b).

Let us consider the forms of functions $\mathcal{R},\mathcal{S}$ generated by some well-known algorithms. To obtain Strongin minimization algorithm (see Strongin (1978)) under the supposition $f\in\mathcal{L}ip(\mathcal{X},\mathcal{L})$, one should choose

$$\mathcal{R}(i) = \mathcal{L}(x_{i+1} - x_i) + \frac{\left(f(x_{i+1}) - f(x_i)\right)^2}{\mathcal{L}(x_{i+1} - x_i)} - 2\left(f(x_i) + f(x_{i+1})\right), \tag{2.3.1}$$

$$\mathcal{S}(j) = (x_j + x_{j+1})/2 - \left(f(x_{j+1}) - f(x_j)\right)/(2\mathcal{L}). \tag{2.3.2}$$

In case of $\mathcal{F}=\mathcal{L}ip(\mathcal{X},\mathcal{L},\rho)$ the functions $\mathcal{R},\mathcal{S}$ for the Strongin algorithm are as follows

$$\mathcal{R}(i) = \mathcal{L}\rho(x_i, x_{i+1}) + \frac{(f(x_{i+1}) - f(x_i))^2}{\mathcal{L}\rho(x_i, x_{i+1})} - 2(f(x_i) + f(x_{i+1})),$$

(2.3.3)

$$\mathcal{S}(j) = \frac{x_j + x_{j+1}}{2} - \frac{1}{2}\,\mathrm{sign}\,(f(x_{j+1}) - f(x_j))\,\eta^{-1}\left(\frac{|f(x_{j+1}) - f(x_j)|}{\mathcal{L}}\right)$$

(2.3.4)

where $\eta(z)=\rho(z,0)$, η^{-1} is the inverse function of η. The polygonal line method of Piyavskii-Shubert (2.2.30) in case of $\mathcal{F}=Lip(\mathcal{X},\mathcal{L})$ is obtained by defining

$$\mathcal{R}(i) = \mathcal{L}(x_{i+1} - x_i)/2 - (f(x_i) + f(x_{i+1}))/2 \tag{2.3.5}$$

and determining $\mathcal{S}$ by (2.3.2). (Note that - as it can be seen easily - if one wants to maximize an objective function f, then it is neccessary to change the sign minus to plus in the last items in (2.3.1) - (2.3.5).) Note also that the one-dimensional algorithms presented later in Section 2.4 can also be represented in the frames of Algorithm 2.3.1.

2.3.2 Dimension reduction in multiextremal problems

There are ways of reducing multidimensional multiextremal optimization problems to one or several optimization problems having smaller dimension (in particular, dimension 1). These approaches will be outlined below.

The simplest way of using one-dimensional global optimization techniques is to utilize the scheme of coordinate-wise optimization similarly to some approaches in local optimization theory, see Mockus (1967). Certainly, a coordinate-wise global optimization algorithm can not guarantee, in general, that its limit point is a global minimizer. As a consequence such algorithms are not popular.

A number of multidimensional global optimization algorithms apply one-dimensional search in randomly chosen directions. Consider e.g. the following.

Algorithm 2.3.2. (Bremermann (1970))

1. Draw a point $x_1 \in \mathcal{X}$ arbitrarily, evaluate $f(x_1)$, set k=1.

2. Choose a random line passing through the point x_k (or choose a random direction emanating from x_k) by generating an isotropic probability distribution with centre x_k (for instance, the uniform distribution on the surface of the unit sphere $\mathcal{S}=\{x \in \mathcal{R}^n: \|x-x_k\|=1\}$ can be used).

3. Choose five equidistant points on the above line with x_k as their middle point. Evaluate the objective function at the above points.

4. Construct fourth-degree polynomial interpolation on the line.

5. Obtain a third-degree polynomial by differentiating the above fourth-degree one. Calculate the zeros of the third-degree polynomial by the Cardano formula. Evaluate the objective function at these calculated zeros which belong to $\mathcal{X}$.

6. Choose x_{k+1} as the point with the smallest objective function value in the point set containing five points from Step 3 and not more than three ones from Step 5.

7. Substitute k+1 for k and return to Step 2.

The convergence of Algorithm 2.3.2 is guaranteed only for the case, when the objective function is a polynomial of order not higher than four. In particular, it may be applied for evaluating roots of any polynomial P(z) by reducing the equation P(z)=0 to a system $Q_i(x)=0$, i=1,...,m, of second-degree simultaneous equations for new variables x and finding the minimizers of the function

$$\sum_{i=1}^{m} Q_i^2(x).$$

Note that there are many algorithms similar to Algorithm 2.3.2. The essential part of the method consists of a fourth-degree polynomial approximation of the function

$$f_k(\alpha) = f((1-\alpha)x_k + \alpha\xi_k)$$

on the interval [-1,1] for α where ξ_k is a realization of a random vector having an isotropic distribution in $\mathbb{R}^n$ with its centre at zero. Gaviano (1975) studied a theoretical version of the above algorithm whose iterations involve the minimization of the one-dimensional function $f_k(\alpha)$ over the interval [0,1]. Computational experiments of some authors showed that better results would be obtained if the range of α is extended. The simplest global random search method (Algorithm 3.1.1) may be referred to as a modification of Algorithm 2.3.2 in which minimization of the function $f_k(\alpha)$ is carried out on the two-point set {0,1}.

The most popular theoretical scheme of dimension reduction is the so-called multistep dimension reduction based on the representation

$$\min_{x\in\mathcal{X}} f(x) = \min_{0\le x(1)\le 1} \dots \min_{0\le x(n)\le 1} f(x(1),\dots,x(n)) \tag{2.3.6}$$

where f is a continuous function on the unit cube $\mathcal{X}=[0,1]^n$ and the point $x\in\mathcal{X}$ has the coordinates x(i), i=1,...,n. In particular, for n=2 the representation (2.3.6) gives

$$\min_{0\le x,z\le 1} f(x,z) = \min_{0\le x\le 1} \varphi(x)$$

where

$$\varphi(x) = \min_{0 \le z \le 1} f(x, z).$$

One may use the formula (2.3.6) for reducing the original optimization problem on a cube to a number of one-dimensional global optimization problems (but usually this number is very large). If $n \ge 3$ and the functional class $\mathcal{F}$ is broad enough (such as $Lip(\mathcal{X},\mathcal{L})$), then the algorithms based on (2.3.6) are cumbersome and inefficient. But for some relatively narrow classes $\mathcal{F}$ the multistep reduction scheme may serve as the base of efficient algorithms. For example, if the objective function f is separable, i.e.

$$f(x) = \sum_{i=1}^{n} f_i(x(i))$$

where $f_1,\ldots,f_n$ are one-dimensional functions, then according to (2.3.6) one has

$$\min_{x \in \mathcal{X}} f(x) = \sum_{i=1}^{n} \min_{0 \le x(i) \le 1} f_i(x(i)).$$

In this case to solve the n-dimensional optimization problem means to solve n one-dimensional ones. In a practically more important case, f is represented as follows

$$f(x) = \sum_{i=1}^{n-1} f_i(x(i), x(i+1)) \tag{2.3.7}$$

where $f_1,\ldots,f_{n-1}$ are two-dimensional functions. Here (2.3.6) reduces to

$$\min_{x \in \mathcal{X}} f(x) = \min_{0 \le x(n) \le 1} \varphi_n(x(n))$$

where

$$\varphi_2(x(2)) = \min_{x(1)} f_1(x(1), x(2)), \quad \varphi_i(x(i)) = \min_{x(i-1)} \left[\varphi_{i-1}(x(i-1)) + f_{i-1}(x(i-1), x(i))\right]$$

for $i \ge 3$. Hence, to solve a global optimization problem for the objective function (2.3.7), one may tabulate one-dimensional functions φ_i ($i=2,\ldots,n$) whose values are solutions of one-dimensional global optimization problems. A more detailed description of the multistep reduction scheme is contained in Strongin (1978).

Another way of reducing multidimensional global optimization problems to one-dimensional ones is the dimension reduction by means of Peano curve type mappings. These are continuous maps φ of interval $[0,1]$ into the cube $\mathcal{X}=[0,1]^n$. Using them one has

$$\min_{x \in \mathcal{X}} f(x) = \min_{t \in [0,1]} f(\varphi(t)).$$

The possibility of using one-dimensional algorithms for optimizing the objective functions $g(t)=f(\varphi(t))$ is followed by the next proposition due to Strongin (1978): if $f \in \mathcal{L}ip(\mathcal{X},\mathcal{L})$, then $g \in \mathcal{L}ip([0,1],M_n(\mathcal{L}),\rho_n)$ where $M_n(\mathcal{L})$ is a constant depending on $\mathcal{L},n,\varphi$ and ρ_n is the metric defined by formula

$$\rho_n(t,t') = |t - t'|^{1/n}.$$

If φ is the Peano curve then

$$M_n(\mathcal{L}) = 4\mathcal{L}\sqrt{n}.$$

Strongin (1978) thoroughly studied the numerical construction of approximate Peano curves and their properties. The author has no convincing data in favour of efficiency of corresponding global optimization algorithms in case of $n>2$. The main difficulty here is that the one-dimensional function $g(t)$ may be so complicated that its minimization may give greater difficulties than the minimization of f. (Already for $n=2$ and a linear function f, the corresponding one-dimensional function g is essentially multiextremal). Nevertheless, it is doubtless that the approach may be useful in a sort of visual analysis of the objective function behavior.

Saltenis (1989) analysed a statistical approach to the study of possibilities of objective function representations, as a sum of functions depending on smaller numbers of variables. This approach is based on using the so-called random balance method which is classical in the theory of screening experiments, see Ermakov and Zhigljavsky (1987). The above approach uses the dispersion of different projections of an objective function f or integrals of f with respect to different subsets of variables as structure characteristics. Although the above approach is heuristic, hardly permitting serious theoretical investigations, there are several examples of successful applications to practical problems.

2.3.3 Reducing global optimization to other problems in computational mathematics

The multidimensional global optimization problem is closely connected with a great number of other problems in computational and applied mathematics. This connection becomes apparent in constructing some algorithms or in immediate reduction of the global optimization problem to other problems. Let us outline principal approaches.

Various global search methods based on multiple use of local algorithms have been decsribed in Section 2.1. Some of them may be generalized for the case, when the objective function is evaluated in presence of random noise. In this case the role of local algorithms is played by stochastic approximation type procedures, see Section 8.1. Section 2.3.2 described multidimensional methods of global optimization based on one-dimensional search methods.

Any global optimization problem, in general, may be reduced to an approximation problem. While doing this, one should obtain a global minimizer of an approximation function and then utilize a local descent routine from the above minimizer as its initial point. If splines are used for approximation in the above methods then they may possess some optimality properties, see Pevnyi (1982, 1988), Ganshin (1977). In spite of this, the mentioned methods have no significant practical importance, because the goodness criteria for optimization and approximation algorithms are different in essence (viz. approximation accuracy has to be assured for all the set $\mathcal{X}$, while in optimization in the vicinity of a global optimizer only). Numerical investigations confirm this. For example, Chujan (1986) searched the global optimizer of some one-dimensional functions several hundred times more economically than the mentioned Pevnyi (1982) algorithm which is based on utilizing splines.

Ideas of approximation are fruitfully applied in global optimization methodology in the following way: a rough approximation of the objective function is constructed (not necessary in an explicit form); *nonpromising* subsets of $\mathcal{X}$ are determined by this approximation and are excluded from further search $\mathcal{X}$; in the remaining subset of $\mathcal{X}$, a more accurate approximation is built, and analogous operations are carried out until a given accuracy is attained. A considerable part of global random search algorithms treated in this book are constructed according to this principle.

Another way of applying approximation in global search is to construct algorithms using multiple approximation of the objective function projections (they may be one-, two- or more dimensional): a typical example is Algorithm 2.3.2.

Global optimization algorithms, consisting of the solution of ordinary differential equations or systems of such equations are important as well: these algorithms were described in Section 2.1.3. The connection of differential equations and global optimization theories is based on the fact that trajectories corresponding to solutions of certain classes of differential equations, contain (or converge to) one or more local optimizers of a given function.

Another type of connection is between stochastic differential equations and global optimization theories. Section 3.3.3 shows for some functional classes $\mathcal{F}$ that if $\varepsilon(t)$ approaches zero slowly enough as t tends to infinity then the trajectory corresponding to the solution of the stochastic differential equation

$$d\,x(t) = -\nabla f(x(t)) + \varepsilon(t) d\,w(t), \qquad x(0) = x_0,$$

converges in probability to a global minimizer; here x_0 is an arbitrary point in $\mathcal{X}$ and w is the n-dimensional Wiener process. This approach has a considerable theoretical interest.

Global optimization and certain integral equations are also related. Section 5.3 deals with probability measures that are solutions of integral equations of some types and concentrated in the vicinity of a global maximizer of the function f: hence, the algorithms of generating random vectors distributed approximately by the above probabilistic solutions may be referred to as global optimization algorithms.

Global search based on using integral representations is a popular approach, especially in theoretical works. Some well-known representations (formulated for the problem of the global maximization) are as follows.

Let $\mathcal{X}$ be a compact subset of $\mathbb{R}^n$, the function f be continuous on $\mathcal{X}$ and attains the global maximum $M = \max f$ at the unique point x^* with coordinates $x^*(j)$, $j=1,\dots,n$. Then, according to Pinkus (1968), we have

$$x^*(j) = \lim_{\lambda \to \infty} \int x(j) \exp\{\lambda f(x)\} dx / \int \exp\{\lambda f(x)\} dx. \tag{2.3.8}$$

If, in addition, f is nonnegative, then

$$x^*(j) = \lim_{\lambda \to \infty} \int x(j) f^{\lambda}(x) dx / \int f^{\lambda}(x) dx \tag{2.3.9}$$

and

$$M = \lim_{\lambda \to \infty} \int f^{\lambda+1}(x) dx / \int f^{\lambda}(x) dx. \tag{2.3.10}$$

General conditions, under which relations (2.3.9) and (2.3.10) hold, will be given in Section 5.2.2.

If f is a continuous nonnegative function then for any $\lambda>0$ the evident inequality

$$\int f^{\lambda+1}(x)\, dx \le M \int f^{\lambda}(x) dx$$

is valid: an equivalent form of it being

$$\int [f(x)/f(x^*)]^{\lambda+1} dx \le \int [f(x)/f(x^*)]^{\lambda} dx \tag{2.3.11}$$

Hence, if for some $\lambda>0$, $x_* \in \mathcal{X}$ the inequality

$$\int [f(x)/f(x_*)]^{\lambda+1} dx > \int [f(x)/f(x_*)]^{\lambda} dx, \tag{2.3.12}$$

(opposite to (2.3.11)), is fulfilled, then the point x_* can not be a global maximizer of f. The condition (2.3.12) is sufficient for a point not being a global maximizer. It is non-constructive and thus seems to be of small practical significance. Namely, if evaluating the integrals in (2.3.12) one finds a point $x_{(0)} \in \mathcal{X}$ such that $f(x_{(0)}) > f(x_*)$ then x_* is surely not a global maximizer and if such a point $x_{(0)}$ is not found then the inequality (2.3.12) for estimators of the integrals will not be valid.

Analogously to (2.3.11) and (2.3.12), a necessary and sufficient condition

$$\lim_{\lambda \to \infty} \sup \int [f(x)/f(x^*)]^{\lambda} dx < \infty$$

for a point x^* to be a global maximizer can be obtained (this condition is also non-constructive).The representations (2.3.8) - (2.3.10) are more constructive: they are basic for some global optimization algorithms (e.g. see Ivanov et al. (1985)) involving simultaneous estimation of several integrals. (Note that the problem of optimal simultaneous Monte-Carlo estimation of several integrals will be studied in Section 8.2.)

Still, it is difficult to understand why a point with approximate limits of (2.3.9) as coordinates should be better than the record point obtained during estimation of the integrals.

It should be noted that the asymptotic representations (2.3.9) and (2.3.10) are useful for theoretical investigations of some global random search algorithms, see Section 5.2. Let us point out also that besides (2.3.8), an equivalent asymptotic representation

$$x^*(j) = \lim_{\lambda \to \infty} \int x(j) \exp\{-\lambda f(x)\} dx / \int \exp\{-\lambda f(x)\} dx$$

for the coordinates of the unique global minimizer x^* of f is widely known.

Section 2.1.6 describes a method of minimization of a smoothed function that is close in spirit to the methods based on the above integral representations.

Ideas of discrete optimization proved to be useful for some global optimization problems. Let us outline some principal ways of their use. The first and the most evident way is the discretization of the set $\mathcal{X}$ (replacing $\mathcal{X}$ by a discrete subset) and using a discrete algorithm for optimizing the corresponding discrete function. In particular, this approach was used by Nefedov (1987) for constructing an optimization algorithm for the case when a set $\mathcal{X}$ is determined by inequality constraints. The second way (see Kleibom (1967), Bulatov (1987)) is based on constructing the convex hull $\mathcal{X}_o$ for $\mathcal{X}$ and the convex envelope f_o for f and minimizing f_o on $\mathcal{X}_o$ with the help of discrete optimization methods. The third and the most useful way is the use of the branch and bound principle that is a powerful general instrument in discrete optimization. This principle is considered in the next section.

2.3.4 *Branch and bound methods*

The main idea of branch and bound strategies is the sequential rejection of those subsets of $\mathcal{X}$ that can not contain a global minimizer and then searching only in the remaining subsets (regarded as *prospective*).

At the k-th iteration of a branch and bound method it is necessary to construct a partition (i.e. branching) of the optimization region (at the first iteration this region is $\mathcal{X}$) into a finite number I_k of subsets $\mathcal{X}_i$, on which lower bounds ℓ_i of

$$m_i = \inf_{x \in \mathcal{X}_i} f(x) \tag{2.3.13}$$

can be given by evaluations of f at some points from $\mathcal{X}_i$, $i \in I_k$.

At each iteration the record $f_k^* = f(x_k^*)$ is also used: this is the smallest objective function value obtained so far, thus being an upper bound for $f^* = \min f$.

Since subsets $\mathcal{X}_i$ for which $\ell_i \geq f_k^*$ can never contain a global minimizer, thay can be excluded from the further search; all subsets $\mathcal{X}_j$ for which $\ell_j < f_k^*$ are left. The partition is then further refined, most naturally by branching the subset $\mathcal{X}_j$ with

$$\ell_j = \min_{i \in I_k} \ell_i$$

into smaller subsets - and the iterations are continued.

The convergence problem and implementation aspects for the above class of (global optimization) methods is investigated under different conditions, see Horst (1986), Horst and Tuy (1987), Pintér (1986a, 1988). Convergence is ensured by the fact that the lower bounding procedure is aymptotically accurate, i.e. ℓ_i converges to m_i, when the volume of $\mathcal{X}_i$ approaches zero.

Of course, for too broad functional classes, finding a lower bound for the global minimum is of similar difficulty as finding the global minimum itself. Thus it is possible to construct efficient branch and bound methods only for sufficiently narrow functional classes $\mathcal{F}$: examples of such classes $\mathcal{F}$ are considered below.

There are many variants of the technique under consideration. As noted, e.g. the majority of covering methods may be referred to as branch and bound procedures, see Horst and Tuy (1987). The same is true for the one-dimensional partition algorithms described in Section 2.4.1. Here we shall deal with other algorithms.

First let us follow McCormick (1983) and assume that $\mathcal{X}$ is a hyperrectangle

$$\mathcal{X} = \{x \in \mathcal{R}^n : a \leq x \leq b\}$$

and f is a factorable function, i.e. f is the last one in a collection of m functions $f^1, f^2, \ldots, f^m$ which is called a factorization sequence and built up as follows

$$f^j(x_1, \ldots, x_n) = x_j \qquad \text{for each } j = 1, \ldots, n, \tag{2.3.14}$$

and for j>n one of the following holds

$$f^j(x) = f^p(x) + f^q(x) \qquad \text{for some } p, q < j, \tag{2.3.15}$$

$$f^j(x) = f^p(x) f^q(x) \qquad \text{for some } p, q < j, \tag{2.3.16}$$

or

$$f^j(x) = \varphi(f^p(x)) \qquad \text{for some } p < j, \tag{2.3.17}$$

where φ belongs to a given class Φ of sufficiently simple functions $\varphi: \mathcal{R} \to \mathcal{R}$ (e.g. $\varphi(t) = t^{\alpha}$, $\varphi(t) = e^t$, $\varphi(t) = \sin t$, etc.).

It is easy to verify that the above factorization is a natural way for representing functions which are given in an explicit algebraic form.

Let the subsets $\mathcal{X}_i$ be hyperrectangles

$$\mathcal{X}_i = \{x \in \mathcal{X} : a_i \leq x \leq b_i\}.$$

A lower bound ℓ_i for m_i will be computed by constructing convex lower bounding functions $\ell_i^j(x)$ and concave upper bounding functions $u_i^j(x)$ with the property

$$\ell_i^j(x) \le f^j(x) \le u_i^j(x) \qquad \text{for all } x \in X_i \tag{2.3.18}$$

and computing

$$\ell_i = \inf_{x \in X_i} \ell_i^m(x).$$

One may use different ways for constructing convex functions $\ell_i^j(x)$ and concave functions $u_i^j(x)$ with the property (2.3.18).

The simplest interval arithmetic methods described later use constant functions $\ell_i^j(x)=\ell_i$ and $u_i^j(x)=u_i^j$. The opposite approach is to find the best possible bounding functions, i.e. taking $\ell_i^j(x)$ as the convex lower envelope of $f^j(x)$ on X_i and $u_i^j(x)=u_i^j$ equal to its concave upper envelope. For functions (2.3.14) we have

$$\ell_i^j(x) = u_i^j(x) = f_i^j(x), \qquad x \in X_i .$$

If (2.3.15) holds, then

$$\ell_i^j(x) = \ell_i^p(x) + \ell_i^q(x), \qquad u_i^j(x) = u_i^p(x) + u_i^q(x)$$

Let (2.3.16) hold, L_i^p, L_i^q and U_i^p, U_i^q be lower and upper bounds for $f^p(x)$ anf $f^q(x)$ over X_i, respectively. If $L_i^p \le 0$, $L_i^q \le 0$ and $U_i^p \ge 0$, $U_i^q \ge 0$ then we may take

$$\ell_i^j(x) = \max\left\{U_i^q \ell_i^p(x) + U_i^p \ell_i^q(x) - U_i^p U_i^q,\ L_i^q \ell_i^p(x) + L_i^p \ell_i^q(x) - L_i^p L_i^q\right\},$$
$$u_i^j(x) = \min\left\{L_i^p u_i^q(x) + U_i^q u_i^p(x) - L_i^p U_i^q,\ L_i^q u_i^p(x) + U_i^p u_i^q(x) - L_i^q U_i^p\right\}.$$

(Analogous results are valid for the other cases.)

In case of (2.3.17) convex lower and concave upper bounding functions (L(t) and U(t), respectively) for φ over the interval $[L_i^p, U_i^p]$ are to be given. Let $\varphi(t)$ attain its minimum and maximum on the interval $[L_i^p, U_i^p]$ at t_o and t_1, respectively, and let *mid* be the operator which selects the middle value. Then in case of (2.3.17) we may take

$$\ell_i^j(x) = L\left(\text{mid}\left\{\ell_i^p(x), u_i^p(x), t_o\right\}\right),$$
$$u_i^j(x) = U\left(\text{mid}\left\{\ell_i^p(x), u_i^p(x), t_1\right\}\right).$$

It has been mentioned that constant functions $\ell_i^j(x)$ and $u_i^j(x)$ satisfying (2.3.18) generate so-called interval methods. Let us consider these methods which are of considerable theoretical (and of increasing practical) importance.

Interval methods are aimed at finding the global extremum of a twice differentiable rational objective function f defined on a hyperrectangle $\mathcal{X}$ and having a gradient ∇f and a Hessian $\nabla^2 f$ with only a finite number of (isolated) zeros. Their essence is in evaluation of images $f(Z)$, $\nabla f(Z)$, and $\nabla^2 f(Z)$ for hyperrectangles $Z \subset \mathcal{X}$ with the purpose of excluding those which can not contain extremal points. Their main drawback seems to be the relatively restricted class of optimization problems which can be solved by these methods, and their substantial computational demand.

Let us introduce some notions which are necessary for describing interval methods. Let $Z_i \subset \mathcal{X}$, i=1,2, be intervals $Z_i=[a_i,b_i]$. We shall call them interval variables and define the interval arithmetic operations by

$$Z_1+Z_2=[a_1+a_2, b_1+b_2], \qquad Z_1-Z_2=[a_1-b_2, b_1-a_2],$$
$$Z_1Z_2=[\min(a_1a_2, a_1b_2, b_1a_2, b_1b_2), \max(a_1a_2, a_1b_2, b_1a_2, b_1b_2)],$$
$$Z_1/Z_2=Z_1[1/b_2, 1/a_2], \qquad \text{if} \quad 0 \notin Z_2.$$

Using these formulas, one may evaluate the interval extension of a rational function f, i.e. the image

$$f(Z)=\{y: y=f(x), x \in Z\}$$

which is an interval for an interval argument Z.

If f is a multidimensional function, then interval values $f(Z)$ may be analogously defined for a multidimensional interval (i.e. hyperrectangle) Z. Algorithms of interval values evaluation are studied e.g. in Moore (1966) or Ratschek and Rohne (1984). Returning to global optimization via interval methods, for simplicity, consider the one-dimensional case. The first step of an interval method is the evaluation of the objective function at one of the interval endpoints. Let $k \geq 1$ steps be already done; f_k^* be a k-step upper bound for f^* (e.g. f_k^* be the minimal value of f obtained so far); $\mathcal{X}_k$ be a subset of $\mathcal{X}$ that surely contains an optimal point, consisting of a union of intervals. The next step of the method is the following: let us choose a subinterval Z of $\mathcal{X}_k$ and divide it into two parts Z_1, Z_2, and calculate the interval values $f(Z_1)$, $f(Z_2)$. If $f_k^* < f(Z_i)$ then subinterval Z_i is excluded from the set $\mathcal{X}_k$, since the above relation guarantees that the objective function can not attain its global minimum on the subinterval Z_i. The subintervals Z_i (i=1,2) may be also excluded from $\mathcal{X}_k$ if $0 \notin f'(Z_i)$ or $f''(Z_i) \subset (-\infty, 0)$ (as they imply that Z_i does not contain a stationary point of f or is concave in Z_i). If the subintervals Z_1 and Z_2 are not excluded from $\mathcal{X}_k$ they should be included into $\mathcal{X}_{k+1}$ replacing Z. Note that to find a zero of f' on an interval Z_1 (or to verify the inclusion

$0 \in f'(Z_1)$) one may use the interval version of the Newton method. If $0 \notin f''(Z_1)$, then the interval Newton step has the form

$$N(Z_j) = z_j - f'(z_j)/f''(Z_j), \qquad Z_{j+1} = Z_j \cap N(Z_j) \qquad j = 1, 2, \ldots$$

where z_j is the midpoint of Z_j. According to Hansen (1979), if $0 \notin f'(Z_1)$, then the set Z_j is empty for some j and if $0 \in f'(Z_1)$, then the point sequence $\{z_j\}$ converges to a stationary point of f quadratically.

In the multidimensional case, the interval methods have the same form but their practical realization is complicated, because of the necessity to store, choose and divide a great number of subrectangles of $\mathcal{X}$.

A detailed description of interval global optimization methods may be found in Hansen (1979, 1980, 1984), Ratschek (1985). Mancini and McCormick (1976) described interval methods for minimizing a convex function. Shen and Zhu (1987) suggest an interval version of the one-dimensional Piyavskii and Shubert algorithm (2.2.30). On the whole, interval methods represent a set of promising global optimization approaches but the class of extremal problems which may be efficiently solved by them is naturally restricted by their analytical requirements.

Another class of problems, in which branch and bound methods are used advantageously, consists of concave minimization problems under convex constraints (for details and references see Pardalos and Rosen (1986, 1987)). Following Rosen (1983), consider the special case where

$$f(x) = \eta' x - \frac{1}{2} x' Q x \tag{2.3.19}$$

(here η is a vector, Q is a positive semidefinite matrix) and

$$\mathcal{X} = \{x \in \mathbb{R}^n : Ax = b,\ x \geq 0\}. \tag{2.3.20}$$

Lower and upper bounds of f^* are needed. To compute an upper bound at the beginning, f is maximized over $\mathcal{X}$. This gives a point $x_0 \in \mathcal{X}$. Then the n eigenvectors $e_1, \ldots, e_n$ of the Hessian at x_0 are determined. To move as far away as possible from x_0, one solves 2n linear programming problems and find vectors

$$v_i = \arg \max_{x \in \mathcal{X}} w_i' x$$

where $w_i = e_i$, $w_{n+i} = -e_i$ for $i = 1, \ldots, n$. An upper bound for f^* is $U = \min\{f(v_1), \ldots, f(v_{2n})\}$.

The vectors v_i define halfspaces

$$\{x \in \mathbb{R}^n : w_i'(x - x_0) \leq w_i'(v_i - x_0)\}$$

whose intersection is a hyperrectangle $\mathcal{X}_o$ containing $\mathcal{X}$. A lower bound L for f^* is the minimum of f over $\mathcal{X}_o$ which is attained at one of the 2^n vertices of $\mathcal{X}_o$.

We also construct a hyperrectangle inscribed into the ellipsoid

$$\{x \in \mathbb{R}^n : f(x) = U\}$$

of the form

$$\bigcap_{i=1}^{2n} \{x \in \mathbb{R}^n : w_i'(x - x_0) \le d_i\}$$

where the constant d_i can easily be computed. Now, it can be seen that x* can not be contained in the interior of this hyperrectangle and the intersection of its exterior with $\mathcal{X}$ defines an appropriate family of subsets in which x* is looked for.

Note that the branch and bound technique was used also for some more general problems than (2.3.19) - (2.3.20), including the case, in which f is the difference of two convex functions and $\mathcal{X}$ is a convex set.

Let us also note that Beale and Forrest (1978) applied the branch and bound technique for minimizing one-dimensional functions of the type

$$f(x) = \sum_{i=1}^{m} f_i(x)$$

where f_i (i=1,...,m) are twice continuously differentiable, the values $f_i(x)$, $f_i'(x)$, $f_i''(x)$ can be calculated for any point $x \in \mathcal{X}$, and the set $\mathcal{X}$ can be *a priori* divided into subsets where all second derivatives are monotone.

Finally it should be pointed out that Chapter 4 presents a generalized branch and bound principle for the case when estimators for the lower bounds (2.3.13) are valid only with a large probability: this generalization will be called the branch and probability bound principle.

2.4 An approach based on stochastic and axiomatic models of the objective function

In the previous sections, many deterministic models of the objective function were considered. Other classes of models are also used for the description of multiextremal functions and construction of global optimization algorithms: the most known of these is the class of stochastic models that uses a set of realizations of a random function as $\mathcal{F}$.

2.4.1 Stochastic models

Let $\varphi(x,\omega)$ be a random function where $x\in\mathcal{X}$ and ω is an element of a probability space Ω; the prior information about f consists in that $f\in\mathcal{F}=\{\varphi: \mathcal{X}\times\Omega\to\mathcal{R}^1\}$. In other words, f is supposed to be a realization of a random function φ for some random element ω: $f(x)=\varphi(x,\omega)$ for all $x\in\mathcal{X}$ and some $\omega\in\Omega$. Such models are not evident in advance, but sometimes they are convenient from the mathematical point of view and can be justified with the help of the axiomatic approach reviewed below.

Frequently, classes of Gaussian random functions are considered, i.e. random functions $\varphi(x)$ such that for each $k\geq 1$ and a collection $\Xi_k=\{x_1,\ldots,x_k\}$ of points in $\mathcal{X}$ the random vector $\varphi(\Xi_k)=(\varphi(x_1),\ldots,\varphi(x_k))'$ has a joint Gaussian distribution with density

$$p(u,\mu,V)=\frac{(\det V)^{1/2}}{(2\pi)^{k/2}}\exp\left\{-\tfrac{1}{2}(u-\mu)'V(u-\mu)\right\}$$

where

$$u\in\mathcal{R}^k, \quad \mu=(\mu(x_1),\ldots,\mu(x_k))', \quad \mu(x)=E\varphi(x),$$

$$\mathcal{R}=\left\|\eta(x_i,x_j)\right\|_{i,j=1}^{k}, \ \eta(x,z)=E(\varphi(x)-\mu(x))(\varphi(z)-\mu(z)),\ V=\mathcal{R}^{-1}.$$

The class of realizations of the classical Wiener process determines the most popular stochastic model of one-dimensional functions $f(x)$, $x\in\mathcal{X}=[0,1]$. It is characterized by the functions

$$\mu(x)=\mu=\text{const}, \qquad \eta(x,z)=\sigma^2\min(x,z) \tag{2.4.1}$$

where σ^2 is a constant. In this case $\varphi(0)=\mu$,

$$\varphi(x)-\varphi(z)\sim N(0,\sigma^2|x-z|).$$

In the case of Gaussian random functions, the marginal distributions conditioned by any number of calculated values of f are still Gaussian and can be computed in the following way. Let $Y_1=(y_1,...,y_k)'$ be the vector of values $y_i=f(x_i)$ $(i=1,...,k)$, $Y_2=(\varphi(z_1),...,\varphi(z_m))'$ be a Gaussian random vector of unknown values of φ at points $z_1,...,z_m$ in $\mathcal{X}$ conditioned by the evaluations Y_1. Set

$$\mu_1=(\mu(x_1),...,\mu(x_k))', \qquad \mu_2=(\mu(z_1),...,\mu(z_m))',$$

$$\operatorname{cov}\begin{pmatrix}Y_1\\Y_2\end{pmatrix}=R=\begin{pmatrix}R_{11}&R_{12}\\R_{21}&R_{22}\end{pmatrix}, \qquad V=R^{-1}=\begin{pmatrix}V_{11}&V_{12}\\V_{21}&V_{22}\end{pmatrix}$$

where V_{11} and R_{11} are of order $k\times k$, V_{22} and R_{22} of order $m\times m$. Then

$$E(Y_2|Y_1)=\mu_2-V_{22}^{-1}V_{21}(Y_1-\mu_1)=\mu_2+R_{21}R_{11}^{-1}(Y_1-\mu_1), \tag{2.4.2}$$

$$\operatorname{cov}(Y_2|Y_1)=V_{22}^{-1}=R_{22}-R_{21}R_{11}^{-1}R_{12} \tag{2.4.3}$$

Formulas (2.4.2) and (2.4.3) are usually applied in the case $m=1$. For some particular cases of covariance function $\eta(x,z)$, they are not very complicated. For example, if $m=1$, $\mu(x)=0$, and

$$\eta(x,z)=\sigma^2/(1+\|x-z\|^2) \qquad \text{for} \qquad x,z\in\mathcal{X} \tag{2.4.4}$$

then (2.4.2) can be simplified to

$$\sum_{i=1}^{k}\alpha_i/(1+\|z_1-x_i\|^2) \tag{2.4.5}$$

where $\alpha_1,..., \alpha_k$ are appropriate constants.

2.4.2 *Global optimization methods based on stochastic models*

Let us use the notations of the beginning of Section 2.2.3, but apply the Bayesian (statistical) approach for defining the accuracy of a method, replacing the minimax approach based on (2.2.35).

The accuracy of an N-point method $d^N=(d_1,...,d_{N+1})$ can be defined in various statistically meaningful ways. For instance, the algorithm defined by

$$\arg\min_{d^N\in\mathcal{D}(N)} E(\varphi(x_{N+1})) \tag{2.4.6}$$

is called optimal with respect to the expected value (E-optimal), and the algorithm

$$\arg \max_{d^N \in \mathcal{D}(N)} \Pr\{\varphi(x_{N+1}) - \min \varphi \leq \varepsilon\} \qquad (2.4.7)$$

is ε-optimal in probability (or P-optimal). The computational difficulties of finding them are tremendous, see Archetti and Betro (1980), Mockus (1988), and so they are practically intractable. Therefore the concept of one-step optimality is often used, instead of optimality with respect to all points of a method. In a one-step optimal algorithm each point is selected as if it were the last one: below we shall present two such methods which are respective modifications of the above E- and P-optimal algorithms.

Let k evaluations of the objective function f at points $x_1,...,x_k$ be performed, $y_i=f(x_i)$ $i=1,...,k$,

$$f_k^* = \min_{1 \leq i \leq k} f(x_i)$$

be the optimum estimate. Then the one-step E-optimal algorithm is defined by

$$x_{k+1} = \arg \min_{x \in \mathcal{X}} E\{\varphi(x) | \Sigma_k\} \qquad (2.4.8)$$

where Σ_k denotes the conditions $\varphi(x_i)=y_i$ for $i=1,...,k$. Furthermore, the one-step analogy of the P-optimal algorithm (2.4.7) is

$$x_{k+1} = \arg \max_{x \in \mathcal{X}} \Pr\{\varphi(x) < f_k^* - \varepsilon_k | \Sigma_k\} \qquad (2.4.9)$$

where $\{\varepsilon_k\}$ is a suitably chosen sequence of positive numbers.

In the case, when φ is a Gaussian random function, the calculations in (2.4.8) and (2.4.9) can be performed applying (2.4.2) and (2.4.3). They are computationally tractable only for some special classes $\mathcal{F}$: the most well-known is the case of the one-dimesional Wiener process model considered in Section 2.4.3. If $\mathcal{X}$ is multidimensional, then it is really hard to find a reasonable stochastic model that would not lead to a tremendous amount of computations. The situation is still worse, if the fact that a reasonable model usually contains unknown parameters is taken into account. The multidimensional Wiener process is already not suitable , from the numerical point of view.

An example of another kind is (2.4.4), in which due to (2.4.5) points (2.4.8) can be found by calculating all roots of a set of polynomial equations (of course, the adequacy of the model (2.4.4) to a given objective function is usually rather questionable).

2.4.3 The Wiener process case

Let $\mathcal{X}=[a,b]$ and $\varphi(x,\omega)$ is the Wiener process with mean and covariation function defined by (2.4.1), where μ and σ^2 may be unknown. It is an acceptable model for the global behavior of a complicated one-dimensional function and leads to not very cumbersome calculations.

If σ^2 is unknown then every algorithm has to start with its estimation. To this end, it is usually recommendable to evaluate f at m equidistant points $x_i=a+(b-a)(i-1)/(m-1)$, $1\leq i\leq m$, and estimate σ^2 by the maximum likelihood estimator

$$\hat{\sigma}^2 = \frac{1}{b-a}\sum_{i=2}^{m}(f(x_i)-f(x_{i-1}))^2.$$

Of course, it can be profitable to reestimate σ^2 during the search.

Let now σ^2 be known and $k\geq m$ evaluations $y_i=f(x_i)$ of f be performed at the ordered points $a=x_1<...<x_k=b$. Then the conditional mean $\mu(x\,|\,\Sigma_k)$ and variance $\sigma^2(x\,|\,\Sigma_k)$ are computed through

$$\mu(x|\Sigma_k) = E(\varphi(x)|\Sigma_k) = y_i\frac{x_{i+1}-x}{x_{i+1}-x_i} + y_{i+1}\frac{x-x_i}{x_{i+1}-x_i}, \qquad (2.4.10)$$

$$\sigma^2(x|\Sigma_k) = \mathrm{var}(\varphi(x)|\Sigma_k) = \sigma^2(x-x_i)(x_{i+1}-x)/(x_{i+1}-x_i) \qquad (2.4.11)$$

for each $x\in\Delta_i=[x_i,x_{i+1}]$, $i=1,...,k-1$. Moreover, the expectation of the minimum value

$$\varphi_i = \min_{x\in\Delta_i}\varphi(x)$$

in an interval Δ_i conditioned on Σ_k is

$$E(\varphi_i|\Sigma_k) = \min(y_i,y_{i+1}) - \left(\frac{\pi\sigma^2(x_{i+1}-x_i)}{2}\right)^{1/2} \times$$

$$\times \exp\left\{\frac{(y_{i+1}-y_i)^2}{2\sigma^2(x_{i+1}-x_i)}\right\}\int_{-\infty}^{0} p\left(t,-|y_{i+1}-y_i|,\sigma^2(x_{i+1}-x_i)\right)dt \qquad (2.4.12)$$

where

$$p(t,a,\delta^2) = (2\pi\delta^2)^{-1/2}\exp\{-(t-a)^2/2\delta^2\}$$

is the Gaussian density. By (2.4.12) one can compute the posterior mean of φ_i for each interval Δ_i and select the interval

$$\Delta_j = \arg\min_{\Delta_i} E(\varphi_i|\Sigma_k).$$

The next point x_{k+1} can be chosen in the interval Δ_j in different ways. The simplest one is $x_{k+1}=(x_j+x_{j+1})/2$, i.e. x_{k+1} is the centre of Δ_j. If we want to confine ourselves to Bayesian techniques then it is natural to select x_{k+1} as the expected location of the minimum φ_j but the corresponding formulas are rather complicated. (Note that this approach was followed by Boender (1984), where favourable numerical test results are also given.)

The one-step P-optimal algorithms (2.4.9) are determined in an easier way:

$$j = \arg\max_{1\le i\le k-1} R(i), \qquad x_{k+1} = a_j \tag{2.4.13}$$

where

$$R(i) = \frac{(f_k^* - \varepsilon_k)(x_{i+1} - x_i) - y_{i+1}(a_i - x_i) - y_i(x_{i+1} - a_i)}{\sigma((a_i - x_i)(x_{i+1} - a_i)(x_{i+1} - x_i))^{1/2}},$$

$$a_i = \frac{x_i + x_{i+1}}{2} + \frac{(x_{i+1} - x_i)(y_{i+1} - y_i)}{2\left(y_{i+1} + y_i - 2(f_k^* - \varepsilon_k)\right)}$$

The efficiency of algorithm (2.4.13) depends to a considerable degree on the choice of ε_k. Zilinskas (1981) proposed to choose

$$\varepsilon_k = (\max_{1\le i\le k} y_i - f_k^*)/2.$$

As a stopping rule for the above algorithms, one may choose the following: reject subintervals $\Delta_i=[x_i,x_{i+1}]$, if the probability of finding a function value in Δ_i better than the current optimum estimate f_k^*, i.e.

$$\Pr\left\{\min_{x\in\Delta_i} \varphi(x) < f_k^* | \Sigma_k\right\} = \exp\left\{-\frac{(f_k^* - y_i)(f_k^* - y_{i+1})}{\sigma^2(x_{i+1} - x_i)}\right\} \tag{2.4.14}$$

is sufficiently small (not greater than a given number $\varepsilon_0>0$), and terminate the algorithm if all subintervals except the one corresponding f_k^* are rejected. Note that if this stopping rule is applied then the algorithm can be regarded also in the class of branch and probability bound methods described in Section 4.3. Besides, the algorithms of this subsection can be incorporated by the general scheme of one-dimensional global optimization algorithms, represented in the form of Algorithm 2.3.1.

Further information about construction and investigation methods based on use of stochastic models may be found in Kushner (1964), Archetti and Betro (1979,1980), Mockus (1988), Mockus et al. (1978), Zilinskas (1978, 1981, 1982, 1984, 1986), Boender (1984).

2.4.4 Axiomatic approach

The stochastic function models described above can be viewed as special cases of a more general axiomatic approach. According to this, the uncertainty about the values $f(x)$ for $x \in X \setminus \Xi_k$ is assumed to be representable by a binary relation $\leq_x$ where $(t,t') \geq_x (\tau,\tau')$ symbolizes that the event $\{ f(x) \in (t,t')\}$ is at least as likely as the event $\{ f(x) \in (\tau,\tau')\}$. Under some reasonable assumptions on this binary relation (e.g., transitivity and comleteness),there exists a unique density function p_x that satisfies the following condition: for every pair (A,A') of countable unions of intervals one has $A \geq_x A'$ if and only if

$$\int_A p_x(t)\,dt \geq \int_{A'} p_x(t)\,dt.$$

For the special case when all densities are Gaussian and hence are characterized by their means $\mu(x)$ and covariances $\sigma^2(x)$ one can suppose that the preference relation $\geq_x$ is defined on the set of estimators of $\mu(x)$ and $\sigma^2(x)$.

Subject again to some reasonable assumptions about this preference, the result is that the unique rational choice for the next point of evaluation of f is the one for which the probability of finding a function value, smaller than $f_k^*-\varepsilon_k$ is maximal. (This result justifies the one-step P-optimal algorithms (2.4.9)). In the case of one-dimensional Wiener process, (2.4.9) together with (2.4.10) and (2.4.11) lead to (2.4.13). In the case of higher dimension, analogies of (2.4.10) and (2.4.11) are not valid, but some approximations for $\mu(x \mid \Sigma_k)$ and $\sigma^2(x \mid \Sigma_k)$ can be axiomatically justified, e.g.

$$\mu_k(x|\Sigma_k) = \sum_{i=1}^{k} y_i W_i(x,\Sigma_k),$$

$$\sigma_k^2(x|\Sigma_k) = c_k \sum_{i=1}^{k} \|x - x_i\| W_i(x, \Sigma_k)$$

where c_k is a normalizing constant and the weights $W_i(x,\Sigma_k)$ have some natural properties, see Zilinskas (1982, 1986).

2.4.5 *Information-statistical approach*

The information-statistical approach is similar to the above described Bayesian one and was mainly developed by Strongin (1978). Its essence is the following: the feasible region $\mathbf{X}$ is discretized, i.e. a finite point collection $\Xi_N=\{x_1,...x_N\}$ is substituted for $\mathbf{X}$ and the N-vector $F=(f(x_1),..., f(x_N))$ approximates the objective function f. So, R^N is substituted for the functional set $\mathfrak{F}$. Setting prior information about f consists in setting up a prior probability density $\varphi(F)$ on R^N which must be successively transformed into a posterior density after evaluating f. The points at which to evaluate f can be determined, for instance, as the maximal likelihood estimators for x^*. This idea leads to extremely cumbersome algorithms which are practically not manageable in the multidimensional case. In the one-dimensional case, however, a slight modification of this idea led Strongin (1978) to the construction of the algorithm (2.3.1) - (2.3.4).

PART 2. GLOBAL RANDOM SEARCH

CHAPTER 3. MAIN CONCEPTS AND APPROACHES OF GLOBAL RANDOM SEARCH

The present chapter contains three sections. Section 3.1 describes and studies the simplest global random search algorithms, outlines the ways of constructing more efficient algorithms, presents a general scheme of global random search algorithms and discusses the connection between local optimization and global random search. Section 3.2 proves some general results on convergence. Section 3.3 is devoted to Markovian algorithms, that are thoroughly theoretically investigated in literature.

3.1 Construction of global random search algorithms : Basic approaches

This section can be referred to as an introduction to the methodology of global random search. It describes a few simple global random search algorithms and ways of increasing their efficiency. The simplest algorithm is considered first.

3.1.1 Uniform random sampling

According to the general concept of global optimization, any global optimization algorithm has to search in all the feasible region $\mathcal{X}$ in some way or another. The simplest of these ways is a *uniform* sampling in $\mathcal{X}$ that can be accomplished in both deterministic (described in Section 2.2.1) and stochastic fashion. The simplest stochastic (global random search) method consists in choosing the points at which f is evaluated randomly, independently and uniformly in $\mathcal{X}$.

Algorithm 3.1.1. (Uniform random search in $\mathcal{X}$).

1. Set $k=1$, $f_0^*=\infty$.
2. Obtain a point x_k by sampling from the uniform distribution on $\mathcal{X}$.
3. Evaluate $f(x_k)$ and set

$$f_k^* = \min\left\{ f_{k-1}^*, f(x_k)\right\}. \tag{3.1.1}$$

4. If $k=N$, then terminate the algorithm; choose the point x_k^* with $f(x_k^*)= f_k^*$ as an approximation for $x^*=\arg\min f$. If $k<N$, then return to Step 2 (substituting $k+1$ for k).

Algorithm 3.1.1 has some different names, viz., crude search, pure random search, random bombardment, Monte Carlo method, etc. It utilizes the simplest stopping rule which terminates the algorithm after a given number N of evaluations of f. Chapter 4 describes and studies a mathematical statistical apparatus that can be used as a basis of various stopping rules of Algorithm 3.1.1 and many others according to which an algorithm terminates by attaining a given accuracy.

While using Algorithm 3.1.1 in practice, it is usually profitable to descend locally from one or several points with the lowest function values obtained (this is true for almost all global random search algorithms, as will be discussed later in Section 3.1.5).

The simplicity of Algorithm 3.1.1 makes possible the direct investigation of its theoretical properties considered below.

Let $x^*=\arg\min f$ be a global minimizer of f. For the set $B(\varepsilon)=B(x^*,\varepsilon)$ we have

$$\mu_n\{B(\varepsilon)\} \le \mu_n\{x \in \mathbb{R}^n: \|x\| \le \varepsilon\} = \pi^{n/2}\varepsilon^n / \Gamma(n/2+1) \tag{3.1.2}$$

where μ_n is the Lebesgue measure, and (3.1.2) becomes an equality if $\{x\in\mathbb{R}^n: \|x-x^*\|\le\varepsilon\}\subset\mathcal{X}$, i.e. if the distance from x^* to the boundary of $\mathcal{X}$ is not less than ε. Using (3.1.2) we obtain for all $\varepsilon>0$, $k=1,2,\ldots$

$$\Pr\{\|x_k - x^*\| \le \varepsilon\} = \Pr\{x_k \in B(\varepsilon)\} = \mu_n\{B(\varepsilon)\}/\mu_n(\mathcal{X}) \le$$

$$\le \pi^{n/2}\varepsilon^n / [\mu_n(\mathcal{X})\Gamma(n/2+1)], \tag{3.1.3}$$

$$\Pr\left\{\min_{1\le i\le k}\|x_i - x^*\| \le \varepsilon\right\} = 1 - (1 - \mu_n\{B(\varepsilon)\}/\mu_n(\mathcal{X}))^k \le$$

$$\le 1 - \left[1 - \pi^{n/2}\varepsilon^n/(\mu_n(\mathcal{X})\Gamma(n/2+1))\right]^k \to 1,\ k\to\infty. \tag{3.1.4}$$

The relation obtained shows that the sequence

$$\min_{1\le i\le k}\|x_i - x^*\|$$

converges in probability to zero for $k\to\infty$. Moreover, there is an estimator of the convergence rate in (3.1.4). The estimator of the expected number of steps before hitting into the set $B(\varepsilon)$ is easily obtained from (3.1.3)

$$E\tau_{B(\varepsilon)} = \frac{\mu_n(\mathcal{X})}{\mu_n\{B(\varepsilon)\}} \ge \mu_n(\mathcal{X})\pi^{-n/2}\varepsilon^{-n}\Gamma(n/2+1) \tag{3.1.5}$$

where τ_A is the moment of first hit of the search sequence $x_1,x_2,\ldots$ into a set $A\subset\mathcal{X}$. These formulas estimate the rate of convergence of Algorithm 3.1.1 with respect to values of the argument. The rate of convergence with respect to function values is estimated analogously:

$$\Pr\left\{f_k^* - f^* \le \delta\right\} = \Pr\left\{x_k^* \in A(\delta)\right\} = 1 - \left(1 - \mu_n\{A(\delta)\}/\mu_n(\mathcal{X})\right)^k, \tag{3.1.6}$$

$$E\tau_{A(\delta)} = \mu_n(\mathcal{X})/\mu_n\{A(\delta)\}. \tag{3.1.7}$$

Note, that to simplify calculations in (3.1.4) and (3.1.6) one can use the approximation $(1-p)^k \approx e^{-kp}$ (valid for $p \approx 0$).

Although Algorithm 3.1.1 converges in various senses, this convergence is slow and greatly depends on the dimension n of the set $\mathcal{X}$. Let us calculate how many evaluations N of f one has to perform, in order to reach a probability not less than $1-\gamma$ (where $\gamma>0$ is a small number) of hitting into $B(\varepsilon)$. Supposing that equality holds in (3.1.2) (that certainly holds for $x^* \in \text{int}\,\mathcal{X}$ and sufficiently small ε) compare the right-hand side of (3.1.4) with $(1-\gamma)$ and solve the obtained equation with respect to $k=N$. We get

$$N = \left[\log\gamma / \log\left(1 - \frac{\pi^{n/2}\varepsilon^n}{\mu_n(\mathcal{X})\Gamma(n/2+1)}\right)\right]^+ \approx -(\log\gamma)\mu_n(\mathcal{X})\frac{\Gamma(n/2+1)}{\pi^{n/2}\varepsilon^n} \tag{3.1.8}$$

Let us take $\mu_n(\mathcal{X})=1$, $\gamma=0.1$, $\varepsilon=0.1$ and consider in Table 2 the dependence $N=N(n)$.

Table 2. The dependence N=N(n)

n	1	2	3	4	5	6	8	10	20	100
N	11	73	549	4666	43744	4.5×10^5	5.7×10^7	9×10^9	9×10^{21}	10^{140}

Some related recent results of Deheuvels (1983) and Janson (1986, 1987) concerning multivariate maximal spacings are of interest in the context of global random search theory. We shall present the principal results below.

Let $\mu_n(x)=1$, A be a cube or a ball in $\mathbb{R}^n$ of volume $\mu_n(A)=1$, $\Xi_N=\{x_1,\dots,x_N\}$ be an independent sample from the uniform distribution on $\mathcal{X}$. Set $\Delta_N=\sup\{\eta$: there exists $x\in\mathbb{R}^n$ such that $x+\eta A\subset \mathcal{X}\setminus\Xi_N\}$ and define the maximal spacing as $V_N=(\Delta_N)^n$, i.e. as the volume of the largest ball (or cube of a fixed orientation) that is contained in $\mathcal{X}$ and avoids all N points of Ξ_N. The result of Janson (1987) states that

$$\lim_{N\to\infty} \frac{NV_N - \log N}{\log \log N} = n - 1 \qquad (3.1.9)$$

almost surely, and for $N\to\infty$ the sequence

$$NV_N - \log N - (n-1) \log \log N + \beta$$

converges in distribution to a random variable with c.d.f. $\exp\{-e^{-t}\}$, where $\beta=0$ if A is a cube and

$$\beta = \log \Gamma(n+1) - (n-1)\log\left[\sqrt{\pi}\ \Gamma(n/2+1)/\Gamma\left(\frac{n+1}{2}\right)\right]$$

for the case, when A is a ball. As the latter value $\beta \geq 0$, spherical spacings are somewhat smaller (for $n \geq 3$) than the cubical ones.

A related result on the multivariate maximal spacing in the particular case $\mathcal{X}=A=[0,1]\times[0,1]$ was presented by Isaac (1988). With respect to the asymptotic study of the c.d.f. $F_{N,n}(t)=\Pr\{V_N<t\}$ of maximal cubic spacings, it states that

$$\lim_{N\to\infty} \frac{1 - F_{N,n}(t)}{N^n (1-t)^N} = c_n(t) \qquad (3.1.10)$$

holds for n=2 and each $t\in(0,1)$, where $c_n(t)$ is a constant depending on n and t. It is widely known that for the univariate case (i.e. for $\mathcal{X}=A=[0,1]$) the relation (3.1.10) holds with $c_n(t)=1$.

Some further investigation of the properties of the uniform random search algorithm can be found in Anderson and Bloomfield (1975), Yakowitz and Fisher (1975).

The uniform random search algorithm finds major applications in global random search theory as a pattern in theoretical and numerical comparison of algorithms and also as a component of many global random search algorithms. It is used also for investigating diverse procedures of mathematical statistics.

The slow convergence rate of Algorithm 3.1.1 has served as a reason for creating a great number of generalizations and modifications discussed below.

3.1.2 General (nonuniform) random sampling

A random independent sampling of points in $\mathcal{X}$ with some given nonuniform distribution is the simplest generalization of Algorithm 3.1.1. It is used in cases, when (i) information concerning the objective function is available (perhaps obtained through some previous sampling) permitting to prefer some subsets of $\mathcal{X}$ more than others, or when (ii) the sampling problem for the uniform distribution on $\mathcal{X}$ is hard or practically unsolvable. The algorithm is as follows.

Algorithm 3.1.2. (General (nonuniform) random sampling of points in $\mathcal{X}$).

1. Choose a probability distribution P on $\mathcal{X}$, set k=1, $f_o^*=\infty$.
2. Obtain a point x_k by sampling from P.
3. Evaluate $f(x_k)$ and determine f_k^* (see (3.1.1)).
4. If k=N, then terminate the algorithm: choosing the point x_k^* with $f(x_k^*)=f_k^*$ as an approximation for x*. If k<N, then return to Step 2 (substituting k+1 for k).

In the case of uniform distribution $P(dx)=dx/\mu_n(\mathcal{X})$: hence Algorithm 3.1.1 is a special case of Algorithm 3.1.2.

For the convergence of Algorithm 3.1.2 the continuity of f in a vicinity of a global minimizer x* and the validity of the inequality $P(B(\varepsilon))>0$ for each $\varepsilon>0$ are sufficient. In this case, the convergence statements concerning Algorithm 3.1.1 are valid, with corresponding modifications for Algorithm 3.1.2. The analogy with (3.1.4) - (3.1.8) are

$$\Pr\left\{\min_{1\le i\le k}\|x_i - x^*\| \le \varepsilon\right\} = 1-(1-P(B(\varepsilon)))^k \to 1, \quad k\to\infty,$$

$$E\tau_{B(\varepsilon)} = 1/P(B(\varepsilon)),$$

$$\Pr\{x_k^* \in A(\delta)\} = 1-(1-P(A(\delta)))^k \to 1, \quad k\to\infty.$$

$$E\tau_{A(\delta)} = 1/P(A(\delta)),$$

$$N = [\log\gamma/\log(1-P(B(\varepsilon)))]^+ \approx -(\log\gamma)/P(B(\varepsilon)).$$

The practical significance of Algorithm 3.1.2 is connected mainly with the fact that it may be used as a component in more complicated algorithms, see later Section 3.1.5. Besides, if the independence condition for $\{x_k\}$ and the identity of their distributions are weakened, then more practical algorithms can be constructed: such methods will be described in Sections 3.3.2, 3.3.5.

3.1.3 Ways of improving the efficiency of random sampling algorithms

The low efficiency of Algorithms 3.1.1 and 3.1.2 is caused largely by their passive character: viz., they do not use the previously obtained information when selecting new points. The ways of improving their efficiency are connected with various modes of evaluation and use of the information obtained during the search and/or given *a priori*.. The corresponding global random search algorithms are sometimes called sequential or adaptive, in order to contrast them with the passive ones.

A simple idea of covering of $\mathcal{X}$ by balls with centres at points generated at random is close to that discussed in Section 2.2. It can be properly realized if e.g. Lipschitzian information about the objective function is available. Section 3.1.4 describes the corresponding algorithms.

A simple manner of including adaptive elements into the global random search technique consists of determining a distribution for x_{k+1} as depending on the previous point x_k and objective function value $f(x_k)$. The corresponding algorithms are called Markovian (since the points $x_1,x_2,...$ generated by them form a Markov chain) and will be studied in Section 3.3. Their theoretical properties are intensively studied nowadays, but prospects of their practical usefulness are still not quite clear.

An important way of improving efficiency in global random search algorithms is connected with the inclusion of local descent techniques, see later Section 3.1.6. A simple algorithm of such a kind is the well-known random multistart consisting of multiple local descents from uniformly chosen random points. It will be theoretically studied in Section 4.5. Its theoretical efficiency is rather low, but some of its heuristic modifications described in Sections 2.1.3 and 3.1.6 can be regarded as fairly efficient for complicated optimization problems.

Another important means of constructing efficient global random search algorithms consists of using mathematical statistics procedures for deriving information about the objective function. Such information can serve, in particular, to check the obtained accuracy for many algorithms and to determine corresponding stopping rules. Chapters 4 and 7 are devoted to various problems connected with the construction of statistical inference procedures and their application in global random search.

A further direction of improving global random search efficiency is to reduce the share of randomness. For instance, Section 2.2.1 shows that the method consisting of evaluation of f at quasirandom points is more efficient than Algorithm 3.1.1. However, nonrandom points are in some sense worse, than random ones, due to the following reasons: (i) generally, statistical inferences for deriving information about f can not be drawn if the points are not random, (ii) if the structure of $\mathcal{X}$ is not simple enough, then the problem of constructing nonrandom grids is usually much more complicated, than that for random grids.

A stratified sample may be regarded as an intermediate between random and quasirandom ones. To construct such a sample of size $N=m\ell$, the set $\mathcal{X}$ is divided into m subsets of equal volume and the uniform distribution is sampled ℓ times in each subset. Already Brooks (1958) pointed out some advantages of stratified sampling as a substitute for pure random search. But only recently the gains caused by this substitution were correctly investigated. (Section 4.4. contains these results as well as suitable statistical inferences.)

Many global random search algorithms are based on the idea of more frequent selection of new points in the vicinity of *good* points, i.e. those earlier obtained ones in which the values of f are relatively small. Corresponding methods (which go under the name of methods of generations) will be considered in Chapter 5. Note that many methods of generations can be used also for the case when a random noise is present in the evaluations of f .

The approaches mentioned do not cover completely the variety of global random search methods. Some of them have been described above (in particular, Algorithm 2.3.1 and the method of Section 2.1.6 based on smoothing the objective function). Many others can be easily constructed by the reader: to do this, it suffices to enter a random element into any deterministic algorithm. (This method was used e.g. by Lbov (1972) who proposed to seek the minimum of the minorant (2.2.30) of the objective function by a random sampling algorithm, and thus transformed the polygonal line method of Section 2.2.2 into a global random search algorithm).

3.1.4 Random coverings

The idea of covering studied in Section 2.2 is used here to improve the efficiency of the simplest random search algorithms.

First, let us introduce the notation P_Z for the uniform distribution on a set Z and consider the algorithm of Brooks (1958, 1959) in which $\mathcal{X}$ is covered by balls of equal radius ε with centres at uniformly chosen random points.

Algorithm 3.1.3. (uniform random covering).

1. Set k=1, $f_o^*=\infty$, $Z_1=\mathcal{X}$.
2. Obtain a point x_k by sampling from P_{Z_k}.
3. Evaluate $f(x_k)$ and determine f_k^*.
4. If k=N or the volume of Z_k is sufficiently small, then terminate the algorithm: choose the point x_k^* with $f(x_k^*)= f_k^*$ as an approximation for x*. If k<N, then set $Z_{k+1}=Z_k \setminus B_\varepsilon(x_k)$ and return to Step 2 (substituting k+1 for k).

Provided that the distributions P_Z are sampled in Algorithm 3.1.3 using the rejection technique (i.e. the distribution $P_\mathcal{X}$ is sampled until a realization occurs in Z), it differs from Algorithm 3.1.1 in the following detail only: if at the k-th iteration of Algorithm 3.1.3 a random point, uniformly distributed in $\mathcal{X}$, falls into a ball $B(x_i,\varepsilon)$ whose centre is a previously accepted point x_i ($1\le i\le k$), then this point is rejected, f is not evaluated at it, and a new random point is generated.

Radius ε of the balls in Algorithm 3.1.3 determines the accuracy of the approximation. Thus, if $f\in Lip(\mathcal{X},L)$ and $\varepsilon=\delta/L$ then the objective function values at the rejected points (each of them belonging to a ball $B(x_i,\varepsilon)$, $1\le i\le N$) can not exceed $f_N^*+\delta$.

Certainly, instead of the balls $B(x_i,\varepsilon)$ one can use other sets $S(x_i)$ containing x_i as the *forbidden* sets for new points in Algorithm 3.1.3. For $\mathcal{X}=[0,1]^n$ it is easy to investigate the algorithm in which $S(x_i)$ are the η-adic cubes (these cubes are obtained by dividing each side of the cube $\mathcal{X}$ into η equal parts, see Section 2.2.1). Indeed, for this algorithm we have

$$\Pr\{x_i \in S(x^*) \text{ for some } i \le k\} =$$

$$= 1 - \prod_{j=1}^{k} \left(1 - \frac{c}{1-(j-1)c}\right) = c\,k$$

for each $k \le \eta^n$ instead of (3.1.4) where $c = \eta^{-n}$ is the volume of each η-adic cube $S(x_i)$.

Algorithm 3.1.3 does not use the information which is contained in the values of f and so its efficiency can not be high. The following method of Devroye (1978) constructed under the supposition $f \in Lip(\mathcal{X}, L, \rho)$ is of higher efficiency.

Algorithm 3.1.4. (nonuniform random covering).

1. Set $k=1$, $f_o^*=\infty$, $Z_1=\mathcal{X}$.
2. Obtain a point x_k by sampling P_{Z_k}.
3. Evaluate $f(x_k)$ and determine f_k^*.
4. Compute

$$\eta_i = \left(f(x_i) - f_k^* + \delta\right)/L \tag{3.1.11}$$

for each $i=1,...,k$ where δ is a given positive number.

5. Set

$$Z_{k+1} = \mathcal{X} \setminus \bigcup_{i=1}^{k} B(x_i, \eta_i, \rho)$$

6. Terminate the algorithm, if Z_{k+1} is empty or $k=N$, otherwise return to Step 2 (substituting $k+1$ for k).

Algorithm 3.1.4 is a typical covering method: it differs from the methods of Section 2.2.2 in the way of obtaining points x_k. According to the results of Section 2.2.2, Algorithm 3.1.4 finds a global minimizer with the accuracy δ with respect to values of f, i.e. $f_k^* - f \le \delta$ where k is the last iteration index for which the set Z_{k+1} becomes empty.

It is not necessary to use just P_{Z_k}, the uniform distribution on Z_k: at each k-th iteration of Algorithm 3.1.4, any distribution P_k on the set Z_k can be used instead of P_{Z_k}. In order to ensure the convergence of the algorithm Devroye (1978) supposed

$$P_k = \alpha_k P_{Z_k} + (1-\alpha_k) G_k \text{ where } \alpha_k \ge 0, \qquad \sum_{k=1}^{\infty} \alpha_k = \infty$$

and G_k is an arbitrary distribution on Z_k (for instance, corresponding to performing some local descent iterations). Algorithm 3.1.4 can be modified for the case when $\mathcal{F}=C(\mathcal{X})$, i.e. f is a continuous function. Devroye (1978) proved for this case that if the sequence $\{\alpha_k\}$ is defined as above and

$$\eta_i = \beta_k\left(f(x_i) - f_k^*\right)$$

instead of (3.1.11), where $\beta_k>0$, $\beta_k \to 0$ for $k\to\infty$, then the algorithm converges almost surely with respect to the values of f. Closely related general convergence investigations can be found e.g. in Solis and Wets (1981) or Pintér (1984).

3.1.5 Formal scheme of global random search

We shall describe here a formal scheme of global random search algorithms that may be useful investigating some of their general properties.

Algorithm 3.1.5.

1. Choose a probability distribution P_1 on $\mathcal{X}$, set $k=1$.
2. Obtain points

$$x_1^{(k)},\dots,x_{N_k}^{(k)}$$

by sampling N_k times from the distribution P_k. Evaluate f (perhaps, with a random noise) at these points.
3. According to a fixed (algorithm-dependent) rule construct a probability distribution P_{k+1} on $\mathcal{X}$.
4. Check some appropriate stopping condition; if the algorithm is not terminated, then return to Step 2 (substituting $k+1$ for k).

In other words, any global random search algorithm involves some iterations; at each iteration, a suitably constructed distribution is sampled. For several classes of algorithms (e.g., described in Sections 3.3. and 5.4) $N_k=1$ for each $k=1,2,\dots$ and their representation in the form of Algorithm 3.1.5 is completely formal and gives nothing. But for some others (e.g., described later in Sections 4.3, 5.1 and 5.3) this representation is essential and helps to understand the algorithm and the possibilities of its theoretical study.

The construction of the distributions $\{P_{k+1}\}$ in Algorithm 3.1.5 determines the way of deriving and using the information given a priori or received during the search concerning the objective function. As a rule, they can be written like

$$P_{k+1}(dx) = \int R_k(dz)\, Q_k(z,dx) \qquad (3.1.12)$$

where R_k is a probability distribution on $\mathcal{X}$ and $Q_k(z,.)$ is a (Markovian) transition probability, i.e. a measurable nonnegative function with respect to the first argument and a probability measure with respect to the second. The distribution (3.1.1) is sampled applying the superposition technique, viz., first R_k and then $Q_k(z,.)$ is sampled, where z is the realization obtained via sampling R_k.

The sense of R_k and $Q_k(z,.)$ in the representation (3.1.12) is different. Namely, the construction of R_k takes into account the information of a global feature of f, but that of $Q_k(z,.)$ considers a local feature of f. In sampling R_k, a point z from all $\mathcal{X}$ is chosen, but in sampling $Q_k(z,.)$ a point from the neighbourhood of z is selected (the word *neighbourhood* is meant here in the probabilistic sense: *with large probability near enough*).

The way of constructing distributions R_k establishes a general structure and originality of the algorithm to a great degree. The description of such ways and the study of corresponding algorithms is the main object of Chapters 3, 4, 5. (Let us remark in advance that in the Markovian algorithms of Section 3.3, and Algorithm 5.1.1 the distributions R_k are concentrated at one of the earlier obtained points; in most methods of generations of Chapter 5 - at the points of a preceding iteration; and in branch and probability bound methods of Section 4.3 the distributions are uniform on subsets of $\mathcal{X}$ which have been recognized as *promising* for further search.

Alternatively, a choice of transition probabilities $Q_k(z,.)$ determines a local behaviour of an algorithm. Let us study now some ways of this choice.

3.1.6 Local behaviour of global random search algorithm

Consider first the interdependence of local and global behaviour of global random search algorithms as presented in Algorithm 3.1.5 with the distribution P_{k+1} in (3.1.12). The algorithm design should take into account the desired solution accuracy: if this accuracy is high, then the algorithm must have good local performance; but if the aim is to hit the set $A(\delta)$ for some not too small $\delta>0$, then the local properties can result in an inefficiency - the better is the local performance, the higher is the amount of additional evaluations of f. Anyway, at the first iterations (for small k) where the main search task is to explore the global properties of f, a simple choice of the transition probabilities is recommandable requiring no additional evaluations of f and being convenient for theoretical study. For instance, such a choice is

$$Q_k(z,dx)=\varphi_k(x-z)\,dx/\int_{\mathcal{X}}\varphi_k(y-z)\,dy \qquad (3.1.13)$$

where φ_k is a distribution density in $\mathbb{R}^n$. In order to obtain a random realization x_k in $\mathcal{X}$ from the distribution (3.1.13), one needs to obtain a realization ξ_k in $\mathbb{R}^n$ from the density φ_k, to check the relation $z+\xi_k\in\mathcal{X}$ (if it does not hold, then a new realization ξ_k is needed), and take $x_k=z+\xi_k$. The choice (3.1.13) is natural only in the case, when a random noise is present in the evaluations of f. Similarly, the transition probabilities are often selected in the form

$$Q_k(z,A)=\int 1_{[x\in A,\, f(x)\le f(z)]}T_k(z,dx)+ \qquad (3.1.14)$$

$$+1_A(x)\int 1_{[f(z)<f(x)]}T_k(z,dx)$$

where $T_k(z,dx)$ is a Markovian transition probability having the form of (3.1.13). Given a fixed z, in order to obtain a realization x_k from the distribution (3.1.14) one needs first to get a realization ζ_k from the distribution $T_k(z,.)$ and put

$$x_k = \begin{cases} \zeta_k & \text{if } f(\zeta_k) \le f(z) \\ z & \text{otherwise}. \end{cases}$$

Attempts to determine the details of objective function behaviour all over the set $\mathcal{X}$, by means of a small number of evaluations, cannot be successful; rough approximations of f can be used for determining the points where many evaluations of f should be carried out for an expected deep local descent only in the simplest cases.

On the other hand, if there are good reasons to believe that some points $x_j^{(k)}$ $(j=1,...,N_k)$ of the k-th iteration of Algorithm 3.1.5 are not very far from a global optimizer, then local descent may be profitable. The depth of the descent is defined by the transition probabilities $Q_k(z,.)$: for a fixed realization z of R_k, the algorithm of choosing the point $x_j^{(k+1)}$ consists in carrying out several (possibly, one) local descent iterations defined by the method of $Q_k(z,.)$ sampling. If Q_k has the form of (3.1.13), then the local descent is not performed and $x_j^{(k+1)}$ is chosen in a vicinity of z; if Q_k has the form of (3.1.14), which is possible only if there is no evaluation noise, then a single local random search iteration is done (with return for unprofitable steps). One can directly see how the method of sampling Q_k should be defined in order to correspond to one or several iterations of either other local random search algorithms (they are thoroughly described e.g. in Rastrigin (1968)), or of any deterministic local descent (in the latter case the distributions $Q_k(z,.)$ are degenerated for each z), or even of stochastic approximation type algorithms (in the case of evaluation noise).

It is also evident how to make the number of local iterations from z to be proportional to the evaluated value of $f(z)$. The extremal case where sampling $Q_k(z,.)$ corresponds to the transition from the point z to a local minimizer, (whose domains of attraction z belongs to) is unlikely to be acceptable, except very simple situation. Optimization problems, of course, occur where evaluation of the derivatives of f is rather simple; in this case, local descent iterations may prove useful already at the first iterations. Corresponding algorithms can be regarded as modifications of the random multistart method, see Sections 2.1.3 and 4.5.

It is worth to note that there is no need to know the analytical form of the transition probabilities $Q_k(z,.)$, as well as the distribution R_k: one needs only an algorithm for their sampling, i.e. for passing from z to the point of the (k+1)-st iteration The efficiency of a global search algorithm, hence, may be improved if the complexity of an algorithm for sampling $Q_k(z,.)$ is increased with k (including also additional evaluations of f). In doing so, it seems natural to take smaller N_k for greater values k than for small ones. The quantities N_k and the indices of iteration, where sampling algorithms for $Q_k(z,.)$ become more complicated can be defined in advance (using the prior knowledge about the objective function bahaviour and the accuracy of resulting approximations) as well as in the course of the search (using the obtained information concerning f).

3.2 General results on the convergence of global random search algorithms

A number of works are devoted to the derivation of sufficient conditions for convergence of general global random search algorithms: notable examples include Devroye (1976, 1978), Marti (1980), Solis and Wets (1981), Pintér (1984). These works contain results similar to Theorem 3.2.1 prsented below as well as consequences for some particular methods (including stochastic programming methods when the objective function is subjected to a random noise and the incidence $x \in \mathcal{X}$ is valid with a probability depending on x). Instead of using the results of the above mentioned works, we shall derive the convergence results relying upon the classical zero-one law.

Consider a general global random search method represented in the form of Algorithm 3.1.5; without loss of generality assume that $N_k=1$ for all k=1,2,.. , that is a separate distribution P_k is constructed for each point $x_k=x_1^{(k)}$.

Theorem 3.2.1. Let f be continuous in the vicinity of a global minimizer x* of f, and assume that

$$\sum_{k=1}^{\infty} q_k = \infty \tag{3.2.1}$$

for any $x \in \mathcal{X}$ and $\varepsilon > 0$ where

$$q_k = q_k(x^*, \varepsilon) = \underset{\Xi_{k-1}}{\text{vrai inf}}\, P_k(B(\varepsilon)), \qquad \Xi_{k-1} = \{x_1, \ldots, x_{k-1}\}.$$

Then for any $\delta>0$ the sequence of random vectors $x_1, x_2, \ldots$ generated by Algorithm 3.1.5 with $N_k=1$ (k=1,2,...) falls infinitely often into the set $A(\delta)$ with probability one.

Proof. Fix $\delta>0$ and then find $\varepsilon>0$ such that $B(\varepsilon) \subset A(\delta)$. Determine the sequence of independent random variables $\{\eta_k)$ on the two point set $\{0,1\}$ so as to obtain for fixed $\varepsilon>0$: $\Pr\{\eta_k=1\}=1-\Pr\{\eta_k=0\}=q_k(x^*,\varepsilon)$. Obviously, the probability of x_k falling into $B(\varepsilon)$ is, for all k=1,2,... , not less than the probability q_k of η_k getting into the state 1, and, therefore, the theorem's assertion will be proved if one demonstrates that the sequence η_k infinitely often takes the value 1. Since the latter follows by (3.2.1) and the first part of Borel's zero-one law, Theorem is proved.

Theorem 3.2.1 is valid also for the general case when function f is subject to random noise. If the noise is absent, then the conditions of the theorem ensure that the point sequence $\{x_k\}$ converges to the set $\mathcal{X}^* = \{\arg\min f\}$ of global minimizers with probability 1.

In virtue of the Borel-Cantelli lemma, one can see that if

$$\sum_{k=1}^{\infty} P_k(B(\varepsilon)) < \infty,$$

then the points $x_1, x_2, \ldots$ fall into $B(\varepsilon)$ at most a finite number of times. Moreover, as is illustrated by the function

$$\min\{1, (x-\varepsilon)^2/\varepsilon^2\} + \min\{1, \varepsilon + (1-\varepsilon-x)^2/\varepsilon\}, \qquad x \in \mathcal{X} = [0,1],$$

where ε is a sufficiently small positive number, (3.2.1) cannot be improved for a rather wide class of functions $\mathcal{F}$ (even for the class of analytical functions with two minima), i.e. if (3.2.1) is not satisfied, there exist $f \in \mathcal{F}$ and $\varepsilon>0$ such that none of the points $x_1, x_2, \ldots, x_N$ falls into $B(\varepsilon)$ with a positive probability where N is an arbitrarily great fixed number.

Since the location of x^* is not known *a priori*, instead of (3.2.1) some more strict, but simpler requirement

$$\sum_{k=1}^{\infty} \operatorname{vrai} \inf_{\Xi_{k-1}} P_k(B(x,\varepsilon)) = \infty \tag{3.2.2}$$

can be prescribed for all $\varepsilon>0$, $x \in \mathcal{X}$.

Let us consider some known ways of selecting probability measures P_k for which the fulfilment of (3.2.2) may be easily guaranteed.

In practice, one of the most popular ways of selection of distributions P_k is

$$P_k = \alpha_k P_{\mathcal{X}} + (1-\alpha_k) G_k \tag{3.2.3}$$

where $0 \le \alpha_k \le 1$, $P_{\mathcal{X}}$ is the uniform distribution on $\mathcal{X}$, G_k is an arbitrary probability measure on $\mathcal{X}$ (for instance, the sampling of G_k may correspond to performing some iterations of a local descent from the point $x_{k-1}{}^*$). The sampling from the distribution (3.2.3) means that of $P_{\mathcal{X}}$ with probability α_k and that of G_k with the complement probability $1-\alpha_k$. In spite of the quite common belief that

$$\liminf_{k \to \infty} \alpha_k > 0$$

should be satisfied for the corresponding algorithm to converge, the weaker requirement

$$\sum_{k=1}^{\infty} \alpha_k = \infty$$

is sufficient for (3.2.2) for ensuring convergence.

In practice, one often chooses P_k also in the form of (3.1.12) and the Markov transition probabilities $Q_k(z,.)$ in the form of (3.1.13) (in the case of evaluations subjected to a noise) or (3.1.14) where $T_k(z,.)$ are chosen following (3.1.13). To satisfy (3.2.2) in the above cases, it is obviously sufficient that for each k=1,2,... the density φ_k of (3.1.13) be represented as $\varphi_k(x)=\beta_k^{-n}\varphi(x/\beta_k)$, where $\{\beta_k\}$ is a nonincreasing sequence of positive numbers and the density $\varphi(x)$ is symmetrical, continuous, decreasing for x>0, decomposes into a product of one-dimensional densities, and that the condition

$$\sum_{k=1}^{\infty}\int_{d_1}^{d_1+\varepsilon}\cdots\int_{d_n}^{d_n+\varepsilon}\beta_k^{-n}\varphi(x/\beta_k)\,dx=\infty \tag{3.2.4}$$

be satisfied where d_i is the diameter of $\mathcal{X}$ with respect to the i-th coordinate, $\varepsilon>0$ is arbitrary. It is, of course, difficult to check (3.2.4) in the general case, but it becomes rather simple for some particular choices of φ. Consider e.g. the following methods of choosing φ:

$$\varphi(x)=2^{-n}\lambda_1\ldots\lambda_n\exp\left\{-\sum_{i=1}^{n}\lambda_i\,|x(i)|\right\}, \tag{3.2.5}$$

$$\varphi(x)=(2\pi)^{-n/2}(\lambda_1\ldots\lambda_n)^{-1/2}\exp\left\{-\sum_{i=1}^{n}\lambda_i x^2(i)\right\} \tag{3.2.6}$$

where x=(x(1),...,x(n)), and $\lambda_i>0$ (i=1,...,n) are arbitrary constants. The coordinates of random vectors distributed with densities (3.2.5) and (3.2.6) are independent and distributed according to the Laplace and Gaussian distributions, respectively.

Lemma 3.2.1. Let φ be defined according to (3.2.5). Then (3.2.4) is satisfied if for any b>0 the following relation holds:

$$\sum_{k=1}^{\infty} b^{1/\beta_k}=\infty. \tag{3.2.7}$$

In particular, β_k ($k\geq 3$) can be chosen as $\beta_k=c_1/\log(c_2\log k)$ where $c_1>0$, $c_2>1$ are arbitrary.

Proof. Set $\lambda=\max\{\lambda_1,\ldots,\lambda_n\}$. The condition (3.2.4) will be satisfied for the density (3.2.5), if it is satisfied for the density

$$\varphi_*(x)=(\lambda/2)^n\exp\left\{-\lambda\sum_{i=1}^{n}|x(i)|\right\}.$$

(3.2.4) is satisfied if for any $\varepsilon>0$

$$\sum_{k=1}^{\infty} \int_{d}^{d+\varepsilon} \cdots \int_{d}^{d+\varepsilon} \beta_k^{-n} \varphi_*(x/\beta_k) dx = \sum_{k=1}^{\infty} \left[\left(1 - e^{-\lambda\varepsilon/\beta_k}\right) e^{-\lambda d/\beta_k} \right]^n = \infty$$

is satisfied, where $d=\max\{d_1,\ldots,d_n\}$. Since

$$1 - \exp(-\lambda\varepsilon/\beta_k) \geq 1 - \exp(-\lambda\varepsilon/\beta_1) > 0$$

for all k=1,2,... , the latter condition does hold if (3.2.7) is met with $b=\exp(-n\lambda d)$. The lemma is proved.

Similar reasoning takes place for (3.2.6), where instead of (3.2.7) we have

$$\sum_{k=1}^{\infty} b^{1/\beta_k^2} = \infty,$$

and a possible choice of parameters β_k is

$$\beta_k = c_1(\log(c_2 \log k))^{-1/2} \quad \text{for} \quad c_1 > 0,\ c_2 \geq 1,\ k \geq 3.$$

To obtain sufficient conditions for convergence of global random search algorithms, Moiseev and Nekrutkin (1987) generalized some results of stochastic approximation theory and proved the next statement (an analogous proof can be found also in Zhigljavsky (1985)).

Theorem 3.2.2. Let f be a bounded function on $\mathcal{X}$ attaining its global minimum at the unique point x^*. Assume, that in the vicinity of x^*, f is continuous, and for any $\varepsilon>0$ there exists $\delta=\delta(\varepsilon)$ such that

$$\sum_{k=0}^{\infty} \operatorname*{vrai\,inf}_{\Xi_k \in (A(\varepsilon))^k} P_{k+1}\left(\left\{ x \in \mathcal{X} : f(x) < f_k^* - \delta \right\}\right) = \infty. \tag{3.2.8}$$

Then $x_k^* \to x^*$ for $k\to\infty$ with probability 1.

It seems that the condition (3.2.8) is a bit less restrictive, then (3.2.1). But it is unknown yet whether the consequences of Theorem 3.2.2 are more interesting than those of Theorem 3.2.1.

One may qualitatively interpret the relationship (3.2.8) as follows. If $x_k^*=x$ and in the vicinity of x the function f varies intensively (it implies that x is far from the local minimizers), then the search has to be local (i.e. x_{k+1} must be located close to x) with a sufficiently high probability. If in the vicinity of x the function f is slowly varying (and thus it is likely that x is close to a local minimizer of f) but it is hardly possible that x is close to x* then the search must be global, i.e. x_{k+1} must be far from x with a high probability. At last, if in the vicinity of x function f is slowly varying and it is possible that x is in the region of attraction of x*, then it is worthwhile to alternate local and global steps: the first ones ensure convergence to x* (if the region of attraction was really attained), while the second ensure divergence of the series (3.2.8) (if the region was not attained).

Finally, let us indicate a general result on the speed of convergence of global random search algorithms. Set

$$\alpha(k,\beta,\varepsilon) = \underset{\Xi_k \in (A(\varepsilon))^k}{\text{vrai inf}} \Pr\left\{ f^*_{k+1} < \beta f(x_1) + (1-\beta) f^*_k \right\},$$

$$\gamma_r(k,\varepsilon) = \sup_{0<\beta<1} \left(1-(1-\beta)^r\right)\alpha(k,\beta,\varepsilon), \quad c_r = E\left(f(x_1) - f(x_2)\right)^r.$$

Then, according to Moiseev and Nekrutkin (1987), the inequality

$$E\left(f(x^*_N) - f(x_1)\right)^r \le \inf_{\varepsilon>0} \left(\varepsilon^r + c_r \prod_{k=1}^{N-1} (1-\gamma_r(k,\varepsilon)) \right)$$

holds for all $r>0$, $N\geq 1$.

It should be noted that the above stated general results on convergence and on convergence speed are mainly of theoretical importance.

3.3 *Markovian algorithms*

Markovian algorithms of global optimization represent an advanced field of random search theory. These are algorithms based on sampling Markov chains, giving a sequence of random points $x_1, x_2, \ldots$ such that x_{k+2} is independent of the collection of points and objective function values

$$\Omega_k = \{x_1, \ldots, x_k, f(x_1), \ldots, f(x_k)\} \tag{3.3.1}$$

for each k=1,2,...

The great theoretical importance of Markovian algorithms is not equivalent to their practical significance. They are inefficient for complicated problems owing to the neglection of the information contained by (3.3.1): this finding is confirmed by various numerical results.

This section summarizes the recent results on Markovian algorithms: its subsections present a general algorithm scheme, describe the well-known simulated annealing method as well as some particular algorithms, and analyse methods connected with the solution of stochastic differential equations.

3.3.1 General scheme of Markovian algorithms

Let y_k (k=1,2,...) be a result of evaluating the objective function f at a point x_k (may be, this evaluation is subject to noise); further let $P_{k+1}(\cdot)=P_{k+1}(\cdot|\, x_1,y_1,\ldots,x_k,y_k)$ be a probability distribution of a point x_{k+1} generated by a global random search algorithm.

The Markovian property of the algorithm means that for all k=1,2,... the distributions P_{k+1} depend only on x_k, y_k, that is

$$P_{k+1}(\cdot|\, x_1, y_1, \ldots, x_k, y_k) = P_{k+1}(\cdot|\, x_k, y_k). \tag{3.3.2}$$

If the evaluations of f are not subject to noise, then (3.3.2) takes the form $P_{k+1}(\cdot|\, \Omega_k)= P_{k+1}(\cdot|\, x_k,y_k)$.

In essence, Markovian global random search algorithms are modifications of local ones, alternating local steps with global ones. If the evaluations of f are subject to noise, these algorithms are sometimes named multi extremal stochastic approximation algorithms due to the analogy with the local case (for instance, see Vaysbord and Yudin (1968)).

Different variants of Markovian global random search algorithms were proposed and studied by many authors, starting from the late 1960's. In spite of abundance of works, many algorithms differ only in secondary details. Below we shall describe the principal algorithms, starting with a general scheme for the case $\mathcal{X}\subset\mathbb{R}^n$.

Algorithm 3.3.1. (General scheme of Markovian algorithms).

1. Sampling a given distribution P_1, obtain a point x_1. Evaluate y_1, the value (may be, subject to noise) of the objective function f at the point x_1. Set k=1.

2. Obtain a point z_k from $\mathbb{R}^n$ by sampling a distribution $Q_k(x_k,.)$ with the density $q_k(x_k,x)$ depending on k and x_k.

3. If $z_k \notin \mathcal{X}$, return to Step 2. Otherwise evaluate η_k, the value of f in z_k (η_k may be subject to noise) and set

$$x_{k+1} = \begin{cases} z_k & \text{with probability } p_k \\ x_k & \text{with probability } 1-p_k \end{cases} \qquad (3.3.3)$$

where $p_k=p_k(x_k, z_k, y_k, \eta_k)$ is the acceptance probability and may depend on k, x_k, z_k, y_k, η_k.

4. Set

$$y_{k+1} = \begin{cases} \eta_k, & \text{if} \quad x_{k+1} = z_k, \\ y_k, & \text{if} \quad x_{k+1} = x_k. \end{cases}$$

5. Check some given stopping criterion. If the algorithm does not stop, then return to Step 2 (substituting k+1 for k).

An ordinary way of realizing (3.3.3) consists of calculating p_k, obtaining a random number α_k, checking the inequality $\alpha_k \le p_k$, and setting

$$x_{k+1} = \begin{cases} z_k, & \text{if} \quad \alpha_k \le p_k, \\ x_k, & \text{if} \quad \alpha_k > p_k. \end{cases}$$

Particular choices of initial probability P_1, transition probabilities $Q_k(x,.)$, and acceptance probabilities $p_k(x,z,y,\eta)$ lead to a concrete Markovian global random search algorithms. To obtain convergence conditions the results of Section 3.2 can be used: however, the simplicity of Markovian algorithms allows to get more specific results on their convergence and convergence speed. These are interesting theoretical results, together with simplicity and plain interpretation which explain the popularity of such algorithms. One of the most well-known is simulated annealing, considered first.

3.3.2 Simulated annealing

The simulated annealing method was recently proposed by Kirkpatrick et al. (1983) and in a short time became popular.

It was called simulated annealing because its similarity to the physical procedure called annealing used to remove defects from metals and crystals by heating them locally near the defect to dissolve the impurity and then slowly recooling them so that they could find a basic state with a lower energy configuration. The simulated annealing method can be referred to as a variation of the classical Metropolis method (see Metropolis et al. (1953), or Bhanot (1988)) which simulates the behaviour of an ensemble of atoms in equilibrium at a given temperature: it is a powerful technique in numerical studies in several branches of science, being useful in situations when it is necessary to generate random variables distributed according to a given multivariate probability distribution.

Let the feasible region $\mathcal{X}$ be a discrete set or a compact subset of $\mathbb{R}^n$ and $f: \mathcal{X} \to (0,\infty)$ be a positive function (not subject to noise). If $\mathcal{X}$ is discrete, then the uniform measure on $\mathcal{X}$ replaces the Lebesgue measure and for any function g defined on $\mathcal{X}$ the symbol $\int g(x)dx$ stands for

$$\sum_{x_i \in \mathcal{X}} g(x_i).$$

According to the physical interpretation, a point $x \in \mathcal{X}$ corresponds to a configuration of atoms of the substance and $f(x)$ determines the energy of the configuration. Because of a very large number of atoms and possible arrangements there are a great number of local energy minimum configurations (that is local minimizers of f).

The standard variant of simulated annealing is as follows.

An initial point $x_1 \in \mathcal{X}$ is chosen arbitrarily. Let x_k be the point of a k-th step ($k \geq 1$), $f(x_k)$ be the corresponding objective function value, β_k be a positive parameter, and ξ_k be a realization of a random vector having some probability distribution Φ_k (if $\mathcal{X} \subset \mathbb{R}^n$ then it is natural to choose Φ_k as an isotropic distribution on $\mathbb{R}^n$). Then one should check the inclusion $z_k = x_k + \xi_k \in \mathcal{X}$ (otherwise return to obtaining a new realization ξ_k) evaluate $f(z_k)$ and set $\Delta_k = f(z_k) - f(x_k)$,

$$p_k = \min\{1, \exp(-\beta_k \Delta_k)\} = \begin{cases} 1 & \text{if } \Delta_k \leq 0 \\ \exp(-\beta_k \Delta_k) & \text{if } \Delta_k > 0, \end{cases} \tag{3.3.4}$$

$$x_{k+1} = \begin{cases} z_k & \text{with probability} \quad p_k \\ x_k & \text{with probability} \quad 1 - p_k. \end{cases}$$

This means that the *promising* new point z_k (for which $f(z_k) \leq f(x_k)$) is accepted unconditionally, but the *non-promising* one (for which $f(z_k) > f(x_k)$) may also be accepted with probability $p_k = \exp\{-\beta_k \Delta_k\}$. As the probability of accepting a point which is worse than the preceding one is always greater than zero, the search trajectory may leave a local and even global minimizer. (Note that the probability of acceptance decreases if the difference $\Delta_k = f(z_k) - f(x_k)$ increases.)

The expression (3.3.4) for the acceptance probability p_k is motivated by the annealing process modelled in the simulated annealing method. In statistical mechanics, the probability that the system will transit from a state with energy E_o to a state with energy E_1, where $\Delta E=E_1-E_o>0$, is $\exp(-\Delta E/KT)$ where $K=1.38\ 10^{-16}$ erg/T is the Boltzmann constant and T is the absolute temperature. Thus $\beta=1/KT$ and the lower the temperature is, the smaller is the probability of transition to a higher energy state.

In a more general case, when z_k is distributed according to a transition probability $Q_k(x_k,.)$, the simulated annealing method is a typical Markovian method (see Algorithm 3.3.1): the only particularity being in the form (3.3.4) of acceptance probabilities.

If $\beta_k=\beta=1/KT$, i.e.

$$p_k(x,z,y,\eta) = p(x,z) = \min\{1,\exp(\beta(f(x)-f(z)))\} \tag{3.3.5}$$

and $Q_k(x,.)=Q(x,.)$ are not dependent on the count k, then the point sequence $\{x_k\}$ constitutes a homogeneous Markov chain, converging (under rather general conditions on Q and f) in distribution on the stationary Gibbs distribution having the density

$$\pi_T(x) = \exp\{-f(x)/KT\}/\int\exp\{-f(z)/KT\}dz \tag{3.3.6}$$

(Density is meant with respect to the Lebesgue measure in the continuous case, and with respect to the uniform measure in the discrete case.) Consequently, as $T\to 0$ (or $\beta\to\infty$), the Gibbs density π_T defined by (3.3.6) tends to concentrate on the set of global minimizers of f (subject to some mild conditions, see Geman and Hwang (1986), Aluffi-Pentini et al. (1985)). In particular, if the global minimizer x^* of f is unique, then the Gibbs distribution converges to the δ-measure concentrated at x^* for $T\to 0$.

Numerically, if T is small (i.e. β is great), then the points x_k obtained by a homogeneous simulated annealing method have the tendency to concentrate *near* to the global minimizer(s) of f. Unfortunately, the time required to approach the stationary Gibbs-distribution increases exponentially with 1/T and may reach astronomical values for small T (as confirmed also by numerical results). This can be explained by the fact that for small T a homogeneous simulated annealing method tends to be like the local random search algorithm that rejects unprofitable steps, and so its global search features are poor.

The homogeneous simulated annealing method is the particular case of the above mentioned Metropolis algorithm that uses the acceptance probabilities

$$p(x,z) = \min\{1, w(z)/w(x)\} = \begin{cases} 1 & \text{if } w(z) \geq w(x) \\ w(z)/w(x) & \text{if } w(z) < w(x) \end{cases} \tag{3.3.7}$$

instead of (3.3.5), where w is an arbitrary summable positive function on $\mathcal{X}$. The expression (3.3.5) comes from (3.3.7) after setting $w(x)=\exp(-\beta f(x))$. The points x_k generated by the Metropolis algorithm converge in distribution to a stationary distribution with the density

$$\varphi_w(x) = w(x)/\int w(z)dz \tag{3.3.8}$$

which is proportional to the function w and generalizes (3.3.6). The conditions guaranteeing the convergence and the convergence speed are thoroughly investigated; for references see Bhanot (1988). The transition probabilities $Q(x,.)$ are usually chosen in the Metropolis algorithm in such a way that the transition density from x to z is equal to the same of z to x for any $x,z \in \mathcal{X}$ (the symmetricity property).

Formulae (3.3.7) are not the unique way of choosing the acceptance probabilities of the above described Markov chain that allows (3.3.8) to be the stationary density of the chain. According to Bhanot (1988), one can use the more general functional form

$$p(x,z) = \begin{cases} g(w(z)/w(x)) & \text{if} \quad w(z) \geq w(x) \\ \dfrac{w(z)}{w(x)} g\left(\dfrac{w(z)}{w(x)}\right) & \text{if} \quad w(z) < w(x) \end{cases}$$

for this purpose, where g is an arbitrary function, $0 \leq g \leq 1$, $g \neq 0$. For instance, for $g(t)=1/(1+t)$ we have

$$p(x,z) = \begin{cases} \dfrac{w(x)}{w(x)+w(z)} & \text{if} \quad w(z) \geq w(x) \\ \\ \dfrac{w(z)}{w(x)+w(z)} & \text{if} \quad w(z) < w(x). \end{cases}$$

Another way of constructing stationary Markov chains having the stationary density (3.3.8) is due to Turchin (1971): it consists in setting $p(x,z)=1$,

$$q(x,z) = \int \rho(x,y)\,\sigma(y,z)\,dy \tag{3.3.9}$$

where $\rho(x,y)$ is an arbitrary transition density and

$$\sigma(x,z) = \rho(z,x)w(z)/\int \rho(z,x)w(z)dz. \tag{3.3.10}$$

Let us turn now to the case of unhomogeneous simulated annealing methods that use $\beta_k \to \infty$ (or $T_k \to 0$) for $k \to \infty$ and, probably, transition probabilities $Q_k(x,.)$ depending on k. It can be seen that the convergence of the distribution of points x_k, generated by the simulated annealing method to a distribution concentrated on the set of global minimizers of f, can be guaranteed if T_k tends to zero slowly enough. For standard variant of the method, the choice $T_k=c/\log(2+k)$ is suitable where c is a sufficiently large number depending on f and $\mathcal{X}$, see Mitra et al. (1986). Anily and Federgruen (1987) proved general results on the above convergence and the rate of convergence of the generalized discrete simulated annealing method described by general transition probabilities $Q_k(x,.)$ and general acceptance probabilities $p_k(x,z)$ tending to either zero, one or a constant for the cases $f(x)<f(z)$, $f(x)>f(z)$, or $f(x)=f(z)$, correspondingly.

A particular case of the above generalized simulated annealing method was proposed and numerically investigated by Bohachevsky et al. (1986): the acceptance probabilities have the form

$$p(x,z)=\begin{cases}1 & \text{if } f(z)\le f(x)\\ \exp\{-\beta[f(x)-f_{min}]^{\ell}(f(z)-f(x))\} & \text{if } f(z)>f(x)\end{cases} \qquad (3.3.11)$$

where ℓ is an arbitrary negative number (for $\ell=0$ (3.3.11) coincides with (3.3.3)) and f_{min} is an estimate of $f^*=\min f$. If, for some x, the value $f(x)-f_{min}$ becomes negative, then it is proposed to decrease f_{min} and continue the search.

The Reader interested in numerical realization of the simulated annealing concept is referred also to Corana et al. (1987), Haines (1987), van Laarhoven and Aarts (1987).

3.3.3 Methods based on solving stochastic differential equations

The simulated annealing method can be interpreted as a discrete approximation to a continuous (vector) process x_t which solves the simultaneous stochastic differential equations

$$dx_t=-\nabla f(x_t)dt+\sqrt{2KT(t)}\,dW_t \qquad (3.3.12)$$

where W_t is the standard multivariate Brownian motion. A detailed study on the solutions of stochastic differential equations of the type (3.3.12) is contained in Geman and Hwang (1986). In particular, it is shown that the trajectory x_t, corresponding to the solution of (3.3.12), has a stationary distribution with the Gibbs density (3.3.6) under some conditions on f and $\mathcal{X}$ and $T(t)\to T=\text{const}$, when $t\to\infty$. Moreover, the following is also shown for the case when the global minimizer x^* of f is unique: if there exists an extension of f to an open set containing $\mathcal{X}$ that is twice continuously differentiable and has no local minimizers outside $\mathcal{X}$, $t\to\infty$, and $T(t)=c/\log(2+t)$, where c is a sufficiently large real number, then the trajectory of the process x_t solving the equation (3.3.12) has limit distribution concentrated at x^*.

A discretized version of the stochastic differential equation (3.3.12), obtained with the help of standard methodology, is the Markov chain

$$x_{k+1}=x_k-\alpha_k\nabla f(x_k)+\alpha_k\sqrt{2KT_k}\,\eta_k \qquad (3.3.13)$$

where $\{\alpha_k\}$ is a sequence of real numbers, $0<\alpha_k\le 1$, and $\{\eta_k\}$ is a sequence of independent Gaussian random vectors with zero mean and unit covariance matrix. The search of the global minimizer in the algorithm (3.3.13) is performed in the direction of the antigradient corrupted by a Gaussian noise. The attributes of the algorithm are similar to those of the simulated annealing method.

General results on the asymptotic behaviour of the solutions of the simultaneous stochastic equations (3.3.12) and their discrete analogues were obtained by Kushner (1987) who studied the simultaneous equations

$$dx_t = -\nabla f(x_t)dt + \sqrt{\varepsilon(t)}\,\sigma(x_t)dW_t$$

and discrete processes

$$x_{k+1} = x_k - \alpha_k v(x_k, \zeta_k) + \alpha_k \sigma(x_k)\eta_k \qquad (3.3.14)$$

where $\{\zeta_k\}$ is a sequence of bounded random variables (possibly, correlated), σ is a nonnegative Lipschitz function on $\mathcal{X}$, function $v(., \zeta)$ satisfies the Lipschitz condition uniformly in ζ, $Ev(x, \zeta_k) = \nabla f(x)$, $\alpha_k = c_o/\log(k+c_1)$, $\varepsilon(t) = c_o/\log(t+c_1)$, $c_o > 1$, $c_1 > 1$. The random vectors ζ_k play the role of random noise arising under evaluation/estimation of the gradient of f. Such a noise occurs e.g. when Monte Carlo estimates of the gradient are used, or the objective function and/or its derivatives are subject to random error. Algorithm (3.3.14) is a generalization of many local optimization type stochastic approximation algorithms (that are determined by setting $\sigma=0$) and so it can be termed as *multiextremal stochastic approximation* . This term was used for a special form of the algorithm (3.3.14), see Vaysbord and Yudin (1968). The properties of (3.3.14) were considered also in Khasminsky (1965), Bekey and Ung (1974), Katkovnik (1976), Aluffi-Pentini et al. (1985) and in some other works. Moreover, the approach of Section 2.1.6 is closely connected with the one presented above, hence that it can be investigated with the help of the above mentioned results.

It should be finally noted that - to the author's knowledge - no versions of the algorithms are known to possess acceptable (competitive) numerical efficiency.

3.3.4 Global stochastic approximation: Zielinski's method

Zielinski (1980) proposed and studied another Markovian algorithm of global optimization of a function f, subject to random noise. It was called global stochastic approximation, choosing the acceptance probabilities p_k of Algorithm 3.3.1 in the form

$$p_k = \begin{cases} r(y_k) & \text{if} \quad y_k > \eta_k \\ 0 & \text{if} \quad y_k \le \eta_k \end{cases} \qquad (3.3.15)$$

where $r: \mathbb{R} \to (0,1]$ is a nondecreasing function, y_k and η_k are results of evaluation of f at points x_k and z_k (the latter point is distributed according to $Q_k(x_k,.)$). Correspondingly, the essence of the algorithm can be expressed by the rule

$$x_{k+1} = \begin{cases} z_k & \text{if} \quad y_k > \eta_k \quad \text{and} \quad \alpha_k \le r(y_k) \\ x_k & \text{otherwise} \end{cases} \tag{3.3.16}$$

where α_k is a random number. Comparing (3.3.4) with (3.3.15) we may hence conclude that in a sense the simulated annealing algorithm and (3.3.16) are opposed. The former method accepts each profitable step and also some unprofitable ones, while the latter rejects all unprofitable steps and even some profitable ones. Of course, the reasonableness of the algorithm (3.3.16) is due to the presence of random noise in the evaluations of f. Zielinski (1980) and Zielinski and Neumann (1983) proved the following.

Proposition 3.3.1. Let the global minimizer x^* of f be unique and f be continuous in a vicinity of x^*. If the random noise present in the evaluations of f is uniformly bounded almost surely, then the sequence of random points x_k determined by (3.3.16) converges to x^* in distribution. If the random noise is not bounded almost surely, then the sequence of distributions of the points (3.3.16) has a limit distribution having a density $\varphi(x)$ which reflects certain features of f (for instance, it has the property: if $f(x) \le f(z)$ for x and z from $\mathcal{X}$ then $\varphi(x) \ge \varphi(z)$).

The above works of R. Zielinski paid attention mainly to the case where the distributions $Q_k(x,.)$ are uniform on $\mathcal{X}$ (in this case the algorithm (3.3.16) is a generalization of Algorithm 3.1.1 for the case of random noise). Besides the mentioned works studied some features of φ and provided recommendations concerning the choice of function r and transition probabilities $Q_k(x,.)$ in the algorithm (3.3.16).

3.3.5 Convergence rate of Baba's algorithm

Let us now return to the case where f is not subject to random noise and consider the variant of Algorithm 3.3.1 in which

$$x_{k+1} = \begin{cases} z_k & \text{if} \quad f(z_k) \le f(x_k) \\ x_k & \text{if} \quad f(z_k) > f(x_k) \end{cases} \tag{3.3.17}$$

and the transition densities q_k have the form $q_k(x_k,x)=\varphi(x-x_k)$ where the density φ is supposed to be given on $\mathbb{R}^n$, continuous in a neighbourhood of zero, and $\varphi(0)>0$. Note that (3.3.17) implies that the acceptance probabilities p_k of Algorithm 3.3.1 are

$$p_k = \begin{cases} 1 & \text{if} \quad f(z_k) \le f(x_k) \\ 0 & \text{if} \quad f(z_k) > f(x_k) \end{cases}$$

The convergence of the above algorithm was studied first by Baba (1981): this is the reason for the algorithm being referred to as Baba's algorithm.

Dorea (1983) proved the following result on the rate of convergence of the algorithm (note that a statement on convergence follows from this result).

Proposition 3.3.2. Set

$$\beta_0=\mu_n(A(\delta))\inf_{\substack{z\in A(\delta),\\ x\in \mathcal{X}\setminus A(\delta)}}\varphi(z-x),\qquad \beta_1=\mu_n(A(\delta))\sup_{\substack{z\in A(\delta),\\ x\in \mathcal{X}\setminus A(\delta)}}\varphi(z-x),$$

$$L_j=\left[\mu_n(A(\delta))+\mu_n(\mathcal{X}\setminus A(\delta))\beta_j\left(1+\beta_{1-j}\right)/\beta_{1-j}^2\right]/\mu_n(\mathcal{X})\ \text{for } j=0,1$$

where μ_n is the Lebesgue measure and let Ev_δ be the mean number of random vectors z_i, obtained in Baba's algorithm, required for a sequence $\{x_k\}$ to attain the set $A(\delta)$. Then $L_0\leq Ev_\delta\leq L_1$ and the quantity L_1 attains its minimal value for the case when the algorithm at hand coincides with Algorithm 3.1.1.

We shall present now a more refined result on the convergence rate of a particular case of Baba's algorithm, revising some unpublished results of V.V. Nekrutkin; other asymptotical properties of the algorithm are investigated in Dorea (1986).

Assume that $\mathcal{X}=[0,1]^n$ is the unit cube, the rule (3.3.17) is applied and the transition density $q_k=q$ has the form

$$q(x_k,x)=(1-\alpha)p_0(x)+\alpha p_a(x_k,x)\tag{3.3.18}$$

where $\alpha\in[0,1)$, $a\in(0,2]$ are parameters of the algorithm, $p_0(x)=1$, $x\in\mathcal{X}$ is the uniform density on $\mathcal{X}$, furthermore

$$p_a(z,x)=\begin{cases} b(a,z) & \text{if}\quad x\in \mathcal{D}_a(z)\\ 0 & \text{otherwise,}\end{cases}\tag{3.3.19}$$

where

$$\mathcal{D}_a(z)=\mathcal{X}\cap\left[z(1)-\tfrac{a}{2},\ z(1)+\tfrac{a}{2}\right]\times\ldots\times\left[z(n)-\tfrac{a}{2},\ z(n)+\tfrac{a}{2}\right],$$

and

$$b(a,z)=1/\mu_n(\mathcal{D}_a(z)),\qquad z=(z(1),\ldots,z(n)).$$

(Under a fixed $z \in \mathfrak{X}$, the function (3.3.19) is the uniform density on the set $\mathfrak{D}_a(z)$ which is the intersection of the cube $\mathfrak{X}$ and the cube with the centre at a point z and a side length a.)

The choice of the transition density (3.3.18) implies that on each iteration of the algorithm the uniform distribution on $\mathfrak{X}$ (with probability $1-\alpha$) or the uniform distribution on $\mathfrak{D}_a(z)$ (with probability α) is sampled. Note that in two particular cases (for $\alpha=0$ or $a=2$) the algorithm at hand coincides with Algorithm 3.1.1, since the density (3.3.18) becomes the uniform density on $\mathfrak{X}$. Note also that from the definition of b(a,z) there follows

$$a^{-n} \le b(a,z) \le (a/2)^{-n}; \tag{3.3.20}$$

here the left-side relation becomes equality, when each coordinate of a point z lies in the interval [a/2, 1-a/2], and the right-side inequality does for the vertices of the cube $\mathfrak{X}$.

Before studying the properties of the algorithm, we formulate an auxiliary assertion well-known in the theory of Markov processes.

Lemma 3.3.1. Let $x_1, x_2, \ldots$ be a homogeneous Markov chain, τ_C be the Markovian moment of first hitting into the set $C \subset \mathfrak{X}$, $E_x \tau_C$ be the expectation of τ_C provided $x_1=x$, A and B be two measurable subsets of $\mathfrak{X}$,

$$\Pr_x\left\{x_{\tau_B} \in dz\right\}$$

be the conditional distribution of the vector x_{τ_B} provided $x_1=x$. Then the relation

$$E_x \tau_A \le E_x \tau_B + \int_B \Pr_x\left\{x_{\tau_B} \in dz\right\} E_z \tau_A \tag{3.3.21}$$

is valid.

The inequality (3.3.21) shows that the mean time before reaching the set A by a trajectory of the Markov chain does not exceed the mean time for reaching this set by those trajectories which have to visit B before reaching A.

Let us take $A=\mathfrak{D}_\varepsilon(x^*)$, where x^* is a global maximizer, and

$$B = \mathfrak{D}_{a-\varepsilon}(x^*) \cap \left\{x \in \mathfrak{X}: f(x) \le \inf_{z \in \mathfrak{X} \setminus \mathfrak{D}_{a-\varepsilon}(x^*)} f(z)\right\},$$

where ε ($0 < \varepsilon < a$) defines a required solution accuracy, and estimate the mean number of iterations for reaching A by Baba's algorithm with transition densities determined via (3.3.18) and (3.3.19).

By virtue of continuity of f in a neighbourhood of x^*, there exists such a constant $\beta > 0$ (depending on f, but independent of a, ε) that $\mu_n(B) \ge \beta(a-\varepsilon)^n$. Applying the inequalities

(3.3.20) and (3.3.21), together with the fact that, for each $x \in B$, the value $f(x)$ does not exceed the values of f outside the set $\mathfrak{D}_{a-\varepsilon}(x^*)$, we obtain

$$E_x \tau_B \le \left[(1-\alpha)\int_B p_0(x)dx\right]^{-1} = [(1-\alpha)\mu_n(B)]^{-1} \le \left[(1-\alpha)\beta(a-\varepsilon)^n\right]^{-1},$$

$$\sup_{z \in B} E_z\{\tau_A | z \in B\} = \sup_{z \in B}[\Pr\{x_{k+1} \in A | x_k = z \in B\}]^{-1} =$$

$$= \left[\inf_{z \in B} [\alpha b(a,z) + (1-\alpha)]\mu_n(A)\right]^{-1} =$$

$$= (1-\alpha+\alpha a^{-n})^{-1} b(\varepsilon, x^*) \le 2^n / [(1-\alpha+\alpha a^{-n})\varepsilon^n],$$

$$E_x \tau_A \le E_x \tau_B + \sup_{z \in B} E_z\{\tau_A | z \in B\}.$$

Define

$$\Phi(\varepsilon, a, \alpha) = \frac{1}{(1-\alpha)\beta(a-\varepsilon)^n} + \frac{2^n}{(1-\alpha+\alpha a^{-n})\varepsilon^n}, \tag{3.3.22}$$

then the obtained estimate for the mean number of iterations to be performed for reaching the set $A = \mathfrak{D}_\varepsilon(x^*)$ can be written as

$$\sup_{x \in \mathfrak{X}} E_x \tau_A \le \Phi(\varepsilon, a, \alpha).$$

Let us investigate now the question concerning the rate of increase of $\Phi(\varepsilon,a,\alpha)$ for $\varepsilon \to 0$. From (3.3.22) it follows that if a and α do not depens on ε, then the order of increase of Φ equals ε^{-n}, $\varepsilon \to 0$, and coincides with the order of increase of the quantity $E\tau_A$ for Algorithm 3.1.1 for which

$$E\tau_A = 1/\mu_n(A) = b(\varepsilon, x^*) \asymp \varepsilon^{-n} \qquad \text{for} \qquad \varepsilon \to 0.$$

Alternatively, selecting a, α in dependence on the desired solution accuracy ε, the order of increase of Φ may be considerably decreased. Indeed, set

$$\alpha(\varepsilon) = \alpha = \text{const}, \qquad 0 < \alpha < 1, \qquad \varphi(\varepsilon) = a(\varepsilon) - \varepsilon,$$
$$\varphi(\varepsilon) \to 0, \qquad \varphi(\varepsilon)/\varepsilon \to \infty \qquad \text{for} \qquad \varepsilon \to 0.$$

Then

$$\Phi(\varepsilon, a(\varepsilon), \alpha) = \frac{1}{(1-\alpha)\beta(\varphi(\varepsilon))^{n}} + \frac{2^{n}}{(1-\alpha)\varepsilon^{n} + \alpha(1+\varphi(\varepsilon)/\varepsilon)^{-n}} \asymp$$

$$\asymp \frac{(\varphi(\varepsilon))^{-n}}{(1-\alpha)\beta} + \frac{2^{n}}{\alpha}(\varphi(\varepsilon)/\varepsilon)^{n} \quad \text{for} \quad \varepsilon \to 0.$$

Hence, for $\varphi(\varepsilon) \asymp e$, $\varepsilon \to 0$, the order of increase of $\Phi(\varepsilon, a(\varepsilon), \alpha)$ is $\varepsilon^{-n/2}$: this is optimal since for $a(\varepsilon) \geq const > 0$, $\alpha(\varepsilon) \to 0$, $\alpha(\varepsilon) \to 1$, as well when omitting the supposition $\varphi(\varepsilon)/\varepsilon \to \infty$ for $\varepsilon \to 0$, the order $\varepsilon^{-n/2}$ is getting worse.

Now set $\alpha(\varepsilon)=\alpha$, $a(\varepsilon)=\varepsilon+d\sqrt{\varepsilon}$ and find the optimal values of the constants α and d. Transforming Φ we obtain

$$\Phi(\varepsilon, a(\varepsilon), \alpha) = \frac{1}{(1-\alpha)\beta d^{n} \varepsilon^{n/2}} + \frac{2^{n}}{(1-\alpha)\varepsilon^{n} + \alpha\left(1+d\varepsilon^{-1/2}\right)^{-n}} \asymp$$

$$\asymp \varepsilon^{-n/2}\left[\frac{1}{(1-\alpha)\beta d^{n}} + 2^{n} d^{n}/\alpha\right] \quad \text{for} \quad \varepsilon \to 0.$$

Thus the problem of (approximately) optimal selection of the constants α and d is reduced to the minimization problem of the function

$$\Psi(\alpha, \gamma) = [\,(1-\alpha)\beta\gamma]^{-1} + 2^{n}\gamma/\alpha,$$

where $\gamma = d^{n}$, on the set $\alpha \in (0,1)$, $\gamma > 0$. Setting to zero the partial derivatives of the function Ψ we get a unique solution of the obtained simultaneous equations: $\alpha=0.5$, $\gamma = 2^{-n/2}\beta^{-1/2}$. Hence, it follows that the *quasi-optimal* values of parameters of the algorithm are: $\alpha=0.5$,

$$a = a(\varepsilon) = \varepsilon + \beta^{-1/2n}\sqrt{\varepsilon/2};$$

under such parametrization the estimation below is valid:

$$\sup_{x} E_{x} \tau_{D_{\varepsilon}(x^{*})} \leq \Phi(\varepsilon, a(\varepsilon), \alpha) \asymp 2^{n/2+2} \beta^{-1/2} \varepsilon^{-n/2}, \qquad \varepsilon \to 0.$$

3.3.6 The case of high dimension

Assume that the dimension n of $\mathcal{X}$ is high. Although the usual versions of random search algorithms are generally in this situation much simpler than the deterministic algorithms, they are, nevertheless, rather labourous because n-dimensional random vectors have to be sampled at each iteration. The relative numerical demand of the algorithms of this section for great n is very modest - only two random variables are to be sampled at each k-th iteration (k>1) of these algorithms.

The approach below can be referred to as a modification of the method considered in Section 3.3.2 and consists of sampling a homogeneous Markov chain with a given stationary density.

Choose a nonnegative function Ψ defined on $\mathcal{X}$ such that the univariate densities proportional to any one-dimensional cross-section of the function are easily sampled and the transition probabilities $Q_k(z,.)$ as follows

$$Q_k(z, dx) = Q(z,dx) = c(z)\int T(z,dt)\Psi((x,t))\,dx \qquad (3.3.23)$$

where T(z,dt) for each $z\in\mathcal{X}$ is a probability distribution on the set of lines passing through the point z and

$$T(z, dt) = \sum_{i=1}^{n} q_i \delta(z,dt - t_i) \qquad (3.3.24)$$

where

$$q_i > 0 \quad (i = 1,\ldots,n), \qquad \sum_{i=1}^{n} q_i = 1, \qquad (3.3.25)$$

$\delta(z,dt-t_i)$ is a distribution concentrated on the line passing through z and parallel to the vector t_i ; $\{t_1,\ldots,t_n\}$ is a set of n-dimensional linearly independent vectors; (x,t) denotes the linear coordinate of a point x on the line t (projection of x on t); c(z) for each z is a normalization constant. The transition probability (3.3.23) is sampled applying the superposition technique: first, a random line is chosen passing through z and parallel to one of the vectors t_i by sampling from the distribution induced by the probabilities (3.3.25) and then a one-dimensional distribution is sampled on the line with the density proportional to Ψ.

Note that the most natural way of choosing q_i is $q_i=1/n$ (i=1,...,n) while the vector set $\{t_1,\ldots,t_n\}$ can be chosen as the set of the coordinate vectors. Note also that the above way of choosing the transition probability Q(z,.) draws on the idea of Turchin (1971).

Proposition 3.3.2. Let the function Ψ be non-negative and piecewise continuous, the set $\mathcal{X}'=\{x\in \mathcal{X}: \Psi(x)>0\}$ be connected, $\Psi(x_1)>0$, and a homogeneous Markov chain be sampled with transition probability (3.3.23) for which (3.3.24) is satisfied. Then the sampled Markov chain has a stationary distribution whose density is proportional to Ψ.

Proof. Under the above formulated conditions the transition probability (3.3.23) meets the Doeblin condition, i.e. there exists a natural number m, two real numbers $\varepsilon_1 \in (0,1)$ and $\varepsilon_2>0$, and probability measure $\nu(dx)$ on $(\mathcal{X},B)$ such that for $x\in\mathcal{X}'$ there follows $Q^m(x,A)\geq\varepsilon_2$ from $\nu(A)\geq\varepsilon_1$, $A\subset X'$. Indeed, if one chooses

$$\nu(dx) = \Psi(x)dx/\int\Psi(z)dz,$$

then the Doeblin condition means that the density $q^m(z,x)$ of the probability of transition from any point $z\in\mathcal{X}'$ to any point $x\in\mathcal{X}'$ in m steps is positive which is true (also) for m=n. Since the Doeblin condition is met, the Markov chain under study has a stationary distribution with the density p(x) which is the unique solution (in the class of all probability densities) of the integral equation

$$p(x) = \int p(y)q(y,x)dy.$$

The fact that Ψ is a solution of this equation follows from Turchin (1971): this proves the proposition.

Sampling of one-dimensional densities will be especially simple if constant or piecewise constant functions are used as Ψ. If $\exp\{-\lambda f(x)\}$ or another function connected with f is used as Ψ, then the stationary distribution density will be proportional to it, but one may face a hard sampling problem for one dimensional distributions: if sampling relies on the rejection technique with a constant majorant, then the efficiency of the resulted optimization algorithm for f will not differ appreciably from that of Algorithm 3.1.1. If one succeeds to employ more efficient sampling methods (e.g. that of inverse functions, if the cross-sections of Ψ along some directions have readily sampled forms, or the rejection technique having good majorants), then the efficiency of the algorithm derived may be significantly superior to that of Algorithm 3.1.1. These devices work only when one knows the analytical form of f; otherwise the procedure described below can be used. This procedure relies upon the fact that under certain conditions an appropriately chosen subsequence of a stationary Markov sequence with stationary density proportional to Ψ may be regarded as a stationary Markov chain with stationary density proportional to w ($\exp\{-\lambda f\}$ or any other function related to f may be used as w).

Before formulating the procedure we shall introduce several notations and prove an auxiliary assertion. Let $(\mathcal{X},\mathcal{B})$ be a measurable space, P_1 be a probability distribution on it, Q(z,dx) be a Markov transition probability, P(z,dx)=(1-g(z))Q(z,dx) where $0\leq g\leq 1$, (g(z) is the probability of termination of the Markov chain with transition probability P(z,dx) in a point $z\in\mathcal{X}$), τ be the termination moment of the Markov chain $x_1,x_2,\ldots$ with initial distribution P_1 and transition probability P(z,dx), i.e.

$$\tau = \min\{j=1,2,\ldots \mid g(x_j)\geq\alpha_j\} \tag{3.3.26}$$

where x_1 has distribution P_1, x_i (i=2,3,...) has distribution $Q(x_{i-1},.)$, α_j are random numbers.

Below the minimal positive solution of an integral equation, i.e. the one obtained using the method of successive approximation, is taken as the solution. For example, for the integral equation

$$G(dx) = \int G(dz)P(z,dx) + P_1(dx) \tag{3.3.27}$$

the solution is represented as the measure

$$G(dx) = \sum_{m=1}^{\infty} P_m(dx)$$

where for m=2,3,...

$$P_m(dx) = \int P_1(dz)P^{m-1}(z,dx),$$

$$P^m(z,dx) = \int \ldots \int P(z,dx_1)P(x_1,dx_2)\ldots P(x_{m-1},dx).$$

Note that the minimal positive solution always exists and is unique.

Theorem 3.3.1. With the above notation, the following statements hold:

1. $\Pr\{\tau < \infty\} = \int g(x)G(dx)$,
2. If $\Pr\{\tau < \infty\} = 1$ then $F(dx) = g(x)G(dx)$

is the probability distribution of the random vector ξ_τ where $\xi_1,\ldots,\xi_\tau$ are random vectors connected into the Markov chain with initial distribution P_1 and transition probability P,

3. If $\Pr\{\tau < \infty\} = 1$ then
   $E\tau = G(\mathcal{X})$, (3.3.28)
4. If $G(\mathcal{X}) < \infty$ then $\Pr\{\tau < \infty\} = 1$.

Proof.

1. We have

$$\Pr\{\tau < \infty\} = \sum_{m=1}^{\infty} p_m$$

where

$$p_m = \int g(x)P_m(dx) \tag{3.3.29}$$

stands for the probability of termination of the Markov chain at the m-th step. Thus

$$\Pr\{\tau < \infty\} = \sum_{m=1}^{\infty} \int g(x) P_m(dx) =$$

$$= \int g(x) \sum_{m=1}^{\infty} P_m(dx) = \int g(x) G(dx).$$

2. For any $A \in \mathcal{B}$ the following holds

$$P\{x_\tau \in A\} = \sum_{m=1}^{\infty} \Pr\{g(x_1) < \alpha_1, \ldots, g(x_{m-1}) < \alpha_{m-1}, g(x_m) \geq \alpha_m, x_m \in A\} =$$

$$= \sum_{m=1}^{\infty} \int_{\mathcal{X}^m} (1 - g(x_1)) \ldots (1 - g(x_{m-1}))\, g(x_m) P_1(dx_1) Q(x_1, dx_2) \ldots Q(x_{m-1}, dx_m) 1_A(x_m) =$$

$$= \sum_{m=1}^{\infty} \int P_m(dx) g(x) 1_A(x) = \int_A g(x) G(dx) = F(A).$$

Since $\Pr\{\tau<\infty\}=1$, it follows from the first assertion and from the fact that G is a measure that F is the probability distribution of the random vector x_τ.

3. By definition $E\tau = \sum_{m=1}^{\infty} m p_m$ where p_m is defined by (3.3.29). Determine now the expression for $b_m = P_m(\mathcal{X})$. We have

$$b_m = \int_{\mathcal{X}^{m-1}} P_1(dx_1) P(x_1, dx_2) \ldots P(x_{m-2}, dx_{m-1})(1 - g(x_{m-1})) = b_{m-1} - p_{m-1}.$$

Since $b_1 = P_1(\mathcal{X}) = 1$, so $b_m = 1 - \sum_{k=1}^{m-1} p_k$, but $\Pr\{\tau < \infty\} = \sum_{k=1}^{\infty} p_k = 1$,

therefore $b_m = \sum_{k=m}^{\infty} p_k$. Compute now

$$G(\mathcal{X}) = \int G(dx) = \sum_{m=1}^{\infty} b_m = \sum_{m=1}^{\infty} \sum_{k=m}^{\infty} p_k = \sum_{k=1}^{\infty} \sum_{m=1}^{k} p_k = \sum_{k=1}^{\infty} k p_k = E\tau.$$

4. Integrating (3.3.27) we have

$$G(\mathcal{X}) = \int\int G(dz)P(z,dx) + 1$$

which is equivalent to

$$G(\mathcal{X}) = \int G(dx) - \int g(x)G(dx) + 1$$

whence taking into consideration the first assertion and the fact that $G(\mathcal{X})<\infty$ we get $\Pr\{\tau<\infty\}=1$. The theorem is proved.

Notably, if the Markov chain $x_1,x_2,...$ starts at a fixed point $x\in\mathcal{X}$, i.e. if $x_1=x$ with probability one, then the measure G corresponds to the potential kernel of the chain in potential theory, cf. Revus (1975). In this theory (3.3.28) was known for $g(x)=\mathbf{1}_A(x)$, $A\subset\mathcal{X}$.

Theorem 3.3.1 can serve (and did serve for the author) as a basis for the creation and investigation of sampling algorithms for a wide range of probability distributions. The approach is thoroughly described in Zhigljavsky (1985).

Denote by $\xi_1, \xi_2,...$ a homogeneous Markov chain with an initial distribution F_o and Markov transition probability $Q(z,.)$. Set $P(z,dx)=(1-g(z))Q(z,dx)$, where g is a function on $\mathcal{X}$, $0\le g\le 1$, $\tau_o=0$,

$$\tau_k = \min\{j=\tau_{k-1}+1, \tau_{k-1}+2,... \mid g(\xi_j) \ge \alpha_j\}$$

where α_j are random numbers. According to Theorem 3.3.1 the random vectors $\xi_{\tau k}$ (k=1,2,...) have distributions F_k representable by $F_k(dx)=g(x)\,G_k(dx)$ in terms of G_k, the solution of the equation

$$G_k(dx) = \int G_k(dz)P(z,dx) + F_{k-1}(dx). \tag{3.3.30}$$

It is proved below that under some assumptions, for any initial distribution F_o the consecutive approximations (3.3.30) weakly converge to the solution of the equation

$$G(dx) = \int G(dz)P(z,dx) + F(dx) \tag{3.3.31}$$

where $F(dx)=g(x)G(dx)$.

Denote by v_o a bounded function that is continuous on $\mathcal{X}$, and introduce a sequence of functions v_k (k=1,2,...) induced by v_o which are the solution of the corresponding integral equation

$$v_k(z) = \int P(z,dx)v_k(x) + g(z)v_{k-1}(z). \tag{3.3.32}$$

Proposition 3.3.4. Let $\mathfrak{X}$ be a compact subset of $\mathfrak{R}^n$, for some $m\geq 1$ the m-step transition probability $P^m(z,dx)$ is strictly positive (i.e. $P^m(z,\mathfrak{D})>0$ for any $z\in\mathfrak{X}$ and each open hyperrectangle $\mathfrak{D}\subset\mathfrak{X}$), the method of successive approximations for (3.3.32) converges in the metric of $C(\mathfrak{X})$ for any function v_{k-1} continuous on $\mathfrak{X}$, the family of functions $\{v_k\}$ is equicontinuous.Then $F(dx)=g(x)G(dx)$, where G is the solution of (3.3.31), is a probability distribution and the sequence of distributions $F_k(dx)=g(x)G_k(dx)$ $(k=1,2,...,$ G_k are the solutions of (3.3.30)) weakly converges to F(dx).

Proof. Assume first that m=1.

Let v_o be an arbitrary continuous function on $\mathfrak{X}$. We shall show that $M_o=\max v_o\geq \max v_1=M_1$. Indeed, by (3.3.32) follows

$$v_1(z)\leq M_1\int P(z,dx)+g(z)v_0(z)=M_1(1-g(z))+g(z)v_0(z)$$

whence $(v_1(z)-M_1)/g(z)+M_1\leq v_o(z)$ for all $z\in\mathfrak{X}$. By passing to maxima obtain that $M_1\leq M_o$ (in doing so, it is evident that if $v_1(x)\neq M_1$ then $g(x_1{}^*)>0$ where $x_1{}^*$ is a maximizer of v_1). From the strict positiveness and continuity of v_o and v_1 obtain that $M_1=M_o$ if and only if $v_1(x)\equiv M_1$ (and, consequently, $v_o(x)\equiv M_1$).

In virtue of the Arzela theorem there exists a subsequence v_{km} of the sequence v_k that uniformly converges to a continuous function u_o. Then v_{km+1} converges to u_1, the transformation (3.3.32) of u_o. Since the numerical sequence max v_k is monotone and bounded,

$$M=\lim_{k\to\infty}\max v_k$$

exists. Since the convergence to u_o and u_1 is uniform, max u_o= max u_1=M and, therefore, $u_o=u_1=M$. The limit, thus is independent of the choice of v_{km}, hence, $v_k\to M$ uniformly for $k\to\infty$.

Now we shall prove that

$$\int v_{k-1}(x)F_k(dx)=\int v_k(x)F_{k-1}(dx) \tag{3.3.33}$$

is valid. Indeed, by definition we have the chain of relations

$$\int v_{k-1}(x)F_k(dx) = \int v_{k-1}(x)g(x)G_k(dx) =$$

$$= \int\left[v_k(x) - \int P(x,dz)v_k(z)\right]G_k(dx) =$$

$$= \sum_{m=1}^{\infty} \int\int F_{k-1}(dz)P^{m-1}(z,dx)v_k(x) - \sum_{m=1}^{\infty} \int\int F_{k-1}(dz)P^{m}(z,dx)v_k(x) =$$

$$= \int F_{k-1}(dz)v_k(z).$$

Let now F_1 be an arbitrary probability distribution on $(\mathcal{X},\mathcal{B})$, v_1 be an arbitrary continuous function on $\mathcal{X}$. From (3.3.33) follows

$$\int v_1(x)F_k(dx) = \int v_k(x)F_1(dx), \qquad k = 1, 2, \ldots$$

but since $v_k \rightarrow M$, therefore

$$\int v_k(x)F_1(dx) \rightarrow M$$

in virtue of the Lebesgue theorem on dominated convergent sequences. Thus, we see that $\int v_1(x)F_k(dx)$ converges for $k\rightarrow\infty$ for any continuous function v_1, whence it follows by Feller (1966), Section 1 of Ch.8 that there exists a probability measure F that is a weak limit of $\{F_k\}$. The validity of (3.3.31) now obviously follows from (3.3.30) and the proposition is proved for the case of m=1.

In order to pass to arbitrary m, apply the above assertion m times and substitute $P^m(z,.)$ and $F_i(dx)$, respectively, for $P(z,.)$ and $F_1(dx)$ $(i=1,\ldots,m)$. The limit of the sequences

$$\{F_{mk+i}(dx)\}_{k=0}^{\infty}$$

will be the same and, therefore, $\{F_k\}$ also will converge to this limit. The proof is complete.

Proposition 3.3.4 is a generalization of the ergodic theorem for Markov chains from Feller (1966) to the case of $m\geq 1$ and function g that is not identically equal to 1.

Rewrite now (3.3.31) as

$$(1 - g(x))G(dx) = \int[(1 - g(z))G(dz)]Q(z,dx).$$

By assumption, the transition probability $Q(z,.)$ is chosen so that all solutions of the equation have the form

$$(1-g(x))G(dx)=\begin{cases} c\Psi(x)dx, & x\in \mathcal{X} \\ 0, & x\notin \mathcal{X} \end{cases} \tag{3.3.34}$$

where c is arbitrary positive constant.

Let us require that the stationary distribution F of the sequence $\xi_{\tau k}$ be proportional to w, i.e. that the relation

$$F(dx)=g(x)G(dx)=\begin{cases} w(x)dx/\int w(z)dz, & x\in \mathcal{X} \\ 0, & x\notin \mathcal{X} \end{cases} \tag{3.3.35}$$

be satisfied. Then from (3.3.34) and (3.3.35) follows

$$g(x)=w(x)/[w(x)+b\Psi(x)]. \tag{3.3.36}$$

Although the constant

$$b=c\int w(z)dz$$

may be taken arbitrarily, numerical calculations indicate that

$$b\approx \int w(z)dz/\int \Psi(z)dz \tag{3.3.37}$$

assures the most rapid convergence of $\xi_{\tau k}$ to the stationary distribution F.

Assume now that exact equality holds in (3.3.37). Then the efficiency of the resulting optimization algorithm depends on how frequently the inequality $g(\xi_i)\geq\alpha_i$ is fulfilled. The probability of this event asymptotically equals

$$p=\frac{\int g(x)\Psi(x)dx}{\int\Psi(z)dz}=\int\frac{[w(x)/\int w(z)dz][\Psi(x)/\int\Psi(z)dz]}{[w(x)/\int w(z)dz]+[\Psi(x)/\int\Psi(z)dz]}$$

If the behaviour of Ψ resembles that of w, then p is close to one and the algorithmic efficiency is high. Otherwise (if Ψ is small where w is large and large where w is small) both p and the efficiency are low. The profile of behaviour of Ψ thus, should be close to that of w. If prior information concerning the behaviour of f is missing, then the function should be estimated in the course of optimization.

The above algorithm can be generalized for the situation where evaluations of w(x) (in particular, we can use $w(x)=-f(x)+\text{const}$) are subject to a random noise $\eta(x)$, $w+\eta\geq 0$ a.s. In this case the analogue of (3.3.36) is as follows

$$g(x)=[w(x)+\eta(x)]/[w(x)+\eta(x)+\Psi(x)]$$

and the subsequence $\xi_{\tau k}$ has a stationary distribution proportional to the function

$$h(x) = \left\{ E[w(x) + \eta(x) + \Psi(x)]^{-1} \right\}^{-1} - \Psi(x)$$

rather than to w(x). The behaviour of this function is related to some extent to that of w(x). Indeed, let the distribution of the random variable $\eta(x)$ is independent of x and two points x, $z \in \mathcal{X}$ be chosen such that $\Psi(x)= \Psi(z)$ and $w(x)>w(z)$, then simple calculations give $h(x)>h(z)$.

CHAPTER 4. STATISTICAL INFERENCE IN GLOBAL RANDOM SEARCH

This chapter considers various approaches to the construction and investigation of global random search algorithms based on mathematical statistics procedures. It appears that many algorithms of this chapter are both practically efficient and theoretically justified. Their main theoretical feature is that convergence of the algorithms studied does not hold in the deterministic sense or even in probability, but is valid with some reliability level only. Thus, the algorithms can miss a global optimizer, but the probability of this unfavourable event is under control and may be guaranteed to be arbitrarily small.

The chapter contains six sections.

Section 4.1 is auxiliary and specifies the ways of applying mathematical statistics procedures for constructing global random search algorithms.

Section 4.2 treats statistical inference concerning the optimal value of a function f, on the base of its values at independent random points: much attention is paid to the issue of accuracy increase by using prior information about the objective function.

Section 4.3 describes a general class of global random search methods which use statistical procedures; furthermore, it generalizes the well-known family of the branch and bound methods, permitting inferences that are accurate only with a given reliability (rather than exactly).

Section 4.4 modifies the statistical inference of Section 4.2 for the case, in which the points where f is evaluated are elements of a stratified sample rather than an independent one. It also demonstrates the gains implied by the stratification and evidences that the decrease of randomness, if possible, generally leads to increased efficiency.

Section 4.5 presents statistical inference in the random multistart method which was described in Section 2.1 and serves as a base of a number of efficient global random search algorithms. The statistical inferences can be applied for its control and modifications.

Finally, Section 4.6 describes statistical testing procedures for distribution quantiles based on using the so-called distributions neutral to the right. They can be optionally applied for checking the accuracy in a number of global random search algorithms, as well as for the construction of a particular algorithms of the branch and probability bound type.

As earlier, the feasible region $\mathcal{X}$ is supposed to be a compact subset of $\mathbb{R}^n$, having a sufficiently simple structure, the objective function f can be evaluated at any point of $\mathcal{X}$ without noise. In contrast with the preceding chapters, the maximization problem of f is considered rather the minimization one. (This decision serves for an easier application of the related results of mathematical statistics; note that the transcription between minimum and maximum problems is obvious.)

4.1 Some ways of applying statistical procedures to construct global random search algorithms

Theoretically, there is a scope of sophisticated results in the field of mathematical statistics, applicable during the construction of global random search algorithms. The number of works containing attractive mathematical results is comparatively small: most of them are related to the topics of Sections 4.2 - 4.5. The aim of this auxiliary section is to review some related results and applications in other fields.

4.1.1 Regression analysis and design

Among the possible applications of design and analysis of experiments in linear regression, of considerable interest are those dealing with the construction of local search algorithms in the case when evaluations of the objective function are subject to a random noise. They concern the theory of extremal experimental design and rely on supposition that evaluations of the objective function f are always made the vicinity of points where the function can be rather well approximated by a first- or second-order polynomial whose unknown parameters are estimated. This operation is aimed at determining the direction of further evaluations of f. Section 8.1 will deal with this problem.

Some attempts (for instance, see Chichinadze (1969), Hartley and Pfaffenberger (1971)) of using linear regression analysis methods for the construction of global search algorithms were based on the formal construction of linear relations between some characteristics (such as the values of the objective function in randomly selected points and the amount of evaluations). The lack of a constructive description of the class of objective functions where the postulated linear relation is valid with an acceptable accuracy is the basic obstacle to the theoretical investigation of these algorithms.

The methods of nonlinear regression analysis are of practical interest, if the objective function is evaluated with a random noise and is acceptably approximated by a regression function that is nonlinearly dependent on some unknown parameters. In principle, these methods are applicable to the situation discussed above when

$$M_Z = \sup_{x \in \mathcal{X}} f(x) \tag{4.1.1}$$

is one of the unknown parameters of the nonlinear regression model describing the cumulative distribution function (c.d.f.)

$$F_Z(t) = \int_{f(x)<t} P(dx) \tag{4.1.2}$$

where P is a probability distribution concentrated on the set $Z \subset \mathcal{X}$. Here practically unsolvable theoretical difficulties are related to the constructive description of proper objective function classes, similarly to the above case.

Applications of non-parametric regression analysis to global random search methodology are interesting, in the presence of random noise when evaluating f. First, the non-parametric estimation of f or some related function(s) underlies the approach developed in Chapter 5. Second, non-parametric regression estimates of f can be used for its thorough investigation, with a view on determining points for subsequent evaluations of f. If, in doing so, the regression function is repeatedly estimates, then regression should be estimated with higher accuracy in those subsets of $\mathcal{X}$, where the previous estimate $\hat{f}$ of f (and, plausibly, f itself) is smaller, i.e. where the probability of locating a global maximizer is larger. To do so, one may take as the loss function for regression estimation a quadratic function with the weight function proportional to

$\exp\{\lambda \hat{f}(x)\}$ where $\lambda > 0$ and $\hat{f}$

is the preceding estimate of f.

Section 8.3 will be devoted to some aspects of the theory of non-parametric estimation of a multivariate regression.

4.1.2 *Cluster analysis and pattern recognition*

The prominent role of the cluster analysis in global random search methodology is connected with the so-called candidate points methods discussed in Section 2.1.3. These are the variants of the famous method named multistart that are practically efficient for a rather wide class of multiextremal problems and use the cluster analysis algorithms to distinguish the regions of attraction of local optimizers.

Another idea underlying the use of pattern recognition and cluster analysis methods in the construction of global random search algorithms is the following. Let us have a sufficiently representative sample $x_1,...,x_N$ and corresponding values $f(x_i)$, $i=1,...,N$. By identifying clusters (or recognizing patterns) in the set $\mathcal{X} \times f(\mathcal{X})$ we want to determine the subsets of $\mathcal{X}$ where the objective function f takes sufficiently high values. These algorithms are essentially intended for identification of promising subsets (represented usually as spheres) where the estimate of the minimal, maximal, or mean value of the objective function is regarded as the *prospectiveness criterion* (see its precise definition later in Section 4.2).

4.1.3 *Estimation of the cumulative distribution function, its density, mode and level surfaces*

The cumulative distribution functions (4.1.2) and the techniques for their estimation are of considerable importance in global random search theory. One should bear in mind the fact that since the behaviour of (4.1.2) is of primary interest for those values of t for which F(t) is close to one, it is unreasonable to estimate F(t) for all t. This fact determines the specific character of the problem: various solution approaches will be considered in Section 4.2 and in Chapter 7.

Kernel (non-parametric) density estimates are useful for global random search theory. Two reasons for this are: the densities that are kernel estimates can be sampled without knowing their values in points and the properties of kernel estimates are well-studied.

Using kernel estimates as an example, we shall demonstrate how the non-parametric density estimates can support the construction of global random search algorithms in the case of possible presence of a random noise and high labour consumption during the function evaluations. Assume that we are given a sample $\{x_1,...,x_N\}$ from a distribution with density $p(x)$ and that $y(x)= f(x)+\xi(x)\geq c_1\geq 0$ with probability 1, where $E\,\xi=0$. Choose another density $\varphi(x)$ on $\mathbb{R}^n$ and consider densities (kernels)

$$\varphi_\beta(x) = \beta^{-n}\varphi(x/\beta), \qquad \beta > 0,$$

induced by it.

Let us demonstrate (for strict proofs see Section 5.2) that for large N the density

$$q_\beta(x) = \sum_{i=1}^{N} \left[y(x_i) / \sum_{j=1}^{N} y(x_j) \right] \varphi_\beta(x - x_i) \tag{4.1.3}$$

is a close estimate of the density

$$r_\beta(x) = \int r(z) \varphi_\beta(x - z) dz \tag{4.1.4}$$

which is the smoothed density of

$$r(x) = f(x)p(x) / \int f(z)p(z)dz. \tag{4.1.5}$$

Indeed, for all $x \in \mathbb{R}^n$ we have

$$q_\beta(x) = \frac{\frac{1}{N}\sum_{i=1}^{N} f(x_i)\varphi_\beta(x - x_i)}{\frac{1}{N}\sum_{i=1}^{N} f(x_i) + \frac{1}{N}\sum_{i=1}^{N} \xi(x_i)} + \frac{\frac{1}{N}\sum_{i=1}^{N} \xi(x_i)\varphi_\beta(x - x_i)}{\frac{1}{N}\sum_{i=1}^{N} f(x_i) + \frac{1}{N}\sum_{i=1}^{N} \xi(x_i)}$$

and for $N \to \infty$ the following relations hold (in virtue of the law of large numbers):

$$\frac{1}{N}\sum_{i=1}^{N} f(x_i) \to \int f(x)p(x)dx,$$

$$\frac{1}{N}\sum_{i=1}^{N} \xi(x_i) \to 0,$$

$$\frac{1}{N}\sum_{i=1}^{N} f(x_i)\varphi_\beta(x - x_i) \to \int f(z)\varphi_\beta(x - z)p(z)dz,$$

$$\frac{1}{N}\sum_{i=1}^{N} \xi(x_i)\varphi_\beta(x - x_i) \to 0.$$

If the points x_i ($i=1,...,N$) with weights

$$y(x_i) / \sum_{j=1}^{N} y(x_j)$$

are regarded as the points z_i of a sample from the distribution with density (4.1.5), which is true asymptotically for $N \to \infty$, then (4.1.3) is the kernel estimate

$$N^{-1}\sum_{i=1}^{N}\varphi_\beta(x-z_i)$$

of the density r. This one reflects the features of f. For example, if p is the density of the uniform distribution on $\mathcal{X}$ then $r(x)= f(x)/\int f(z)dz$, i.e. r is proportional to f. The smoothed density (4.1.4) also has this property under a reasonable choice of β and the sufficient smoothness of f. The choice of β is discussed in the corresponding literature (c.f. for instance, Devroye and Györfi (1985)). Essentially the less β is, the higher is the kernel estimate accuracy near to the points x_i and the less regular is the curve corresponding to this estimate. Roughly speaking, β should not be unduely small and its choice should be coordinated with the choice of N.

Therefore the density (4.1.3) may be expected to reflect the basic behavioural aspects of f (under *good behaviour* of f, large N, and reasonable values of β), i.e. to be large whenever f itself is large and small whenever f is small. By studying (4.1.3) we, thus, are studying (approximately) the behaviour of the objective function. The density (4.1.3) can be studied without evaluating it at points (the latter may be rather difficult because of the prohibitive number of points to be evaluated), but rather by making a suitable sample from the distribution with this density. The distribution with density (4.1.3) can be generated by means of the superposition method and presents no principle difficulties.

Having a sample $\{x_1,...,x_N\}$ from a distribution with density p, one can estimate the mode (i.e. the maximizer of p) and the level surfaces of this density, i.e. the sets

$$\{x \in \mathcal{X}: p(x)=h\} \qquad (4.1.6)$$

One of the most convenient techniques of the mode estimation is the following. Define a numerical sequence $\{m(N)\}$ so that the asymptotic relations

$$m(N)=o(N), \qquad (N\log N)^{1/2}/m(N)=o(1)$$

hold for $N\to\infty$. Select the least-radius ball among those having their centres at the sample points and containing at least m(N) points. Then the mode of the distribution with density p is estimated by the vector consisting of the average coordinates of those points that are inside the ball. For estimating the mode domain $\mathcal{M}=\{x\in\mathcal{X}: p(x)\le h\}$, bounded by the surface (4.1.6), one should take in the above procedure the convex minimal-volume envelope of m(N) sample points instead of the minimal-radius ball. The above and similar procedures have been thoroughly investigated by Sager (1983) for the unimodal density case. It seems that some of the results can be generalized for the case, when the density belongs to some classes of multiextremal functions (in particular, small local maxima far from the global maximizer do not influence). To reduce possible errors and information losses, the above procedures should be simultaneously applied to several subdomains, say, by separating clusters at the beginning.

4.1.4 Statistical modelling (Monte Carlo method)

There are several basic connections between the theory and methods of statistical modelling and global random search. First, since the global random search algorithms are based on sequential sampling from probability distributions, sampling algorithms are their substantial part: the theory of global random search generates in this case specific sampling problems. For example, the construction of random search algorithms for extremal problems with equality constraints is possible only, if there exist suitable sampling algorithms for probability distributions over surfaces. The problem of sampling on surfaces will be discussed in Section 6.1. Other sampling problems arise e.g. when optimizing in functional spaces, see Section 6.2.1.

Second, that part of statistical modelling theory which deals with optimal organization of random sampling for estimation of multidimensional integrals is of importance also in the theory of global random search. Of special interest in this case is the problem of simultaneous estimation of several integrals which is exemplified as follows. Let the distribution sequence $\{P_k\}$ weakly converge for $k\to\infty$ to a probability measure concentrated at the global maximizer x^* (the construction of such $\{P_k\}$ is described in Section 2.3.3 and Chapter 5). For sufficiently large k, the estimates of $\int f(x)P_k(dx)$ and $\int x(j)P_k(dx)$ $(j=1,\dots,n)$ are, respectively, the estimates of $\max f$ and $x^*(j)$ where $x(j)$ and $x^*(j)$ are the coordinates of the points x and x^*, respectively. One more integral, the normalization constant of the distribution P_k is added to the above n+1 integrals. Moreover, simultaneous estimation of several integrals using the same sample underlies a number of non-parametric regression estimation methods (see Section 8.3). Finally, many useful characteristics of the objective function (e.g. mean values on subsets of $\mathcal{X}$) are representable as linear integral functionals of either the objective function itself, or functions/measures closely related to it (functionals of the form of (5.3.6) or functionals of estimates of f).

Third, various procedures created within the framework of Monte Carlo theory in an effort to reduce the variance of linear integral functional estimates are used for construction of optimization algorithms. For example, Section 4.4 considers stratified sampling, while Section 8.2 will analyse importance sampling. Let us demonstrate that dependent sampling procedures may also be used in the search algorithms.

Let $\mathcal{X}\subset\mathbb{R}$, the objective function f be sufficiently smooth and subject to random noise, $f(x)=E f(x,\omega)$ where $f(x,\omega)$ is the result of a simulation trial conducted at a point $x\in\mathcal{X}$ under the random occurance ω that in practice is a collection of random numbers (i.e. realizations of the random variable uniformly distributed over the interval [0,1]). Let the derivative $f'(x_0)$ at a point $x_0\in\mathcal{X}$ be estimated and the approximations

$$f'(x_0)\approx\big(f(x_0+h)-f(x_0+h)\big)/2h, \tag{4.1.7}$$

$$f(x_0+h)\approx\frac{1}{N}\sum_{i=1}^{N} f(x_0+h,\omega_i), \tag{4.1.8}$$

$$f(x_0-h)\approx\frac{1}{N}\sum_{i=1}^{N} f(x_0-h,\omega_i') \tag{4.1.9}$$

be used for the estimation of $f'(x_o)$ where h is a small value, N is an integer (resp. the step size and sample size).

In ordinary simulation experiments, all random elements ω_i and ω_i' are different and the error of the estimator of $f'(x_o)$ arises from the error of the approximation (4.1.7) (that asymptotically equals $h^2|f'''(x_o)|$ for $h\to 0$), and the error due to randomness of (4.1.8) and (4.1.9). The latter error is estimated by the value

$$3\gamma\sqrt{D_o}\,(2h^2N)^{-1/2},$$

where γ is determined by the significance level of the estimate and

$$D_0 = \int f^2(x_0, \omega)P(d\omega) - f^2(x_0).$$

is the variance of the estimate $f(x_o, \omega)$ of $f(x_o)$. Thus, one can refer to

$$h^2|f'''(x_0)| + 3\gamma\sqrt{D_0}/\sqrt{2h^2N} \tag{4.1.10}$$

as the error of the approximation (4.1.7) - (4.1.9) under different random occurance ω_i and ω_i'. If $N\to\infty$, $h=h(N)\to 0$, $h(N)N^{1/2}\to 0$ then (4.1.10) approaches zero while $N\to\infty$ and the optimal sequence $h_{opt}(N)$ minimizing (4.1.10) is determined by

$$h_{opt}(N) = (3\gamma)^{1/3}|f'''(x_0)|^{-1/3}D_0^{1/2}(2N)^{-1/6}.$$

The minimum of (4.1.10) is $cN^{-1/3}$ where

$$c = \left(36\gamma^2|f'''(x_0)|D_0\right)^{1/3}.$$

While using the dependent sample procedure one uses $\omega_i' = \omega_i$ for i=1,...,N in (4.1.8), (4.1.9), i.e. the simulations for x_o+h and x_o-h are accomplished under the same collection of realizations of random numbers. Choosing h arbitrary small, one obtains $3\gamma D_1^{1/2}N^{-1/2}$ instead of (4.1.10), where

$$D_1 = \int\left(f_x'(x_0, \omega)\right)^2 P(d\omega) - \left(f'(x_0)\right)^2.$$

This way, the order of convergence of the approximation as $N\to\infty$ is higher for dependent sample. Analogous results are valid for the multidimensional case and the case of higher derivatives as well. Recently an improvement of the dependent sampling technique was created for optimization of discrete simulation models of particular classes, for references see e.g. Suri (1987).

Note that some of the approaches to global optimization (e.g. see later Section 6.2.1) need the drawing of a random function g from a parametric functional set G: setting randomness is equivalent to setting a probability measure on a parameter set. The prior information is often insufficient for a reasonable choice of this probability measure, and frequently a uniform distribution on the parameter set is used. Such a choice of randomness on the parameter set, with great chance, might lead to *exotic* random functions from G. Kolesnikova et al. (1987) proposed to construct the probability distribution by defining uniformity properties on the set of realizations of random functions from G, see also Zhigljavsky (1985).

It should be also noted that global random search theory can be regarded as a part of the theory of Monte Carlo method, if the latter concept is interpreted in a broad sense.

4.1.5 Design of experiments

The following basic kinds of experiments are distinguished (see Ermakov and Zhigljavsky (1987)): regression, factorial, extremal, simulation (i.e. Monte Carlo), and discrimination (including screening) for hypotesis testing. In the above subsections, possible applications of procedures for design of regression, extremal and simulation experiments have been discussed. The theory of screening experiment is also important for global search theory: it consists in determination of the basic potential for construction of algorithms, in the construction itself, and in separation from many factors those several ones that define the relation at issue. Some applications of screening experiment theory in global search were investigated by Szltenis (1989), see also Section 2.3.2.

4.2 Statistical inference concerning the maximum of a function

This is one of the principal sections of this chapter (often referred to later): its results can be widely used in global random search theory and its applications.

Subsection 4.2.1 states the problem of drawing statistical inference for $M=\max f$ and presents the main conditions applied in the section. Subsection 4.2.2 outlines the ways of constructing statistical inferences for M. Subsections 4.2.3 and 4.2.4 describe statistical inferences for M, borrowing them to some extent from Chapter 7. Subsection 4.2.5 deals with the problem of estimating the c.d.f. F(t) defined by (4.1.2). Subsection 4.2.6 contains a number of results concerning the prior determination of the value of the tail index of the extreme value distribution: these results are of great importance for global random search theory. Finally, Subsection 4.2.7 deduces from previous parts of the section a result on the exponential complexity of the uniform random sampling algorithm.

Due to their significance for the global random search theory, the main findings of the section are formulated not only for the maximization problem, but also for the minimization one.

4.2.1 *Statement of the problem and a survey of the approaches for its solution*

At each step of most global random search algorithms, there is a sample of points from subsets $Z\subset\mathcal{X}$ and values of the objective function f at these points; the distributions for subsequent points are constructed after drawing certain statistical inferences concerning the objective function behaviour. The parameters

$$M_Z = \sup_{x\in Z} f(x)$$

and the behaviour of the c.d.f. (4.1.2) in the vicinity of the parameter M_Z (i.e. for t such that $F_Z(t)\approx 1$), which is unambigously related to the behaviour of the objective function near the point

$$x^*_Z = \arg\max_{x\in Z} f(x),$$

are of primary importance in making a decision about the prospectiveness of $Z\subset\mathcal{X}$ for further search.

Since for all sets Z and at various iterations statistical inferences are drawn in a similar manner, for $M=M_{\mathcal{X}}=\max f$ and $F=F_{\mathcal{X}}$ they only will be drawn through independent sample $\Xi=\{x_1,...,x_N\}$ from distribution P on $\mathcal{X}$ satisfying the condition $P(B(\varepsilon))>0$ for all $\varepsilon>0$. It will also be assumed that $\mathcal{X}\subset\mathcal{R}^n$ and f is evaluated without a noise. The elements $x_1,...,x_N$ of the independent sample Ξ are mutually independent random vectors (a way of their generation can either consist of the direct sampling of a distribution or include iterations of a local ascent starting at random points, see Section 3.1.6). Some generalizations to the cases of random noise and of dependence for elements of Ξ be found in Sections 4.2.8 and 4.4.

Having a sample Ξ, pass to the sample

$$Y = \{y_1, \ldots, y_N\} \tag{4.2.1}$$

where $y_i = f(x_i)$ for $i=1,\ldots,N$ are independent realizations of the random variable η with c.d.f.

$$F(t) = \int_{f(x)<t} P(dx) = P(\{x \in \mathcal{X}: f(x) < t\}) \tag{4.2.2}$$

and to the order statistics $\eta_1 \leq \eta_2 \leq \ldots \leq \eta_N$ corresponding to the sample (4.2.1). The parameter $M = \max f$ is the upper bound of the random variable η (M=vrai sup η), i.e.

$$\Pr\{\eta \leq M\} = 1, \qquad \Pr\{\eta \leq M - \varepsilon\} < 1$$

holds for any $\varepsilon > 0$. Now, the problem of drawing statistical inferences for $M = \max f$ is stated as follows: having a sample (4.2.1) of independent realizations of the random variable η with c.d.f. (4.2.2), statistical inferences for the upper bound

$$M = \text{vrai sup}\,\eta \tag{4.2.3}$$

are to be made. The statistical inference described below can both provide supplementary information on the objective function, at each iteration of numerous global random search algorithms (in particular, in order to construct suitable stopping rules) and support the construction of various branch and probability bound methods (see later Section 4.3).

For convenience, the main conditions to be applied in Section 4.2 are collected below.

(a) The function

$$V(v) = 1 - F(M - 1/v), \quad v > 0,$$

is regularly varying at infinity with some exponent $(-\alpha)$, $0 < \alpha < \infty$, i.e.

$$\lim_{v \to \infty} [V(tv)/V(v)] = t^{-\alpha} \tag{4.2.4}$$

for all $t > 0$.

(b) The representation

$$F(t) = 1 - c_0 (M - t)^{\alpha} + o((M - t)^{\alpha}), \qquad t \uparrow M, \tag{4.2.5}$$

is valid, where $c_0 > 0$ is an appropriate constant.

(c) The global maximizer $x^* = \arg\max f$ is unique and f is continuous in a vicinity of x^*.

(d) There is an $\varepsilon_o>0$ such that the sets

$$A(\varepsilon)=\{x\in \mathcal{X}: |f(x)-f(x^*)|\le\varepsilon\}$$

are connected for all ε, $0<\varepsilon\le\varepsilon_o$.

(e) There is such $c_1>0$ that there holds the relation

$$\lim_{\varepsilon\to 0}\varepsilon^{-n}P(B(\varepsilon))=c_1. \tag{4.2.6}$$

(f) In a vicinity of x^* the representation

$$|f(x)-f(x^*)|=w(\|x-x^*\|)H(x-x^*)+o(\|x-x^*\|^{\beta}),\quad \|x-x^*\|\to 0, \tag{4.2.7}$$

is valid, where H is a positive homogeneous function on $\mathcal{R}^n\setminus\{0\}$ of order $\beta>0$ (for H the relation $H(\lambda z)=\lambda^{\beta}H(z)$ holds for all $\lambda>0$, $z\in\mathcal{R}^n$) and w is a positive, continuous, one-dimensional function.

(g) There exists a function U: $(0,\infty)\to(0,\infty)$ such that for any $x\in R^n$, $x\neq 0$, the limit

$$\lim_{t\downarrow 0}|f(x^*)-f(x^*+tx)|U(t) \tag{4.2.8}$$

exists and is positive.

(h) There exist positive numbers ε_1, c_1 and c_2 such that for all ε, $0<\varepsilon\le\varepsilon_1$, the two-sided inequality

$$c_1\varepsilon^n\le P(B(\varepsilon))\le c_2\varepsilon^n \tag{4.2.9}$$

is valid.

(i) There exist positive numbers ε_2, β, c_3 and c_4 such that for all $x\in B(\varepsilon_2)$ the inequality

$$c_3\|x-x^*\|^{\beta}\le|f(x)-f(x^*)|\le c_4\|x-x^*\|^{\beta} \tag{4.2.10}$$

holds.

If the minimization problem of f is considered, then one has to deal with the lower bound vrai inf $\eta=\min f$ of the random variable η, make the evident substitution in the definition of x^* and M, and change the conditions a) and b) into

a') the function $U(u)=F(M-1/u)$, where $u<0$ and $M=\text{vrai inf}\,\eta>-\infty$, regularly varies at infinity with some exponent $(-\alpha)$, i.e.

$$\lim_{u \to -\infty} U(tu)/U(u) = t^{-\alpha}$$

for all $t > 0$,

b') the representation

$$F(t) = c_0(t - M)^{\alpha} + o((t - M)^{\alpha}), \qquad t \downarrow M = f^*$$

is valid, where $c_0 > 0$ is a constant.
All other conditions are unchanged.

4.2.2 *Statistical inference construction for estimating M*

Several basic approaches to estimate the optimum value $M = \max f$ will be outlined below.

An approach involving the construction of a parametric (e.g. quadratic in x) regression/approximation model of $f(x)$ in the vicinity of a global maximizer is often used in the case, when evaluations of f are subject to a random noise and it is likely that the vicinity of x^* is reached where the objective function behaves *well* (for example, f is locally unimodal and twice continuously differentiable), see later Section 8.1.

Another approach (see Mockus (1967), Hartley and Pfaffenberger (1971), Chichinadze (1967, 1969), Biase and Frontini (1978)) is based on the assumption that the c.d.f. (4.2.2) is determined (identified) to a certain number of parameters that are estimated via (4.2.1) by means of standard mathematical statistics tools. Essentially, the results of Betro (1983, 1984) (described later in Section 4.6) can be also classified to this general approach. It has three main drawbacks:
(i) for many commonly used classes of objective functions, the adequacy of the parametric models is not valid and there do not exist yet constructive descriptions of functional classes for which these models have acceptable accuracy;
(ii) the presence of redundant parameters decreases the estimation accuracy of M and increases the computational efforts; and
(iii) construction of a confidence interval for M and testing statistical hypothesis about M are frequently impossible, while that is of primary importance in optimization problems (an exception being the method of Hartley and Pfaffenberger (1971) designed especially for confidence interval construction).

Certainly, a non-parametric approach for estimating c.d.f. (4.2.2) together with M can be used but it is hard to expect its high efficiency. Generally, the semi-parametric approach described below is more efficient than the both mentioned. Its another advantage is that it is theoretically justified (if the sample size N is large enough). This semi-parametric approach is based on the next classical result of the asymptotic theory of the extreme order statistics, see e.g. Galambos (1978).

Theorem 4.2.1. Let $M = \text{vrai sup}\ \eta < \infty$ where η is a random variable with c.d.f. $F(t)$ and the condition a) of Section 4.2.1. be fulfilled. Then

$$\lim_{N \to \infty} F^N(M + \chi_N z) = \Phi_\alpha(z) \qquad (4.2.11)$$

where

$$\chi_N = M - \theta_N + o(1) \qquad \text{for} \qquad N \to \infty, \tag{4.2.12}$$

$$\theta_N = \inf \{ v | 1 - F(v) \le N^{-1} \}, \tag{4.2.13}$$

$$\Phi_\alpha(z) = \begin{cases} \exp\{-(-z)^\alpha\}, & z < 0 \\ 1 \quad , & z \ge 0. \end{cases} \tag{4.2.14}$$

The distribution having the c.d.f. (4.2.14) will be called an extreme value distribution and α - its tail index. The c.d.f. $\Phi_\alpha(z)$, including the limit case $\Phi_\infty(z)=\exp\{-e^{-z}\}$, is the unique nondegenerate limit of c.d.f. sequences for $(\eta_N-a_N)/b_N$ in the case $M<\infty$, where $\{a_N\}$ and $\{b_N\}$ are arbitrary numerical sequences. The condition a) is a regularity condition for the c.d.f. F in the vicinity of M: only some *exotic* functions fail to satisfy it, see Gumbel (1958), Galambos (1978). In particular, it is fulfilled if the natural condition b) is met.

The asymptotic representation (4.2.11) means that the sequence of random variables $(\eta_N-M)/\chi_N$, where η_N is the maximal order statistic from the sample (4.2.1), converges in distribution to the random variable with c.d.f. (4.2.14).

As shown below, the convergence rate of F(t) to 1 for $t\uparrow M$ is mainly defined by the tail index α of the extreme value distribution. The main difficulties in applying the results of the extreme order statistics theory are due to the fact that α may not be known in practice. For this reason, e.g. Clough (1969) suggested $\alpha=1$ that is, of course, generally false. Dannenbring (1977), Makarov and Radashevich (1981) and some others suggested - following Gumbel's (1958) recommendation to estimate jointly α, M, and θ_N via taking a sample of order N of independent maximal order statistics corresponding to samples for the r.v. η. This approach requires an order of N^2 evaluations of f and has the drawbacks (ii) and (iii) detailed above.

The approach below also relies on the theory of extreme order statistics, but is free of the drawbacks mentioned. It is based on some recent advances of mathematical statistics described later in Chapter 7, enabling for a prior determination of the value of the tail index α for a wide range of objective function classes.

The present approach was developed by Zhigljavsky (1981, 1985, 1987), independently of other authors (cited below) who obtained similar results. The main similarity is between Theorem 4.2.2 and the outstanding result of de Haan (1981) that will be properly discussed.

4.2.3 *Statistical inference for M, when the value of the tail index α is known*

Let us cite from Chapter 7 some results concerning statistical inference for M: the corresponding proofs, references, comments as well as additional details and information will be given in Chapter 7.

We shall suppose below that condition a) defined in Section 4.2.1 is satisfied and that the value of the tail index $\alpha \geq 1$ of the extreme value distribution is known. (This is a realistic assumption for many global random search problems, as will be seen in Section 4.2.6.)

First, let us consider some estimators of M. Linear estimators have the form

$$M_{N,k} = \sum_{i=0}^{k} a_i \eta_{N-i} \tag{4.2.15}$$

where k is an integer, k may depend on N, i.e. $k=k(N)$, $k^2(N)/N \to 0$ if $N \to \infty$, a_i $(i=0,1,\ldots,k)$ are real numbers (coefficients). The estimator (4.2.15) is consistent for $N \to \infty$ if and only if

$$a'\lambda = \sum_{i=0}^{k} a_i = 1 \tag{4.2.16}$$

where $a=(a_0,a_1,\ldots,a_k)'$ $\lambda=(1,1,\ldots,1)'$.

The quality of the estimator (4.2.15) satisfying (4.2.16) is measured by the mean squared error that is asymptotically represented as

$$E(M_{N,k} - M)^2 \sim (M - \theta_N)^2 a'\Lambda a, \qquad N \to \infty \tag{4.2.17}$$

where Λ is the symmetrical matrix of order $(k+1)\times(k+1)$ with elements

$$\lambda_{ij} = \Gamma(2/\alpha + i + 1)\Gamma(1/\alpha + j + 1)/[\Gamma(1/\alpha + i + 1)\Gamma(j+1)], \; j \leq i.$$

The optimal coefficient vector

$$a^* = \arg \min_{a:a'\lambda=1} a'\Lambda a$$

corresponds to the asymptotically optimal linear estimator $M_{N,k}{}^*$ and equals

$$a^* = \Lambda^{-1}\lambda/(\lambda'\Lambda^{-1}\lambda). \tag{4.2.18}$$

The explicit form for the components a_i^* $(i=0,1,\ldots,k)$ of the vector a^* is as follows:

$$a_0^* = c(\alpha + 1)/\Gamma(2/\alpha + 1),$$
$$a_j^* = c(\alpha - 1)\Gamma(j + 1)/\Gamma(2/\alpha + j + 1) \qquad \text{for} \qquad j = 1, \ldots, k - 1,$$
$$a_k^* = -\, c(k\alpha + 1)\Gamma(k + 1)/\Gamma(2/\alpha + k + 1)$$

where

$$c = \left[\frac{\alpha\Gamma(k+2)}{(\alpha-2)\Gamma(2/\alpha+k+1)} - \frac{2}{(\alpha-2)\Gamma(2/\alpha+1)}\right]^{-1} \qquad (\alpha \neq 2)$$

is the normalization constant. (c can also be computed from (4.2.16).)

Under some supplementary conditions namely, $\alpha>2$, $k=k(N)\to\infty$, $k^2(N)/N\to 0$ for $N\to\infty$) the estimator $M_{N,k}^*$, defined by (4.2.15) and (4.2.18), is asymptotically Gaussian and asymptotically optimal. Its mean squared error is asymptotically equal to

$$E\left(M_{N,k}^* - M\right)^2 \sim (1 - 2/\alpha)(c_0 N)^{-2/\alpha} k^{-(1-2/\alpha)}, \qquad N \to \infty \qquad (4.2.19)$$

where c_0 is a constant that can be estimated by the estimator

$$\hat{c}_0 = k(\eta_N - \eta_{N-k})^{-\alpha}/N \qquad (4.2.20)$$

which is consistent, asymptotically unbiased, and

$$\lim_{k\to\infty} kE\left(\hat{c}_0 - c_0\right)^2 = c_0^2.$$

Some other linear estimators are also asymptotically Gaussian and asymptotically optimal (for them an analogy of (4.2.19) is valid). For instance, the estimator (4.2.15) with coefficients

$$\begin{cases} a_i = (k+1)^{(2-\alpha)/\alpha}(\alpha - 1)\left[(i+1)^{1-2/\alpha} - i^{1-2/\alpha}\right] & \text{for } i = 0, 1, \ldots, k-1 \\ a_k = (k+1)^{(2-\alpha)/\alpha}(\alpha - 1)\left[(k+1)^{1-2/\alpha} - k^{1-2/\alpha}\right] - (\alpha - 2) & \end{cases} \qquad (4.2.21)$$

has these properties.

The same property is possessed by the maximal likelihood estimator

$$\hat{M}$$

that is the solution of the equation

$$(\alpha-1)\sum_{j=0}^{k-1}\beta_j(\hat{M})=k+1 \tag{4.2.22}$$

under the condition $M \geq \eta_N$; here

$$\beta_j(\hat{M})=(\eta_{N-j}-\eta_{N-k})/(\hat{M}-\eta_{N-j}). \tag{4.2.23}$$

The maximum likelihood estimator

$\hat{M}$

is nonlinear and it cannot be represented in explicit form.

Consider the ways of constructing confidence intervals for M with asymptotic confidence level $1-\gamma$, where a small positive number γ is fixed.

If k is large, then to construct a confidence interval one may use the above mentioned asymptotic normality property and the expression (4.2.19) for the optimal linear estimator, for the linear estimator defined by (4.2.15) and (4.2.21), and for the maximal likelihood estimator. If N is not sufficiently large, say $N<500$, then k cannot be chosen sufficiently great and the suggested method is inapplicable.

Instead of the asymptotical normality property, one may also use the Chebyshev inequality that gives for linear estimators the approximate relation

$$\Pr\left\{\left|M_{N,k}-M\right|\geq\varepsilon\right\} \lesssim a'\Lambda a(c_0N)^{-2/\alpha}\varepsilon^{-2}, \qquad N\to\infty, \tag{4.2.24}$$

where c_0 can be estimated by (4.2.20). This approach leads to less precise confidence intervals, than the preceding one, but is applicable for any $k\geq 1$.

The one-sided confidence interval

$$[\eta_N, \eta_N + r_{k,1-\gamma}(\eta_N-\eta_{N-k})] \tag{4.2.25}$$

where

$$r_{k,1-\gamma}=1/\left[\left(1-\gamma^{1/k}\right)^{-1/\alpha}-1\right]$$

is usually narrower than the one-sided interval based on (4.2.24). It should be noted that the one-sided confidence intervals for M - with η_N as a natural lower bound - are of more interest for global random search theory, than the usual two-sided ones.

Statistical hypothesis testing procedures for M are of great importance in the present context, too. The standard situation here is as follows. Let there be a record value M_0 and an independent sample (4.2.1) from values of $\eta=f(\xi)$ where ξ is a random vector given on a subset Z of $\mathcal{X}$. Suppose also that

$$\eta_N=\max\{f(x_1),\ldots,f(x_N)\}<M_0$$

and it is to be decided whether the function f can have values in Z greater than M_0 (i.e. whether f can achieve its global maximum in Z). In other words, one has to test the statistical hypothesis H_0: $M_Z > M_0$ under the alternative H_1: $M_Z \leq M_0$.

For testing the hypothesis H_0, one can use ways similar to those stated for confidence intervals construction. According to the approach leading to (4.2.25), the rejection region for H_0 is

$$W = \{Y: (M_0 - \eta_N)/(\eta_N - \eta_{N-k}) \geq r_{k,1-\gamma}\} \tag{4.2.26}$$

where $r_{k,1-\gamma}$ is given by (4.2.25). In this test the power function

$$\beta_N(M,\gamma) = \Pr\{Y \in W\}$$

is a decreasing function of M having the asymptotic representations

$$\beta_N(M_0,\gamma) \to \gamma \qquad \text{for} \qquad N \to \infty$$

$$\beta_N(M,\gamma) \sim \frac{1}{k!} \int_0^\infty t^k e^{-t} \left(1 - \left(\frac{r_{k,1-\gamma} - t^{-1/\alpha}(M_0 - M)/(M - \theta_N)}{1 + r_{k,1-\gamma}}\right)_+^\alpha\right)_+^k dt$$

where $N \to \infty$ and $(a)_+ = \max\{0,a\}$.

A related numerical study shows that for moderate values of N (say $N \leq 100$), it is advantageous to choose k accelerating to the *rule* $N/20 \leq k \leq N/10$; for large N ($N>100$) it is often sufficient to take k not greater than 10. For such a choice of k the precision of the statistical inferences (4.2.25) and (4.2.26) is almost as good as for the asymptotically optimal choice $k(N) \to \infty$, $k^2(N)/N \to 0$ for $N \to \infty$.

If the minimization problem of f is stated then the condition a') is to be substituted for a) and in the suitable places of the formulas of this subsection η_{i+1} is to be substituted for η_{N-i} for all $i=0,1,\ldots,k$. Thus, one has to test the hypothesis

$$H_0: \min_Z f < M_0$$

instead of H_0: $M_Z > M_0$, and should use the formulas

$$M_{N,k} = \sum_{i=0}^{k} a_i \eta_{i+1}, \tag{4.2.27}$$

$$\hat{c}_0 = k(\eta_{k+1} - \eta_1)^{-\alpha}/N,$$

$$\beta_j(\hat{M}) = (\eta_k - \eta_j)/(\eta_j - \hat{M}),$$

$$[\eta_1 - r_{k,1-\gamma}(\eta_{k+1} - \eta_1), \eta_1], \tag{4.2.28}$$

and

$$W = \{Y: (\eta_1 - M_0)/(\eta_{k+1} - \eta_1) \geq r_{k,1-\gamma}\} \tag{4.2.29}$$

replacing (4.2.15), (4.2.20), (4.2.23), (4.2.25) and (4.2.26), respectively: the other formulas of the subsection, given above are unchanged.

We turn now to another approach to construct confidence intervals for M which may be termed as the improved Hartley-Pfaffenberger method. Let us start by quoting an auxiliary result from Hartley and Pfaffenberger (1971).

It is well known that $t_i = F(\eta_i)$, i=1,...,N, form an order statistics corresponding to an independent sample from the uniform on [0,1] distribution and their means and covariances are $E\, t_i = \mu_i$, $cov(t_i, t_j) - v_{ij}$ (for i,j=1,...,N) where

$$\mu_i = i/(N+1), \qquad v_{ji} = v_{ij}, \qquad v_{ij} = \mu_i(1-\mu_j)/(N+2) \qquad \text{for } i \leq j.$$

Note that $v_{ij} = v_{N-j+1,N-i+1}$ for $i \leq j$. Define the k-vectors

$$t = (t_N, \ldots, t_{N-k+1})', \qquad \mu = (\mu_N, \ldots, \mu_{N-k+1})'.$$

The covariance matrix V of t is symmetric, has the order k×k and elements v_{ij} for $1 \leq i \leq j \leq k$. According to Hartley and Pfaffenberger (1971), the quadratic form

$$s^2(k,N) = (t-\mu)'V^{-1}(t-\mu)$$

can be written as

$$s^2(k,N) = (N+1)(N+2)\left[\frac{t^2_{N-k+1}}{N-k+1} + \sum_{i=N-k+1}^{N+1} (t_i - t_{i-1})^2\right] - (N+2)$$

where $t_{N+1}=1$. The exact distribution of $s^2(k,N)$ is rather complicated: a table of numerically evaluated γ-quantiles s_γ of this distribution is given in the work mentioned.

The dependence of s_γ on N is rather mild: one may approximately select s_γ for γ=0.05, 0.01 and k=5, 10, 15, 20 as follows $s_{0.05}\approx$15, 25, 33, 40, $s_{0.01}\approx$28, 40, 50, 55.

Hartley and Pfaffenberger (1971) used the unreasonable supposition that the c.d.f. F(t) is represented by a Taylor expansion at t=M, rather, than a condition of type (a). We shall suppose that (b) is fulfilled and k is such that the extreme order statistics $\eta_N,\ldots,\eta_{N-k+1}$ fall into the vicinity A_M of M, where the approximation

$$F(t) \approx 1 - c_0(M-t)^\alpha, \qquad t \in A_M,$$

is valid with a rather high accuracy. Thus, we may set

$$t_{N-i} = F(\eta_{N-i}) = 1 - c_0(M - \eta_{N-i})^\alpha \qquad \text{for } i = 0, 1, \ldots, k-1$$

and put them into the expression for $s^2(k,N)$ that depends on c_0 and M as well as k and N. Denoting $s^2(k,N)= s^2(k,N,M, c_0)$ and

$$s^2(k,N,M) = \min_{c_0} s^2(k,N,M,c_0),$$

we come to the upper confidence bound $M_{1-\gamma}$ for M of a level $1-\gamma$ defined as the solution of the equation $s^2(k,N,M_{1-\gamma})=s_\gamma$ under the condition $M_{1-\gamma}\geq\eta_N$. The corresponding confidence interval for M is $[\eta_N, M_{1-\gamma}]$.

Since $s^2(k,N,M, c_0)$ as a function of c_0 is a polynomial of degree two, the expression for $s^2(k,N,M)$ is easily derived:

$$s^2(k,N,M) = s^2(k,N,M,c_*)$$

where

$$c_* = \frac{(M-\eta_{N-k+1})^\alpha}{N-k+1} \Big/ \left[\sum_{i=N-k+1}^{N} \left((M-\eta_i)^\alpha - (M-\eta_{i-1})^\alpha \right)^2 + \right.$$

$$\left. + \frac{(M-\eta_{N-k+1})^{2\alpha}}{N-k+1} + (M-\eta_N)^{2\alpha} \right].$$

In principle, instead of $s^2(N,k,M,c*)$ one may use $s^2(N,k,M,\hat{c}_0)$ - where $\hat{c}_0$ is an estimator of c_0 (cf. (4.2.20)) - as $s^2(N,k,M)$.

Roughly speaking, the use of $c*$ corresponds to the max-min approach to a confidence interval construction, but the use of $\hat{c}_0$ - to a Bayesian one.

Note that the improved Hartley-Pfaffenberger approach described above is simpler than the original one (and, unlike the original method, it is correct). The author knows nothing, however, about the accuracy of the confidence intervals constructed by applying the improved Hartley-Pfaffenberger approach.

4.2.4 *Statistical inference, when the value of the tail index α is unknown*

We continue to cite results from Chapter 7.

Suppose that the condition (a) of Section 4.2.1 holds, but the value of the parameter α is unknown (practically this is quite typical).

In this case, one may apply the presented statistical inferences for M, using an estimator for α instead of its true value.

Many estimators for α are known: for instance, the estimator

$$\hat{\alpha} = \left(\log \frac{k}{m}\right) / \log[(\eta_N - \eta_{N-k})/(\eta_N - \eta_{N-m})] \tag{4.2.30}$$

can be used in which $m<k$. If $k \to \infty$, $m \to \infty$, $k/N \to 0$ (for $N \to \infty$) then $\hat{\alpha}$ is consistent and asymptotically unbiased. For the practically advantageous case $k=5m$, there holds

$$\lim_{k \to \infty} k(\hat{\alpha} - \alpha)^2 = 1.544 \ \alpha^2.$$

The estimator

$$\tilde{\alpha} = \left(\log \frac{k}{m}\right) \Big/ \left[\log \frac{\psi(k+1) - \psi(1)}{\psi(m+1) - \psi(1)} - \log \frac{\psi(k+1) + \hat{\alpha}}{\psi(m+1) + \hat{\alpha}} + \frac{\log(k/m)}{\hat{\alpha}}\right], \tag{4.2.31}$$

where $\hat{\alpha}$ is determined by (4.2.30) and $\psi(k)=\Gamma'(k)/\Gamma(k)$ is the ψ-function, is more precise than $\hat{\alpha}$.

Set

$$F_k(u) = \begin{cases} u^{m+1} \sum_{i=0}^{k-m-1} C^i_{m+i} (1-u)^i, & \text{for } 0 \le u \le 1 \\ 1 \quad , & \text{for } u > 1 \end{cases}$$

and $u_{k,\delta}$ is such that $F_k(u_{k,\delta}) = \delta$. Then the confidence region

$$\left[\hat{\alpha}\left(\log \frac{k}{m}\right) \log u_{k,1-\gamma/2}, \hat{\alpha}\left(\log \frac{k}{m}\right) \log u_{k,\gamma/2}\right] \tag{4.2.32}$$

for α has asymptotical confidence level 1-γ.

The results of the next subsection show that the procedures of testing the statistical hypothesis H_0: α=n/2 under the alternative H_1: α=n are of importance in global random search theory. To construct a procedure for testing the above hypothesis, one may choose

$$W = \left\{Y: \hat{\alpha} > n\left(\log \frac{k}{m}\right)\left[2 \log u_{k,1-\gamma}\right]^{-1}\right\} \tag{4.2.33}$$

as the rejection region. The power of the test is asymptotically equal to

$$\Pr\{Y \in W | \alpha = n\} \sim 1 - F_k\left(u^2_{k,1-\gamma}\right), \qquad N \to \infty.$$

In this case the maximal likelihood estimator

$\hat{M}$

for M is the solution of the equation

$$\left[\sum_{j=0}^{k-1} \log\left(1 + \beta_j(\hat{M})\right)\right]^{-1} - \left[\sum_{j=0}^{k-1} \beta_j(\hat{M})\right]^{-1} = 1/(k+1)$$

under the condition

$\hat{M} \geq \eta_N$ where $\beta_j(\hat{M})$

is determined by (4.2.23). Under some auxiliary conditions (including α>2, N$\to\infty$, k$\to\infty$, k/N$\to$0), this estimator is asymptotically Gaussian with mean M and variance,

$$(1 - 2/\alpha)(\alpha - 1)^2 c_0^{-2/\alpha} N^{-2/\alpha} k^{-(1-2/\alpha)}.$$

For large values of N and k, this result may be used for constructing confidence intervals and statistical testing procedures for M.

If k can not be large (for the case of moderate values of N) then an other way of constructing confidence intervals and hypothesis testing for M is more attractive. Its essence is in the result of de Haan (1981) that if k$\to\infty$, k/N$\to$0, N$\to\infty$ then the test statistic

$$\frac{(\log k)\log\left[(M - \eta_{N-1})/(M - \eta_N)\right]}{\log\left[(\eta_{N-2} - \eta_{N-k})/(\eta_{N-2} - \eta_{N-1})\right]} \tag{4.2.34}$$

asymptotically follows standard exponential distribution with the density e^{-t}, t>0.

If the minimization problem of f is stated, then the formulas (4.2.30) and (4.2.34) are to be transformed into

$$\hat{\alpha} = \left(\log \frac{k}{m}\right)/\log[(\eta_k - \eta_1)/(\eta_m - \eta_1)], \tag{4.2.35}$$

and

$$\frac{(\log k)\log[(\eta_2 - M)/(\eta_1 - M)]}{\log[(\eta_k - \eta_3)/(\eta_3 - \eta_2)]}$$

All other formulas are unchanged.

4.2.5 *Estimation of F(t)*

Estimators of the c.d.f. F(t) defined by (4.2.2) can be widely used in global random search algorithms for the prediction of further search, construction of stopping rules or rules for change to local search (if such a search change is designed). Again, the behavior of F(t) in the vicinity of $M = \text{vrai sup}\, \eta$ (i.e. for t such that $F(t) \approx 1$) is of most interest.

The ways of estimating F(t) for $t \approx M$ are based on the asymptotic representation (4.2.11) which follows by supposition (a) and can be rewritten in the form

$$F(t) \approx \exp\left\{-\frac{1}{N}((M-t)/\chi_N)^{\alpha}\right\} \qquad \text{for } t < M,\ N \to \infty. \tag{4.2.36}$$

This asymptotic representation is valid for all $t<M$, but the more close is t to M, the more close the right and left hand sides of (4.2.36) are.

The simplest way of using (4.2.36) for estimating F(t) consists of replacing M, χ_N, and α by their estimators. Thus, if one uses (4.2.15), (4.2.30) and $M_{N,k} - \eta_N$ as an estimator for χ_N, the estimator

$$F(t) \approx \exp\left\{-\frac{1}{N}((M_{N,k} - t)/(M_{N,k} - \eta_N))^{\hat{\alpha}}\right\} \tag{4.2.37}$$

is obtained for $t \le M_{N,k}$.

If the condition (b) is met, then

$$1 - \frac{1}{N} = F(\theta_N) \sim 1 - c_0(M - \theta_N)^{\alpha} \qquad \text{for } N \to \infty$$

that implies

$$M - \theta_N \sim (c_0 N)^{1/\alpha} \qquad \text{for } N \to \infty. \tag{4.2.38}$$

So one may use

$$\chi_N = (c_0 N)^{1/\alpha} \tag{4.2.39}$$

or substitute here

$\hat{\alpha}$

for α if α is unknown.

Combining (4.2.36) and (4.2.39) one obtains

$$F(t) \sim \exp\left\{-c_0(M-t)^{\alpha}\right\} \quad \text{for } t \uparrow M \tag{4.2.40}$$

and the corresponding estimator

$$F(t) \approx \exp\left\{-\hat{c}_0(M_{N,k}-t)^{\hat{\alpha}}\right\} \tag{4.2.41}$$

where $\hat{c}_0$ is determined from (4.2.20).

Of course, if an exact value of the tail index α can be determined (see Section 4.2.6), then this value is to be used in (4.2.37) and (4.2.41) instead of

$\hat{\alpha}$.

If the minimization problem is considered, then the conditions (a') and (b') are to be substituted for (a) and (b) and the formulas (4.2.36), (4.2.37), (4.2.40) and (4.2.41) are to be modified as follows, viz.,

$$F(t) \sim 1-\exp\left\{-\tfrac{1}{N}\left((t-f^*)/\chi_N\right)^{\alpha}\right\}, \qquad t > f^* = \min f,\ N \to \infty,$$

$$F(t) \approx 1-\exp\left\{-\tfrac{1}{N}\left((t-M_{N,k})/(\eta_1-M_{N,k})\right)^{\hat{\alpha}}\right\},$$

$$F(t) \sim 1-\exp\left\{-c_0\left(t-f^*\right)^{\alpha}\right\},$$

$$F(t) \approx 1-\exp\left\{-\hat{c}_0\left(t-M_{N,k}\right)^{\hat{\alpha}}\right\}, \qquad t \le M_{N,k},$$

where χ_N is such that

$$F(\chi_N) \sim N^{-1} \qquad (N \to \infty),$$

the estimator $M_{N,k}$ for f^* is determined by (4.2.27) and

$\hat{\alpha}$

using (4.2.35); the other formulas and notions are the same as above.

4.2.6 *Prior determination of the value of the tail index α*

Sometimes the value of the tail index α of the extreme value distribution can be determined, due to the following obvious result : if a c.d.f. F(t) is a sufficiently smooth in a vicinity of M=vrai sup $\eta<\infty$ and there exists an integer ℓ such that $F^{(i)}(M)=0$ for $0<i<\ell$ and $0<|F^{(\ell)}(M)|<\infty$ then (a) is fulfilled and $\alpha=\ell$, see Gumbel (1958). The above sufficient condition for (a) is a particularity of (b) for the case of integer ℓ. If one would apply the fractional differentiation concepts (see Ariyawansa and Templeton (1983) for references) then the integrality requirement can be omitted for ℓ, and the above condition coincides with (b). Note passing by that the condition (b) is somewhat stricter than (a), because the latter permits for c_0 from (4.2.5) to be a slowly varying function.

The distinguishing feature of using the above described statistical procedures in global random search lies in the fact that the c.d.f. F(t) has the form of (4.2.2). Using this fact and prior knowledge about the objective function behavior in the vicinity of a global maximizer, the tail index α of the extreme value distribution can be often exactly determined as is demonstrated below. The basic result in this direction is as follows.

Theorem 4.2.2. Let the conditions (c) through (f) (see Section 4.2.1) be satisfied. Then the condition (a) holds and $\alpha=n/\beta$.

Proof. In terms of the notation used, the condition (a) is represented as follows: the limit of $P(A(t\varepsilon))/P(A(\varepsilon))$ for $\varepsilon\downarrow 0$ exists for all $t>0$ and equals t^{α}. It is well known (see de Haan (1970)) that it suffices to require the existence of the limit only for $t\in[0.5, 2]$.

Put $z=x-x^*$,

$$g(z)=f(x^*)-f(x^*+z)-w(\|z\|)H(z),$$

$$g_\varepsilon=\sup_{0.5\le t\le 2}\ \sup_{x\in A(t\varepsilon)} g(z)/w(\|z\|).$$

The assumptions (c), (d) and (f) together with the continuity and positivity of w imply $g_\varepsilon/\varepsilon\to 0$ for $\varepsilon\to 0$. Further, for each $t\in[0.5, 2]$ and $\varepsilon>0$ there holds

$$P(A(t\varepsilon))=P(\{x\in \mathcal{X}: H(z)+g(z)/w(\|z\|)\le t\varepsilon\})\le P(\{x: H(z)\le t\varepsilon+g_\varepsilon\}).$$

Analogously,

$$P(A(t\varepsilon))\ge P(\{x: H(z)\le t\varepsilon-g_\varepsilon\}),$$
$$P(A(\varepsilon))\le P(\{x: H(z)\le \varepsilon+g_\varepsilon\}),$$
$$P(A(\varepsilon))\ge P(\{x: H(z)\le \varepsilon+g_\varepsilon\}).$$

The homogeneity of H gives

$$\mu_n\{z \in \mathcal{R}^n: H(z) \le u\} = \mu_n\{z: H(u^{-1/\beta} z) \le 1\} = cu^{n/\beta}$$

for all u>0 where $c=\mu_n\{z: H(z)\le 1\}$.

From (e) and the relations obtained above, one can deduce for fixed $t\in[0.5, 2]$

$$\limsup_{\varepsilon\to 0} P(A(t\varepsilon))/P(A(\varepsilon)) \le$$

$$\le \limsup_{\varepsilon\to 0} P(\{x: H(z) \le t\varepsilon + g_\varepsilon\})/P(\{x: H(z) \le \varepsilon - g_\varepsilon\}) =$$

$$= \limsup_{\varepsilon\to 0} \mu_n\{z \in \mathcal{R}^n: H(z) \le t\varepsilon + g_\varepsilon\}/\mu_n\{z \in \mathcal{R}^n: H(z) \le \varepsilon - g_\varepsilon\} =$$

$$= \limsup_{\varepsilon\to 0} (t\varepsilon + g_\varepsilon)^{n/\beta}/(\varepsilon - g_\varepsilon)^{n/\beta} = t^{n/\beta}.$$

Analogously,

$$\liminf_{\varepsilon\to 0} P(A(t\varepsilon))/P(A(\varepsilon)) \ge$$

$$\le \liminf_{\varepsilon\to 0} P(\{x: H(z) \le t\varepsilon + g_\varepsilon\})/P\{x: H(z) \le \varepsilon - g_\varepsilon\} =$$

$$= \liminf_{\varepsilon\to 0} (t\varepsilon - g_\varepsilon)^{n/\beta}/(\varepsilon + g_\varepsilon)^{n/\beta} = t^{n/\beta}.$$

These relations give the assertion of the theorem.

Let us make some comments on the conditions of the theorem. Conditions (c) and (d) are typical for studying global random search algorithms. The requirements of the uniqueness of x* and for the corresponding condition (d) fulfilment are imposed only for convenience and may be relaxed (see Theorem 4.2.4 below).

The condition (e) is satisfied if the measures P and μ_n are equivalent in a vicinity of x* which is either an interior point of $\mathcal{X}$ or a boundary point situated outside an *appendix* of the set $\mathcal{X}$ of zero Lebesgue measure μ_n, i.e. the following should be satisfied

$$\mu_n(B(\varepsilon))/\mu_n\{z \in \mathcal{R}^n: \|z\| \le \varepsilon\} = \text{const} + o(1), \qquad \varepsilon \to 0.$$

The condition (f) characterizes the behavior of the objective function near to x*. There are two important special cases of (f). Assume that all the components of $\nabla f(x^*)$ (the gradient of f in x*) are finite and non-zero which usually happens if the extremum is attained on the boundary. Then $H(z)=z'\,\nabla f(x^*)$, w=1, and therefore $\beta=1$ and $\alpha=n$. If it is assumed that f is twice continuously differentiable in some vicinity of x*, $\nabla f(x^*)=0$, and the Hessian $\nabla^2 f(x^*)$ is nondegenerate then w=1, $H(z)=-z'[\nabla^2 f(x^*)]z$, $\beta=2$, and $\alpha=n/2$. These two special cases (under somewhat stricter conditions on P and x*) were known mainly due to de Haan (1981) who deduced them (as well as (a)) from the condition (g) rather than (f). The former case was investigated also by Patel and Smith (1983) for the problem of concave minimization under linear constraints.

In cases where it is not evident in advance that (4.2.7) or (4.2.8) holds, but the growth rate of f near x* is known, then the following assertion may be useful.

Theorem 4.2.3. Let the conditions (a), (c), (d), (h) and (i) of Section 4.2.1 be met. Then $\alpha=n/\beta$.

Proof. Put $\varepsilon_3=\min\{\varepsilon_1,\varepsilon_2\}$,

$$\varepsilon_4=\sup\{\varepsilon \mid A(\varepsilon)\subset B(\varepsilon_3)\},\qquad \varepsilon_5=\frac{1}{2}\min\{\varepsilon_3,\varepsilon_4\}.$$

It follows from (c) that $\varepsilon_5>0$ and from (h) and (i) that for all u, $0<u\leq\varepsilon_5$, and $t\in[0.5, 2]$ the followings holds

$$P(A(tu))\leq P\left(B\left((tu/c_3)^{1/\beta}\right)\right)\leq c_2(tu/c_3)^{n/\beta},$$

$$P(A(tu))\geq P\left(B\left((tu/c_4)^{1/\beta}\right)\right)\geq c_1(tu/c_4)^{n/\beta},$$

$$P(A(u))\leq c_2(u/c_3)^{n/\beta},\qquad P(A(u))\geq c_1(u/c_4)^{n/\beta}.$$

Hence,

$$c\,t^{n/\beta}\leq P(A(tu))/P(A(u))\leq c^{-1}t^{n/\beta} \tag{4.2.42}$$

where

$$c=c_1(c_3/c_4)^{n/\beta}/c_2.$$

In virtue of (a) the limit P(A(tu))/P(A(u)) exists for $u\downarrow 0$ and any $t\in[0.5, 2]$, and it equals t^{α}. It follows from (4.2.42) that any value of α different from $\alpha=n/\beta$ is inadmissible. The proof is complete.

The conditions of Theorem 4.2.3 are slightly weaker than those of Theorem 4.2.2, except for the requirement of meeting (a), but Theorem 4.2.2 not only infers the expression for the tail index α, but also substantiates the fact that the convergence to the extreme value distribution takes place.

The following statement is aimed at relaxing the uniqueness requirement for the global maximizer.

Theorem 4.2.4. Let P be equivalent to the Lebesgue measure μ_n on $\mathfrak{X}$ and let the function f attain its global maximum at a finite number of points x_i^* ($i=1,2,\dots,\ell$) in whose vicinities the tail indexes α_i can be estimated or determined. Then the condition (a) for c.d.f. F(t) is met and $\alpha=\min\{\alpha_1,\dots,\alpha_\ell\}$.

The assertion follows from the fact that f is represented under our assumptions as

$$f=\sum_{i=1}^{\ell} f_i$$

where

$$\overline{\operatorname{supp} f_i}=\mathfrak{X}_i\in\mathfrak{B},\qquad \mu_n(\mathfrak{X}_i\cap\mathfrak{X}_j)=0\qquad \text{for } i,j=1,\dots,\ell,\ i\neq j$$

and f_i ($i=1,\dots,n$) is a measurable function attaining the global maximum M at the unique point x_i^*, as well as from two following lemmas.

Lemma 4.2.1. If

$$\mathfrak{X}_i=\overline{\operatorname{supp} f_i},\qquad f=\sum_{i=1}^{\ell} f_i,$$

$$P(\mathfrak{X}_i\cap\mathfrak{X}_j)=0\qquad \text{for } i\neq j,\qquad P(\mathfrak{X}_i)>0\qquad (i,j=1,\dots,\ell)$$

then

$$F(t)=\sum_{i=1}^{\ell} p_i F_i(t)$$

where

$$p_i=P(\mathfrak{X}_i)>0,\qquad \sum_{i=1}^{\ell} p_i=1,\qquad F_i(t)=\int_{f_i(x)<t} P_i(dx),$$

$$P_i(A)=P(A\cap\mathfrak{X}_i)/p_i\qquad \text{for}\qquad A\in\mathfrak{B}.$$

The proof is evident:

$$F(t)=\int 1_{[f(x)<t]}P(dx)=\sum_{i=1}^{\ell}\int_{\mathfrak{X}_i} 1_{[f(x)<t]}P(dx)=$$

$$= \sum_{i=1}^{\ell} \int_{X_i} 1_{[f_i(x)<t]} P(dx) = \sum_{i=1}^{\ell} p_i F_i(t).$$

Lemma 4.2.2. If the condition (a) with parameters $\alpha = \alpha_i$ is fulfilled for c.d.f. $F_i(t)$, $i=1,...,\ell$, then it is also met for c.d.f.

$$F(t) = \sum_{i=1}^{\ell} p_i F_i(t) \qquad (\text{where } p_i > 0 \text{ for } i = 1,\dots,\ell, \qquad \sum_{i=1}^{\ell} p_i = 1)$$

with the value $\alpha_m = \min\{\alpha_1,..., \alpha_\ell\}$ for the tail index.

Proof. Represent the functions

$$V_i(v) = 1 - F_i(M - 1/v) \qquad (v > 0, i = 1,\dots,\ell)$$

as

$$V_i(v) = L_i(v) v^{-\alpha_i}$$

where $L_i(v)$ are slowly varying functions, i.e. such functions that the limit of $L_i(tv)/L_i(v)$ exists and equals 1 for any $t>0$ and $v \to \infty$.

It suffices to demonstrate that under $v \to \infty$ the ratio

$$A(t,v) = t^{\alpha_m}(1 - F(M - 1/(tv))/(1 - F(M - 1/v)))$$

has limit 1. We have

$$A(t,v) = t^{\alpha_m} \sum_{i=1}^{\ell} p_i V_i(tv) / \sum_{j=1}^{\ell} p_j V_j(v) = \sum_{i=1}^{\ell} p_i B_i(t,v)$$

where

$$B_i(t,v) = \frac{t^{\alpha_m} V_i(tv)}{\sum_{j=1}^{\ell} p_j V_j(v)} = \left[\sum_{j=1}^{\ell} p_j t^{-\alpha_m} \frac{V_j(v)}{V_i(tv)} \right]^{-1} =$$

$$= \left[\sum_{j=1}^{\ell} p_j t^{\alpha_i - \alpha_m} \frac{v^{\alpha_i - \alpha_j} L_j(v)}{L_i(tv)} \right]^{-1}.$$

If $\alpha_i>\alpha_m$ then $B_i(t,v)\to 0$ $(v\to\infty)$ in virtue of the properties of slowly varying functions (see de Haan (1970)). If $\alpha_i=\alpha_m$ and $\alpha_j>\alpha_i$ for $j\neq i$ then $B_i(t,v)\to 1/p_i$ $(v\to\infty)$. Existence of limits in these cases is obvious. It remains to consider the case where several α_i's are simultaneously equal to α_m.

We shall prove the following: let the c.d.f.'s $F_i(t)$, $i=1,\ldots,r$ meet the condition (a) with index α, then the c.d.f.

$$F_0(t)=\sum_{j=1}^{r} q_j F_j(t) \qquad \left(\text{where } q_j>0 \text{ for } j=1,\ldots,r,\ \sum_{j=1}^{r} q_j=1\right)$$

meets (a) with index α as well. If this assertion is true and if

$$\alpha_m=\alpha_{i_1}=\alpha_{i_2}=\ldots=\alpha_{i_r} \qquad (1\le i_1\le\ldots\le i_r\le \ell)$$

then, having assumed that

$$F_0(t)=\sum_{j=1}^{r}\left[p_{i_j}\Big/\sum_{k=1}^{r}p_{i_k}\right]F_{i_j}(t),$$

we reach the case where $\alpha_j>\alpha_m$ for all $j=1,\ldots,\ell$ except $j=i$.

The assertion desired will be proved for $r=2$ and by induction can be extended to an arbitrary r.

Through a transformation similar to the above ones, we obtain

$$A_0(t,v)=t^{\alpha}\left[1-F_0(M-1/(tv))\right]/\left[1-F_0(M-1/v)\right]=$$

$$=q_1\Big/\left[q_1\frac{L_1(v)}{L_1(tv)}+q_2\frac{L_2(v)}{L_1(tv)}\right]+q_2\Big/\left[q_1\frac{L_1(v)}{L_2(tv)}+q_2\frac{L_2(v)}{L_2(tv)}\right]$$

Putting

$$c=\lim_{v\to\infty}L_1(v)/L_2(v), \qquad 0\le c\le\infty$$

(the limit exists, because V_1 and V_2 are monotone) obtain

$$\lim_{v\to\infty}\frac{L_1(v)}{L_2(tv)}=\lim_{v\to\infty}\frac{L_1(v)}{L_2(v)}\lim_{v\to\infty}\frac{L_2(v)}{L_2(tv)}=c.$$

Similarly

$$\lim_{v\to\infty}\left[L_2(v)/L_1(tv)\right] = 1/c.$$

Therefore, the limit of $A_0(t,v)$ for $v\to\infty$ exists and equals

$$\frac{q_1}{q_1+q_2/c}+\frac{q_2}{cq_1+q_2}=1.$$

This completes the proof of Lemma.

Let us present some generalizations of the above results permitting to extend the applicability domain of the apparatus of this subsection. We shall omit further detailed proofs and discussions restricting ourself to short comments. Note that an alternative approach to the exact evaluation of the value of the tail index α was treated in Dorea (1987), mainly for the univariate case (i.e. n=1).

Proposition 4.2.1. Assume that the set of global maximizers $\mathcal{X}^*=\{\arg\max f\}$ is a continuously differentiable m-dimensional manifold ($0\le m\le n-1$), f is continuous in the vicinity of the set $\mathcal{X}^*$, and the conditions (d), (e) and (f) are satisfied for all $x^*\in\mathcal{X}^*$. Then (a) is satisfied and $\alpha=(n-m)/\beta$.

The proof is similar to that of Theorem 4.2.2.

From Proposition 4.2.1 and Lemma 4.2.2 directly follows the next.

Proposition 4.2.2. Assume that

$$\mathcal{X}^*=\bigcup_{j=1}^{\ell}\mathcal{X}_j^*$$

where $\mathcal{X}_j^*$ is a continuously differentiable m_j-dimensional manifold ($0\le m_j\le n-1$, $j=1,\dots,\ell$); f is continuous in the vicinity of $\mathcal{X}^*$ and (d), (e) and (f) are satisfied for all $x^*\in\mathcal{X}_j^*$ ($j=1,\dots,\ell$) if $c_1(j)$ and β_j are substituted, respectively, for c_1 and β. Then (a) is fulfilled and

$$\alpha=\min_{1\le j\le\ell}\left[(n-m_j)/\beta_j\right].$$

Another generalization of Theorem 4.2.2 is the next.

Proposition 4.2.3. Let all the conditions of Theorem 4.2.2 be fulfilled but H(z) rather than being a homogeneous function of order β, be representable as

$$H(z) = H_1(z^{(1)}) + H_2(z^{(2)}) + \ldots + H_\ell(z^{(\ell)})$$

where

$$H_j: \mathcal{R}^{n_j} \rightarrow \mathcal{R}$$

are homogeneous functions of order β_j, $\ell \leq n$, $z=(z^{(1)},\ldots,z^{(\ell)}) \in \mathcal{R}^n$, $z^{(j)}$ are disjoint groups of n_j variables of vector z, $n_1+\ldots+n_\ell=n$. Then (a) is satisfied and

$$\alpha = \sum_{j=1}^{\ell} n_j/\beta_j$$

For the simplest case of n=2, ℓ=2, the proof differs from that of Theorem 4.2.2 in the following: for $u \rightarrow 0$ there holds

$$P(\{z: H_1(z^{(1)}) + H_2(z^{(2)}) \leq u\}) =$$

$$= P(\{z: H_1(u^{-1/\beta_1} z^{(1)}) + H_2(u^{-1/\beta_2} z^{(2)}) \leq 1\}) \sim$$

$$\sim (c_1/\pi)\mu_2\{z: H_1(u^{-1/\beta_1} z^{(1)}) + H_2(u^{-1/\beta_2} z^{(2)}) \leq 1\} =$$

$$= \frac{c_1}{\pi} \iint_{x+y\leq 1} u^{1/\beta_1+1/\beta_2} D(x,y)dxdy = \frac{c_1}{\pi} u^{1/\beta_1+1/\beta_2} \iint_{x+y\leq 1} D(x,y)dxdy$$

where D stands for the Jacobian corresponding to the change of variables $x=H_1(z^{(1)})$, $y=H_2(z^{(2)})$. For the general case the proof is analogous.

The case of $\ell=2$, $\beta_1=1$, $\beta_2=2$ typifies situations where the above result proves useful: this case corresponds to non-zero first derivatives of f at x* with respect to the variables $z^{(1)}$ and to zero first derivatives and non-degenerate Hessian matrix with respect to the variables $z^{(2)}$.

The following generalization of the situation, where Theorem 4.2.2 and similar assertions can be used, is based on the possibility of the exact determination of the tail index α in the case where the objective function values $f(x)$ are evaluated with a random noise whose distribution is not influenced by the location of $x \in \mathcal{X}$ and lies within a finite interval. The generalization relies upon the following assertion that can readily be proved.

Proposition 4.2.4. Let the condition (a) be satisfied for distributions F_1, F_2 and F_1*F_2, the tail index of the extreme value distribution for F_i being α_i (i=1,2). Then for the distribution F_1*F_2 the tail index is $\alpha=\max\{\alpha_1, \alpha_2\}$.

It follows from Proposition 4.2.4 that if the tail index corresponding to the distribution of the random variable $f(\xi)$ is α_1 and if that of the noise ζ distribution is α_2, then the sum $\max f$+vrai sup ζ is the bound of the random variable $y=f(\xi)+\zeta$, and for y only $\max\{\alpha_1, \alpha_2\}$ can be the tail index.

Further generalization of the situation to which the above apparatus is applicable is possible, if one abandons the independence assumption for realizations x_i of the random vector ξ having distribution P(dx). To this end, it should be noted that the apparatus of statistical inferences for the random variable bounds is based on the fact that the distributions of random variables $(\eta_N-M)/(M-\theta_N)$ converge to that of extreme values for $N\to\infty$. This fact was shown (see Resnick (1987), Lindgren and Rootzen (1987)) to be true not only in the case of independent random variables $y_1, y_2, \ldots$, but be generalizable to cases where this sequence is a homogeneous Markov chain exponentially converging to a stationary distribution, or strictly stationary m-dependent sequence, or that of symmetrically dependent random variables. The apparatus described in this section is, thus, applicable to the majority of algorithms presented in Part 2. Section 4.4 will investigate another case of dependent sample $Y=\{y_1,\ldots, y_N\}$ for which the above apparatus is generalizable.

4.2.7 Exponential complexity of the uniform random sampling algorithm

As a consequence of the results of the present section, we shall obtain a result on the exponential complexity of Algorithm 3.1.1. The mean length of the one-sided confidence interval (4.2.25) for $M=\max f$ of fixed level $1-\gamma$ is taken as the measure of algorithm accuracy; we shall study the growth rate of the number N of evaluations of f required for reaching a given accuracy under $n\to\infty$.

Theorem 4.2.5. Let assumption (b) and the conditions of Theorem 4.2.2 or 4.2.3 be fulfilled. Then, in order to make the asymptotic mean length of the confidence interval (4.2.25) be equal to ε, the number N of objective function evaluations in the algorithm of uniform random sampling of points in $\mathcal{X}$, has to grow with rate $N\sim c_o c^n$ $(n\to\infty)$ where the parameter c_o has the same sense as in (b) and

$$c=\left[(\psi(k+1)-\psi(1))/\left(-\varepsilon\log\left(1-\gamma^{1/k}\right)\right)\right]^{1/\beta}, \qquad \psi(u)=\Gamma'(u)/\Gamma(u)$$

is the Ψ-function. The theorem's assertion follows from

Lemma 4.2.3. Let the conditions of Theorem 4.2.5 be fulfilled. Then the mean length of one-sided confidence interval (4.2.25) is asymptotically equal (for $N\to\infty$, $n\to\infty$) to

$$\ell(\gamma,\beta\, n,k,N) \sim (\psi(k+1)-\psi(1))/\left[(c_0N)^{\beta/n}\left(-\log\left(1-\gamma^{1/k}\right)\right)\right].$$

Proof. It follows from Lemma 7.1.2 (given later) that the mean length of the confidence interval (4.2.25) for $N\to\infty$ equals $\ell_1\ell_2\ell_3$ where

$$\ell_1 = n[\Gamma(1/\alpha+k+1)/\Gamma(k+1)-\Gamma(1/\alpha+1)],$$

$$\ell_2 = 1/\left[n\left(\left(1-(1-\gamma)^{1/k}\right)^{-1/\alpha}-1\right)\right], \qquad \ell_3 = M-\theta_N.$$

If the conditions of Theorems 4.2.2 or 4.2.3 are satisfied then $\alpha=n/\beta$ and therefore for $n\to\infty$

$$\ell_1 = n\left[\frac{\Gamma(\beta/n+k+1)-\Gamma(k+1)}{\Gamma(k+1)}-\frac{\Gamma(\beta/n+1)-\Gamma(1)}{\Gamma(1)}\right]\sim$$

$$\sim n\left[\frac{\Gamma'(k+1)}{\Gamma(k+1)}\frac{\beta}{n}-\frac{\Gamma'(1)}{\Gamma(1)}\frac{\beta}{n}\right]=\beta(\psi(k+1)-\psi(1)),$$

$$\ell_2 = \frac{\left(1-\gamma^{1/k}\right)^{\beta/n}}{n\left(1-\left(1-\gamma^{1/k}\right)^{\beta/n}\right)} \sim \frac{1}{\beta\log\left(1-\gamma^{1/k}\right)}.$$

It follows from the condition (b) that the representation (4.2.38) is applied with $\alpha=n/\beta$ and thus

$$\ell_3 = M-\theta_N \sim (c_0N)^{-\beta/n}, \qquad N\to\infty.$$

Combining the limit relations for ℓ_1, ℓ_2 and ℓ_3, one obtains the desired result: this completes the proof of the lemma (i.e. also that of Theorem 4.2.5).

4.3 Branch and probability bound methods

This section describes a class of global random search algorithms applying the mathematical apparatus developed in Chapter 7 and in the preceding section. These algorithms are closely related, in terms of their philosophy, to the branch and bound methods that were reviewed in Section 2.3 and the key concept in their description is that of prospectiveness.

4.3.1 Prospectiveness criteria

A sub-additive set function $\varphi:\mathcal{B}\rightarrow\mathcal{R}$ resulting from processing the outcomes of previous evaluations of the objective function and reflecting the possibility of locating the global optimizer in subsets is referred to as prospectiveness criterion. If for two sets $Z_1, Z_2 \in \mathcal{B}$ the inequality $\varphi(Z_1) \geq \varphi(Z_2)$ holds, then the location of the global optimizer in Z_1 is at least as probable as in Z_2, according to the prospectiveness criterion φ.

A number of set functions φ (assuming $Z\in\mathcal{B}$) can be used as prospectiveness criteria; examples are:

a) $\varphi(Z)$ is an estimate of the maximum

$$M_Z = \max_{x\in Z} f(x),$$

b) $\varphi(Z)$ is an estimate of the mean value

$$\int_Z f(x)\nu(dx)$$

c) $\varphi(Z)$ is an estimate of the minimum

$$\min_{x\in Z} f(x)$$

d) $\varphi(Z)$ is an upper confidence bound of a fixed confidence level for M_Z,

e) $\varphi(Z)$ is an estimate of the probability $P\{M_Z \geq M_k^*\}$, where M_k^* is the maximal among the already determined values of f.

Chapter 7 and the preceding section consider the construction of prospectiveness criteria a), d) and e); criteria b) and c) are mentioned in Section 4.1. The author's order of preference is as follows: e), d), a), b) and c): thus, criteria e) or d) should be used if possible. In complicated situation (for instance, in case of random noise presence), however, they cannot be usually constructed, and one has to rely on a b)-type criterion that can be constructed in rather general situation.

The estimates listed above can be constructed either via the results of evaluating the objective function, or after investigation of estimates or approximations of f. As a rule, the estimates derived are probabilistic; deterministic estimates can be constructed only for certain (e.g. Lipschitz-type) functional classes and correspond to the standard branch and bound approach treated earlier in Section 2.3.4.

4.3.2 The essence of branch and bound procedures

Branch and bound methods, used to advantage in various extremal problems, consist in rejecting some of the subsets of $\mathcal{X}$ that can not contain an optimal solution and searching only among the remaining subsets regarded as promising. Branch and bound methods may be summarized, as successive implementation at each iteration of the following three stages:

i) branching (decomposition) of the (one or several) sets into a tree of subsets and evaluating the objective function values at points from the subsets;

ii) estimation of functionals that characterize the objective function over the obtained subsets (evaluation of subset prospectiveness criteria); and

iii) selection of subsets, that are promising, for further branching.

In standard versions of branch and bound methods, deterministic upper bounds of the maximum of f on subsets are used as subset prospectiveness criteria. By doing so, all the subsets Z are rejected whose upper bounds for M_Z do not exceed the current optimum estimate. Prospectiveness criteria in these methods, thus, can be either 1 or 0 which means that branching of a subset should go on or be stopped.

In the following consideration will be given to non-standard variants of branch and bound methods that will be referred to as branch and probability bound methods. Their distinctive feature is that the maximum estimates on the subsets are probabilistic (i.e. these estimates are valid with a high probability) rather than deterministic.

4.3.3 Principal construction of branch and probability bound methods

The methods under consideration are distinguished by the (i) organization of set branching, (ii) kinds of prospectiveness criteria, and (iii) rules for rejecting unpromising subsets.

Set branching depends on the structure of $\mathcal{X}$ and on the available software and computer resources. If $\mathcal{X}$ is a hyperrectangle, then it is natural to choose the same form for branching sets Z_{kj} where k is the iteration count, and j is a set index. In the general case, simplicial sets, spheres, hyperrectangles and, sometimes, ellipsoids can be naturally chosen as Z_{kj}. Two conditions are imposed on the choice of Z_{kj}: their union should cover the domain of search and the number of points where f is evaluated should be, in each set, sufficiently large for drawing proper statistical inference. There is no need for the sets Z_{kj} to be mutually disjoint, for any fixed k.

Branching/decomposition of the search domain can be carried out either *a priori* (i.e. independent of the values of f), or *a posteriori* . Numerical experience indicates that the second approach provides more economical algorithms. For example, the following branching technique has proved to be advantageous. At each iteration k, first select in the search domain $\mathcal{X}_k$ a subdomain Z_{k1} with the centre at the current optimum estimate. The point corresponding to the record of f over $\mathcal{X}_k / Z_{k1}$ is the centre of subdomain Z_{k2}.

Similar subdomains Z_{kj} $(j=1,...,l)$ are isolated until either $\mathcal{X}_k$ is covered, or the hypothesis is rejected that the global maximum can occur in the residual set

$$\mathcal{X}_k \setminus \bigcup_{j=1}^{l} Z_{kj}$$

(the hypothesis can be verified by the procedure described in Section 4.2.3). This way, the search domain $\mathcal{X}_{k+1}$ of the (k+1)-th iteration is

$$Z^{(k)} = \bigcup_{j=1}^{l} Z_{kj},$$

or a hyperrectangle covering $Z^{(k)}$, or a union of disjoint hyperrectangles covering $Z^{(k)}$. In the multidimensional cases, the latter two ways induce more conveniently realizable variants of the branch and probability bound method, than the first one.

In contrast to the standard branch and bound algorithms, in these methods the prospectiveness criterion may take any real value (e.g. for some criteria, the interval [0,1] can be the set of values) rather than only two values, say 0 and 1. For example, an estimate of M_Z can be used as the prospectiveness criterion; it is natural to take the upper confidence bound for M_Z of a fixed level $1-\gamma$, as the prospectiveness criterion of Z (this bound can be computed through (4.2.25)). The prospectiveness criterion that will be discussed at the beginning of the following subsection (relying upon the procedure for testing statistical hypothesis about M_Z) is both natural and easily computable.

Rules for rejecting unpromising sets may also be diverse. Under reasonable organization of branching and use of the apparatus of the preceding section for constructing the prospectiveness criterion, there is no absolutely unpromising set Z_o. Narrowing of the search domain (i.e. rejection of subsets) therefore may occur only, if a lower prospectiveness bound δ is defined; if $\varphi(Z) \leq \delta$, then the set Z may be regarded as unpromising and be rejected. Furtherly, it is intuitively evident that the more promising is a subset, the greater number of sample points should be located in it. This can be assured by taking, for example, the number of points in a subset to be directly proportional to the value of the prospectiveness criterion. Note that the extremal approach in which all points are always located in the most promising set, is not very good and basically does not differ much from Algorithm 5.1.2. The following note can also be taken into consideration when constructing the rejection rule. A reasonable prospectiveness criterion may depend not only on function values over a given subset but also on those over the whole set $\mathcal{X}$. A subset of a mean prospectiveness can, therefore, become an unpromising one (and be rejected), due to the increase of prospectiveness of other subsets. The following numerical organization of the search algorithm is, thus, natural: if a subset Z was recognized at the k-th iteration as being of medium prospectiveness (i.e. $\delta < \varphi(Z) \leq \delta_*$, for a suitable pair $\delta_* > \delta$), then one shall not evaluate function f values in Z at several subsequent iterations and wait; may be, Z will become unpromising and be rejected.

4.3.4 Typical variants of the branch and probability bound method

Below we shall consider one of the most natural and readily computable prospectiveness criteria.

Let M_k^* be the largest value of f obtained so far and let the search domain $\mathcal{X}_k$ be covered at each iteration k by the subsets Z_{kj} $(j=1,...,l_k)$:

$$\mathcal{X}_k \subset \bigcup_j Z_{kj}.$$

The prospectiveness criterion value on Z_{kj} is defined as follows:

$$\varphi_k(Z_{kj}) = p_{kj} = 1 - \left[1 - \left(\frac{M_k^* - \eta_m}{M_k^* - \eta_{m-i}}\right)^{\alpha}\right]^i \tag{4.3.1}$$

where η_m and η_{m-i} are respective elements of the order statistics corresponding to the sample $\{y_\ell = f(\xi_\ell),\ \ell=1,...,m\}$, i is much smaller than m, ξ_ℓ are independent realizations of a random vector on $\mathcal{X}$ such that fall into Z_{kj}. The value p_{kj} can be treated in two ways: in the asymptotically worst case (for i=const, $m\to\infty$) it is more than or equal to, first, the probability that

$$M_{Z_{kj}} = \sup_{x\in Z_{kj}} f(x) \geq M_k^*$$

and, second, to the probability of accepting the hypothesis

$$H_0: M_{Z_{kj}} \geq M_k^*$$

under the alternative

$$H_1: M_{Z_{kj}} < M_k^*$$

and provided that the hypothesis is true and that the hypothesis testing procedure is that of Section 7.3. In order to obtain (4.3.1) from (4.2.25) and (4.2.26) in which m, i, and M_k^* are substituted for N, k, and M_o, it suffices to solve with respect to $1-\gamma$ the inequality

$$(M_k^* - \eta_m)/(\eta_m - \eta_{m-i}) \geq r_{i,1-\gamma},$$

which is basic in the formulas, and to substitute it by an equality.

In the algorithm below the number of points at each k-th iteration in the promising subset Z_{kj} is assumed to be (in the probabilistic sense) proportional to the value of the prospectiveness criterion $\varphi_k(Z_{kj})$; further branching is performed over those sets Z_{kj} whose values (4.3.1) are not less than the given numbers δ_k.

Algorithm 4.3.1.

1. Set $k=1$, $\mathcal{X}_1=\mathcal{X}$, $M_o^*=-\infty$. Choose a distribution P_1.
2. Generating N_k times the distribution P_k, obtain a random sample

$$\Xi_k = \left\{ x_1^{(k)}, \dots, x_{N_k}^{(k)} \right\}.$$

3. Evaluate f at the points of Ξ_k and put

$$M_k^* = \max\left\{ M_{k-1}^*, f\left(x_1^{(k)}\right), \dots, f\left(x_{N_k}^{(k)}\right) \right\}.$$

4. Organize the branching of $\mathcal{X}_k$ by representing this set as

$$\mathcal{X}_k \subset \bigcup_j Z_{kj}$$

where Z_{kj} are measurable subsets of $\mathcal{X}$ having a sufficient number of points from Ξ_k for statistical inference drawing.

5. For each subset Z_{kj}, compute (not necessarily through (4.3.1)) the values of the prospectiveness criterion $\varphi_k(Z_{kj})$.
6. Put

$$\mathcal{X}_{k+1} = \bigcup_j Z_{kj}^*$$

where

$$Z_{kj}^* = \begin{cases} Z_{kj} & \text{if} \quad \varphi_k(Z_{kj}) > \delta_k, \\ \varnothing & \text{if} \quad \varphi_k(Z_{kj}) \le \delta_k, \end{cases}$$

i.e. those subsets Z_{kj} for which $\varphi_k(Z_{kj}) \le \delta_k$ are rejected from the search domain $\mathcal{X}_k$. Let l_k be a number of remaining subsets Z_{kj}^*.

7. Put

$$P_{k+1}(dx) = \sum_{j=1}^{\iota_k} p_j^{(k)} Q_{kj}(dx) \qquad (4.3.2)$$

where $p_j^{(k)}=1/\iota_k$ and Q_{kj} are the uniform distributions on sets Z_{kj}^* ($j=1,\ldots,\iota_k$). If φ_k is a nonnegative criterion, then we also may take

$$p_j^{(k)} = \varphi_k\left(Z_{kj}^*\right) / \sum_{\ell=1}^{\iota_k} \varphi_k\left(Z_{k\ell}^*\right)$$

where

$$\varphi_k\left(Z_{kj}^*\right) = \begin{cases} \varphi_k(Z_{kj}) & \text{if} \quad Z_{kj}^* = Z_{kj}, \\ \emptyset & \text{if} \quad Z_{kj}^* = 0. \end{cases}$$

8. Return to Step 2 (substituting k+1 for k.

The closeness of M_k^* to an estimate of M or to the upper bound of confidence interval (4.2.26) is a natural stopping rule for Algorithm 4.3.1. Another type of stopping rule is based on reaching a small (fixed) volume of the search domain $\mathcal{X}_k$.

Of course, after terminating the algorithm, one may use a local ascent routine to make more precise the location of the global maximizer.

The distributions P_k in Algorithm 4.3.1 are sampled by means of the superposition method: first, the discrete distribution concentrated at $\{1,2,\ldots,\iota_k\}$ with probabilities $p_j^{(k)}$ is sampled, this is followed by distribution sampling $Q_{k\tau}$ where τ is the realization obtained by sampling the discrete distribution. If the first procedure for choosing the probabilities $p_j^{(k)}$ (i.e. $p_j^{(k)}=1/\iota_k$) is used at Step 7 of Algorithm 4.3.1 and the sets Z_{kj}^* ($j=1,\ldots,\iota_k$) are disjoint, then (4.3.2) is the uniform distribution over the set $\mathcal{X}_{k+1}$.

All the points obtained at previous iterations and falling into the set $\mathcal{X}_k$ can be included into the collection Ξ_k at Step 2 of Algorithm 4.3.1: this improves the accuracy of statistical procedures for determination of the prospectiveness criterion value. If all the distributions P_k are uniform on $\mathcal{X}_k$, then all the resulting samples are uniform on $\mathcal{X}_k$ as well; if the form of P_{k+1} is (4.3.2), then, in general, the distributions of the resulting samples are not uniform on $\mathcal{X}_{k+1}$, but this fact is of no importance for drawing statistical inference (see Chapter 7). It is not necessary, of course, to store all the previous points and objective function values, because statistical inferences are made only using the points where f is relatively large; one even does not need to know the number of points within the domain.

After completion of Algorithm 4.3.1, one can apply (4.3.1) to the union of all rejected subsets in order to determine the probability of not missing the global maximum. The following should be taken into consideration. Let $\gamma_{kj}=1-p_{kj}$ be the probabilities of missing global maximum in the rejected sets Z_{kj} as computed via (4.3.1). The total probability of missing the global maximum point, as determined by (4.3.1)is, then at most max γ_{kj}. Indeed, let us take the set $Z_{kj}{}^{o}= Z_{kj}$ which contains the point corresponding to the statistics η_m of the set

$$Z = \bigcup_{k,j} Z_{kj}.$$

Then η_{m-i} for Z is not less than η_{m-i} for $Z_{kj}{}^{o}$ and

$$\max_k M_k^*$$

(it plays the role of $M_k{}^*$ for Z) is not less than $M_k{}^*$. But p_{kj} defined via (4.3.1) is an increasing function of both η_{m-i} and $M_k{}^*$ for fixed α, i, and η_m.

The philosophy of constructing the branch and probability bound method is closely related to that of those algorithms which are based on objective function estimation. However, in the methods under consideration only functionals

$$M_{Z_{kj}}$$

are estimated rather than the function f itself. But if one assumes that in Algorithm 4.3.1 $\delta_k=-\infty$ for all k=1,2,... , then one arrives at a variant of adaptive random search, in which more promising domains are looked through more thoroughly. Algorithm 4.3.1 is then a special case of Algorithm 5.2.1, where $f_k(x)$ for all the points of Z_{kj} is equal to $\varphi_k(Z_{kj})$ (if the subsets Z_{kj} are disjoint for fixed k).

If the evaluations of f are costly, then one should be extremely cautious in planning computations. After completing a certain number of evaluations of f, one should increase the amount of auxiliary computations with the aim of extracting and exploiting as much information about the objective function as possible. To extract this information one should: compute the probabilities (4.3.1); construct various estimates of

$$M_{Z_{kj}};$$

check the hypothesis about the value of the tail index α of the extreme value distribution which can provide information whether a vicinity of the global maximizer is reached or not; estimate the c.d.f. F(t) for the values of t close to M (this will enable one to draw a conclusion about the approximate effort related to the remaining computations); and, in addition, one can estimate f (and related functions of interest) in order to recheck and update the information. Decisions about prospectiveness of subsets should be made applying suitable statistical procedures. It is natural that these procedures can be taken into consideration only if the algorithms are realized in an interactive fashion.

The major part of assertions made in this section have precise meaning only if for the corresponding c.d.f. F(t) the condition (a) of Section 4.2 is met and the parameter α of the extreme value distribution is known. As for the condition (a), practice shows that it may always be regarded as met, if the problem at hand is not too specific. In principle, statistical inference about α can be made via the procedure of Section 4.2.4 that is to be carried out successively as points are accumulated. However, it is recommendable to use the results of Section 4.2.6 if possible, since the accuracy of procedures for statistical inference about α is high only for large sample sizes. According to these results for the case when $\mathcal{X}\subset\mathbb{R}^n$ and the objective function f is twice continuously differentiable and approximated by a non-degenerate quadratic form in the vicinity of a global maximizer x^* one can always set $\alpha=n/2$. While doing so one may be confident that statistical inferences are asymptotically true for subsets $Z\subset\mathcal{X}$ containing x^* (together with subsets Z containing maximizers

$$x^*_Z = \arg\max_{x\in Z} f(x)$$

as interior points). The prospectiveness of other subsets Z may be lower than for the case of using their true values of α, but this will not crucially affect the methods in question.

Let us finally describe a variant of the branch and probability bound method that is convenient for realization, uses most of the above recommendations, and proved to be efficient for a wide range of practical problems.

Algorithm 4.3.2.

1. Set $k=1$, $\mathcal{X}_1=\mathcal{X}$, $M_0^*=-\infty$.

2. Sample N times the uniform distribution on the search domain $\mathcal{X}_k$, obtain $\Xi_k=\{x_1^{(k)},\dots,x_N^{(k)}\}$.

3. Evaluate $f(x_j^{(k)})$ for $j=1,\dots,N$. Put

$$M_k^* = \max\left\{M_{k-1}^*, f\left(x_1^{(k)}\right),\dots,f\left(x_N^{(k)}\right)\right\}.$$

Check the stopping criterion (closeness of M_k^* and the optimal linear estimator (4.2.15) for M with $\alpha=n/2$).

4. Set $Y_{k,0}=\mathcal{X}_k$, $j=1$.

5. Set $Y_{k,j}=Y_{k,j-1}\setminus Z_{kj}$ where Z_{kj} is a cube (or a ball) of volume $\rho\mu_n(\mathcal{X}_k)$ centered at the point having the maximal value of f among the points from the set $Y_{k,j-1}$ in which the objective function is evaluated.

6. If the number m of points in $Y_{k,j}$ with known objective function values is insufficient for drawing statistical inference (i.e. $m\le m_0$), then set $\mathcal{X}_{k+1}=\mathcal{X}_k$ and go to Step 9. If $m>m_0$, then take

$$\varphi_k(Y_{k,j}) = \left[1-\left(\frac{M_k^*-\eta_m}{M_k^*-\eta_{m-i}}\right)^{\alpha}\right]^i$$

where the order statistics η_m, η_{m-i} correspond to values of f at

$$Y_{k,j} \cap \bigcup_{\ell=1}^{k} \Xi_\ell .$$

If $\varphi_k(Y_{k,j}) \leq \delta$, then go to Step 8.

7. Substitute j+1 for j and go to Step 5.
8. Choose $\mathcal{X}_{k+1}$ as the union of disjoint hyperrectangles covering the set $Z_{k1} \cup \ldots \cup Z_{kj}$.
9. Substitute k+1 for k and return to Step 2.

Under the condition $m_o \to \infty$ Algorithm 4.3.2 converges with probability 1- δ, i.e. it misses a global maximizer with probability not larger than δ: this follows from the results of Section 4.2.

Based on numerical experiments, the author proposes to use N=100, m_o=15, ρ=0.1, and i=min{5,[m/10]}, where [.] is the integer part operation, as the standard collection of parameters of Algorithm 4.3.2. (Other choices of the parameters, corresponding to the recommendations of Section 4.2, are possible as well.)

4.4. Stratified sampling

As it was pointed out in Section 3.1, the global random search algorithms consists of iterations which involve random points sampled from some distributions. Independent random sampling is the simplest standard way of obtaining random points: naturally it is equivalent to the use of an independent sample. The aim of the present section is to establish the possibility and advantage of using a stratified sample - instead of an independent one - at each iteration of the algorithm in question.

The essence of stratified sampling is that the feasible region is divided into a number of disjoint subsets and random points are generated independently on each one. The way of stratified sample construction will be considered in Subsection 4.4.1: there the case of a hypercube feasible region is studied in details.

Subsection 4.4.2 concerns statistical inference procedures for the maximum of a function based on its values at the points of a stratified sample. The procedures are, essentially, analogous to those considered in Section 4.2 for independent samples. The use of stratified sampling is a standard way of decreasing the variance of Monte Carlo estimators of integrals. While solving this problem, stratified sampling outperforms the corresponding independent one and is admissible for a wide range of functions. Subsection 4.4.3 establishes similar results for the global optimization problem.

The results of this section enable us to draw conclusions on a whole series of global random search algorithms. The main conclusion is that decreasing the randomness leads to improved efficiency (implying, of course, increased algorithmic complexity).

4.4.1 *Organization of stratified sampling*

Let P be a probability measure on $(\mathcal{X},\mathcal{B})$ that is absolutely continuous with respect to the Lebesgue measure μ_n on $\mathcal{X}$ (in the important particular case $P(dx)=\mu_n(dx)/\mu_n(\mathcal{X})$ it is the uniform distribution on $\mathcal{X}$). Define $\mathcal{D}=\mathcal{X}^N$, $\Xi=(x_1,\ldots,x_N)$ where $x_i\in\mathcal{X}$ for $i=1,\ldots,N$. Assume that Ξ is a random vector on $\mathcal{D}$, denote its distribution by $Q(d\Xi)$, and suppose that $N=m\ell$ where m and ℓ are integers.

Divide $\mathcal{X}$ into m disjoint subsets $\mathcal{X}_1,\ldots,\mathcal{X}_m$ of the same P-measure:

$$\mathcal{X}=\bigcup_{j=1}^{m}\mathcal{X}_j,\qquad \mathcal{X}_j\in\mathcal{B},\qquad P(\mathcal{X}_j)=1/m,\qquad P(\mathcal{X}_i\cap\mathcal{X}_j)=0 \text{ for } i\neq j$$

The distribution P induces the distribution P_j on $\mathcal{X}_j$ as defined by $P_j(A)=mP(A\cap\mathcal{X}_j)$ for $A\in\mathcal{B}$. These distributions are probability measures on $(\mathcal{X},\mathcal{B})$, and if P is the uniform distribution on $\mathcal{X}$, then P_j is the uniform distribution on $\mathcal{X}_j$.

A stratified sample is a sample $\Xi=\{x_1,\ldots,x_N\}$ that can be divided into m subsamples $\Xi=\Xi_1\cup\ldots\cup\Xi_m$ where each Ξ_j $(j=1,\ldots,m)$ consists of ℓ independent realizations of the random vector with the distribution P_j (i.e. Ξ_j is an independent sample from P_j). If the distributions P_j $(j=1,\ldots,m)$ are sampled sequentially, ℓ times each, then the distribution of the stratified sample Ξ is

$$Q(d\Xi) = \prod_{j=1}^{m} P_j(dx_{\ell j-\ell+1}) P_j(dx_{\ell j-\ell+2}) \dots P_j(dx_{\ell j}).$$

In the particular case m=1, the stratified sample is independent.

Sometimes it is convenient in practice to organize a stratified sample so that the arrangement of elements of Ξ is random. Under sequential sampling of random points $x_i \in \Xi$, this is attained by means of uniform random choice of the distribution to be sampled among the distributions $P_1, \dots, P_m$ sampled less than ℓ times.

We shall describe an economical way of stratified sample organization when the feasible region $\mathcal{X}$ is a cube $\mathcal{X}=[0,1]^n$ and the stratification consists of dividing $\mathcal{X}$ into hyperrectangles $\mathcal{X}_j$.

Let us represent the number m as a product $m=m_1 \dots m_n$ where m_i is a number of intervals into which the cube $\mathcal{X}$ is divided by the i-th coordinate. Suppose that $m=d^{p_i}$, where d, $p_1, \dots, p_n$ are some integers. In the case the intervals have length $1/m_i$, it is convenient to correspond them to the ordered collections of p_i figures in the d-adic representations (i.e. figures 0,1,...,d-1): this is connected with the fact that all points of each interval have the same first p_i figures. A random point from an interval is easily obtained by adding on the right-hand side some more figures corresponding to the d-adic representation of a random number to the given collection of p_i figures.

Hyperrectangles $\mathcal{X}_j$ of volume 1/m correspond to the ordered collections θ_j from $p=p_1+\dots p_n$ d-adic figures: each of them $\theta=(u_1,\dots,u_p)$ is naturally identified with a number $\theta=u_1 d^{p-1}+u_2 d^{p-2}+\dots+u_p$. Let us show how multiplicative random number generators should be used for obtaining pseudorandom numbers so that a stratified sample would possess the mentioned property of the random arrangement of $\mathcal{X}_j$.

If $m=2^p$, i.e. d=2, then to get a number θ_i we may take e.g. the generator

$$\theta_j \equiv \lambda \theta_{j-1} \qquad (\text{mod } 4m)$$

where $\lambda \equiv 5$ (mod 8), $\theta_o \equiv 1$ (mod 4) and in each number θ_j only its first p digits are to be used. As it follows from Ermakov (1975), p. 405, this generator has period length m. By means of this generator the sets $\mathcal{X}_1, \dots, \mathcal{X}_m$ are chosen pseudorandomly until they all are chosen. If $\ell>1$ and it is required to choose pseudorandomly the sets $\mathcal{X}_1, \dots, \mathcal{X}_m$ again, then the same generator can be used, with a new initial value θ_o.

If $m=10^p$, i.e. d=10, then in order to obtain the numbers θ_i we may proceed as follows. Take a multiplicative generator $v_j \equiv \lambda v_{j-1}$ (mod K) where λ is a primitive root with respect to the modulus K, $K \geq m$, K is the prime number nearest to m, and one can take $v_o=K-1$. This generator has the period K-1, i.e. pseudorandomly gives K-1 different numbers from the collection {1,2,...,K-1}. Those numbers v_j which exceed m should be omitted and for the remaining ones set $\theta_j=v_j-1$. The amount of omitted numbers equals K-m-1 and does not exceed m-1 (as it follows from the so-called Bertran postulate).

4.4.2 *Statistical inference for the maximum of a function based on its values at the points of stratified sample*

Let $\mathcal{X}$ be divided into m disjoint subsets $\mathcal{X}_{jm}$ (j=1,...,m) of equal volumes $\mu_n(\mathcal{X}_{jm})=\mu_n(\mathcal{X})/m$ and let there be given ℓ uniformly distributed points $x_{j1},...,x_{j\ell}$ in each subset $\mathcal{X}_{jm}$. Then the sample

$$\Xi_{m,\ell} = \{x_{ji}, j = 1,\ldots,m,\ i = 1,\ldots,\ell\} \tag{4.4.1}$$

of size $N=m\ell$ is stratified.

For the sake of simplicity we suppose that the above introduced measure P is uniform, the objective function f is continuous and attains its global maximum at a unique point $x^*=\arg\max f$. Under these conditions, we shall construct procedures of statistical inference for $M=\max f$ knowing the values of f at points of a stratified sample (4.4.1). The procedures are similar to those for independent samples, given in Section 4.2.

Denote by $\mathcal{X}_{*m}$ that (unknown) set from the collection of sets $\{\mathcal{X}_{1m},...,\mathcal{X}_{mm}\}$ which contains the global maximizer x^*. P_{*m} is the uniform distribution on $\mathcal{X}_{*m}$; P_o is the uniform distribution on $\mathcal{X}$, i.e. $P_o(dx)=\mu_n(dx)/\mu_n(\mathcal{X})$; $x_{*1},...,x_{*\ell}$ are elements of $\Xi_{m\ell}$ belonging to $\mathcal{X}_{*m}$; $\eta_{m,1}\leq\ldots\leq\eta_{m,\ell}$ are the order statistics corresponding to the sample $\{y_{m,i}=f(x_{*i}),\ i=1,...,\ell\}$; further on

$$F_{*m}(t) = \int_{f(x)<t} P_{*m}(dx) \tag{4.4.2}$$

is the c.d.f. of random variables $y_{m,i}$, $i=1,...,\ell$;

$$F(t) = \int_{f(x)<t} P_0(dx) \tag{4.4.3}$$

is the c.d.f. (4.2.2) corresponding to the uniform distribution P_o on $\mathcal{X}$; $\theta_{m,\ell}$ is the $(1-1/\ell)$-quantile of $F_{*m}(t)$ determined by the condition $F_{*m}(\theta_{m,\ell})=1-1/\ell$; $\theta_{m,\ell}$ is the $(1-1/m\ell)$-quantile of c.d.f. F(t) determined by $F(\theta_{m,\ell})=1-1/(m\ell)$.

From the theoretical point of view, the case when the number m of stratification of the set $\mathcal{X}$ tends to infinity but the number of points ℓ in each subset $\mathcal{X}_{jm}$ is constant, is of most interest and will be considered below.

The main asymptotic properties of the order statistics $\eta_{m,\ell-i}$, $i=0,1,\ldots,\ell-1$ are presented in the following.

Theorem 4.4.1. Let a functional set $\mathfrak{F}$ consist of continuous function f given on $\mathcal{X}$ such that the condition (a) from Section 4.2 for the c.d.f. (4.4.3) holds and the unique global maximizer x* has a certain distribution R on $(\mathcal{X},\mathfrak{B})$ equivalent to the Lebesgue measure μ_n on $(\mathcal{X},\mathfrak{B})$. Then for $m\to\infty$, ℓ=const, with R-probability one, the following statements are valid:

a) the limit distribution of the random variable sequence

$$\left(\eta_{m,\ell}-M\right)/\left(M-\theta_{m\ell}\right) \tag{4.4.4}$$

has the c.d.f.

$$\Phi_{\alpha,\ell}(u)=\begin{cases}\left(1-(-u)^{\alpha}/\ell\right)^{\ell} & \text{for } u<0,\\ 1 & \text{for } u\geq 0;\end{cases} \tag{4.4.5}$$

b)

$$F_{*m}^{-1}(v)\sim M-\left(M-\theta_{m\ell}\right)[\ell(1-v)]^{1/\alpha} \tag{4.4.6}$$

for each v, 0<v<1;

c)

$$M-E\eta_{m,\ell-i}\sim\Psi(\ell,1/\alpha)\left(M-\theta_{m\ell}\right)b_i \tag{4.4.7}$$

for all $i=0,1,\ldots,\ell-1$ where

$$b_i=\Gamma(1/\alpha+i+1)/\Gamma(i+1), \tag{4.4.8}$$

$$\Psi(\ell,u)=\ell^{u}\Gamma(\ell+1)/\Gamma(\ell+1+u),\qquad u>0; \tag{4.4.9}$$

d)

$$E\left(\eta_{m,\ell-i}-M\right)\left(\eta_{m,\ell-j}-M\right)\sim\Psi(\ell,2/\alpha)\left(M-\theta_{m\ell}\right)^{2}\lambda_{ij} \tag{4.4.10}$$

for all $0\leq j\leq i\leq\ell-1$ where

$$\lambda_{ij} = \Gamma(2/\alpha + i + 1)\Gamma(1/\alpha + j + 1)/[\Gamma(1/\alpha + i + 1)\Gamma(j + 1)],\ i \geq j \quad (4.4.11)$$

Proof. It follows from the above listed conditions that with R-probability one, x* does not fall into the boundary of the sets $\mathcal{X}_{jm}$ for all $j \leq m$, m=1,2,... : therefore we shall deal with such events only during the proof.

Consider the ℓ greatest values

$$\eta_N = f(x_{(N)}), \ldots, \eta_{N-\ell+1} = f(x_{(N-\ell+1)})$$

from the collection

$$\{y_i = f(x_i),\ x_i \in \Xi_{m,\ell}\}.$$

Since f is continuous and its global maximum is attained at the unique point x*, thus for $m \to \infty$ the points $x_{(N)}, \ldots, x_{(N-\ell+1)}$ belong only to the sets $\mathcal{X}_{*m}$. Therefore under the given conditions the c.d.f. (4.4.2) and (4.4.3) are connected by the relationship

$$F_{*m}(t) = P_{*m}\{x\colon f(x) < t\} =$$

$$= m\left[P_0\{x\colon f(x) < t\} - (m-1)/m\right] = mF(t) - m + 1. \quad (4.4.12)$$

Hence and from the definition of the quantities $\theta_{m,\ell}$ and $\theta_{m\ell}$ there follows

$$\theta_{m,\ell} = \theta_{m\ell}. \quad (4.4.13)$$

Indeed, from the equality $1 - F_{*m}(\theta_{m,\ell}) = 1/\ell$ and (4.4.12) we have

$$1 - mF\left(\theta_{m,\ell}\right) + m - 1 = 1/\ell, \qquad F\left(\theta_{m,\ell}\right) = 1/(m\ell)$$

that is equivalent to (4.4.13).

Since F(t) satisfies the condition (a) from Section 4.2, thus by virtue of Theorem 4.2.1, we have

$$F\left(M + \left(M - \theta_{m\ell}\right)u\right) \sim \exp\left\{-(-u)^{\alpha}/(\ell m)\right\} \qquad \text{for } u < 0.$$

Whence there follows the chain of relations

$$\Pr\left\{\left(\eta_{m,\ell}-M\right)/\left(M-\theta_{m\ell}\right)<u\right\}=F^{\ell}_{*m}\left(M+\left(M-\theta_{m\ell}\right)u\right)\sim$$

$$\sim\left[m\exp\left\{-(-u)^{\alpha}/(\ell m)\right\}-m+1\right]^{\ell}\sim$$

$$\sim\left[m\left[1-(-u)^{\alpha}/(\ell m)\right]-m+1\right]^{\ell}=\left(1-(-u)^{\alpha}/\ell\right)^{\ell}.$$

The asymptotic relation obtained is equivalent to the statement a) of Theorem 4.4.1. It may be also rewritten like

$$F_{*m}(t)\sim 1-\left[(M-t)/\left(M-\theta_{m\ell}\right)\right]^{\alpha}/\ell \qquad \text{for } t\le M. \qquad (4.4.14)$$

But this implies (4.4.6), i.e. the statement b) of Theorem 4.4.1.

The statement c) can be proved analogously to the proof of Lemma 7.1.2. The distribution density of the statistic $\eta_{m,\ell-i}$ equals

$$p_{m,\ell-i}(t)=\ell C^{i}_{\ell-1}F^{\ell-i-1}_{*m}(t)\left(1-F_{*m}(t)\right)^{i}F'_{*m}(t).$$

Thus

$$E\eta_{m,\ell-i}=\int_{-\infty}^{\infty}t\,p_{m,\ell-i}(t)dt=$$

$$=\ell C^{i}_{\ell-1}\int_{-\infty}^{\infty}tF^{\ell-i-1}_{*m}(t)\left(1-F_{*m}(t)\right)^{i}dF_{*m}(t).$$

Introducing the variable $u=F_{*m}(t)$, we obtain

$$E\eta_{m,\ell-i}=\ell C^{i}_{\ell-1}\int_{0}^{1}F^{-1}_{*m}(u)\,u^{\ell-i-1}(1-u)^{i}du.$$

Furthermore, applying (4.4.6) we get

$$E\eta_{m,\ell-i}\sim$$

$$\sim\ell C^{i}_{\ell-1}\left[M\int_{0}^{1}u^{\ell-i-1}(1-u)^{i}du-\left(M-\theta_{m\ell}\right)1^{1/\alpha}\int_{0}^{1}u^{\ell-i-1}(1-u)^{i+1/\alpha}du\right]=$$

$$=M-\left(M-\theta_{m\ell}\right)\ell^{1/\alpha}\Gamma(\ell+1)\Gamma(i+1+1/\alpha)/[\Gamma(i+1)\Gamma(\ell+1+1/\alpha].$$

The statement c) has been proved. Finally, statement d) is proved in analogy with Lemma 7.1.3. The joint distribution density of the random variables $\eta_{m,\ell-i}$ and $\eta_{m,\ell-j}$ for $i \geq j$ equals

$$p_{m,\ell-i,1-j}(t,s) =$$

$$= AF_{*m}^{\ell-i-1}(t)F'_{*m}(t)[F_{*m}(s) - F_{*m}(t)]^{i-j-1} F'_{*m}(s)(1 - F_{*m}(s))^{j}, \ t \leq s,$$

where for brevity

$$A = \Gamma(\ell+1)/[\Gamma(\ell-i)\Gamma(i-j)\Gamma(j+1)].$$

This way,

$$E(\eta_{m,\ell-i} - M)(\eta_{m,\ell-j} - M) =$$

$$= \int_{-\infty}^{\infty} ds \int_{-\infty}^{s} (t-M)(s-M)p_{m,\ell-i,\ell-j}(t,s)dt = I.$$

Changing variables by $u=F_{*m}(t)$, $v=F_{*m}(s)$ and applying (4.4.6) we get

$$I = A\int_0^1 dv \int_0^v \left(F_{*m}^{-1}(u) - M\right)\left(F_{*m}^{-1}(v) - M\right) u^{\ell-i-1}(v-u)^{i-j-1}(1-v)^j du \sim$$

$$\sim A\left(M - \theta_{m\ell}\right)^2 \ell^{2/\alpha} I_1 \tag{4.4.15}$$

where

$$I_1 = \int_0^1 dv \int_0^v (1-u)^{1/\alpha}(1-v)^{j+1/\alpha} u^{\ell-i-1}(v-u)^{i-j-1} du .$$

Replacing the integration order in the integral I_1 and using the beta-function property

$$\int_0^a t^{m-1}(a-t)^{k-1}dt = a^{m+k-1}B(m,k), \qquad a,m,k > 0,$$

we have

$$I_1 = \int_0^1 du\,(1-u)^{1/\alpha} u^{\ell-i-1} \int_u^1 (1-v)^{j+1/\alpha} (v-u)^{i-j-1} dv =$$

$$= \int_0^1 du\,(1-u)^{1/\alpha} u^{\ell-i-1} \int_0^{1-u} ((1-u)-r)^{j+1/\alpha} r^{i-j-1} dr =$$

$$= B(j+1+1/\alpha, i-j) \int_0^1 u^{\ell-i-1} (1-u)^{1/\alpha} (1-u)^{i+1/\alpha} du =$$

$$= B(j+1+1/\alpha, i-j) B(i+1+2/\alpha, \ell-i) =$$

$$= \frac{\Gamma(j+1+1/\alpha)\Gamma(i-j)}{\Gamma(i+1+1/\alpha)} \frac{\Gamma(i+1+2/\alpha)\Gamma(\ell-i)}{\Gamma(\ell+1+2/\alpha)}.$$

Substituting A and the expression obtained for I_1 into (4.4.15), we get (4.4.10). The theorem is proved completely.

It follows from the theorem that the objective function records corresponding to stratified samples are asymptotically subject to probabilistic laws similar to those which hold for the case of independent samples. It is easily seen that for $\ell \to \infty$ the mentioned asymptotic laws coincide. In particular, $\Phi_{\alpha,\infty}(u) = \Phi_\alpha(u)$ where Φ_α and $\Phi_{\alpha,\ell}$ are determined by (4.2.14) and (4.4.5), respectively. This implies that the limit distributions of record values for independent and stratified samples for $m \to \infty$, $\ell \to \infty$ coincide.

The conditions of Theorem 4.4.1 are slightly stricter, than the condition (a) of Section 4.2: under condition (a), the record values for an independent sample are subjected to analogous asymptotic relations. The additional condition requires the existence of a probability distribution R for x* which is equivalent to the Lebesgue measure on $\mathcal{X}$. The distribution can be regarded to as a prior distribution for the maximizer, and thus the set of objective functions $\mathcal{F}$ is regarded as being stochastic. This requirement is necessary in order to assure that only such events be considered when x* does not hit the boundary of any set $\mathcal{X}_{*m}$. In practice, the number m is finite: thus the requirement is not very restricted, so much the more that the distribution R does not occur in the formulas.

The availability condition for a prior distribution R for x* may be replaced by the more explicit, but more restrictive one:

$$\inf_{x \in \mathcal{X}_{*m}} f(x) \geq \sup_{x \in \mathcal{X} \setminus \mathcal{X}_{*m}} f(x).$$

In general, this condition can be satisfied with some error only. It is satisfied with a high accuracy for the case where Ξ is a Π_τ–grid (see Section 2.2.1). Although the hitting of elements of a Π_τ-grid into the set $\mathcal{X}_{*m}$ cannot be considered as uniformly distributed in the probabilistic sense, a stratified sample pattern is a good approximation for a Π_τ-grid.

To estimate the parameter M, we shall use linear estimators of the type

$$M_{m,k} = \sum_{i=0}^{k} a_i \eta_{m,\ell-i} \tag{4.4.16}$$

where $k \leq \ell-1$ and $a_0, a_1, \ldots, a_k$ are some real numbers defining the estimate.

From (4.4.7) it follows that under the fulfilment of the conditions of Theorem 4.4.1,

$$EM_{m,k} \sim Ma'\lambda - \left(M - \theta_{m\ell}\right)\Psi(\ell, 1/\alpha)a'b, \qquad m \to \infty, \tag{4.4.17}$$

is valid where

$$a = (a_0, \ldots, a_k)', \qquad \lambda = (1, 1, \ldots, 1)', \qquad b = (b_0, \ldots, b_k)',$$

the values b_i are defined by (4.4.8) and the function Ψ by (4.4.9). Now, it follows from (4.4.10) that if the mentioned conditions are met, then

$$E\left(M_{m,k} - Ma'\lambda\right)^2 \sim \left(M - \theta_{m\ell}\right)^2 \Psi(\ell, 2/\alpha)a'\Lambda a \tag{4.4.18}$$

holds where Λ is a symmetric matrix of order $(k+1)\times(k+1)$ with elements λ_{ij} defined by formula (4.4.11) for $0 \leq j \leq i \leq k$.

Since f is continuous, the c.d.f. $F(t)$ and $F_{*m}(t)$ are continuous and therefore $\theta_{m\ell} \neq M$, $M - \theta_{m\ell} \to 0$ for $m \to \infty$ and each integer ℓ. Applying now the Chebyshev inequality

$$\Pr\left\{\left|M_{m,k} - Ma'\lambda\right| \geq \varepsilon\right\} \leq E\left(M_{m,k} - Ma'\lambda\right)^2 / \varepsilon^2$$

and (4.4.18), we can conclude that under the fulfilment of the conditions of Theorem 4.4.1 the estimators (4.4.16) converge to $Ma'\lambda$ for $m \to \infty$. Hence, the sequence of estimators (4.4.16) is consistent if and only if the relation

$$a'\lambda = \sum_{i=0}^{k} a_i = 1 \tag{4.4.19}$$

coinciding with (4.2.16) is satisfied.

If the conditions of Theorem 4.4.1 are satisfied and $\alpha > 1$ then for consistent estimators $M_{m,k}$ formulas (4.4.17) and (4.4.18) have the form

$$EM_{m,k} \sim M - \left(M - \theta_{m\ell}\right)\Psi(\ell, 1/\alpha)a'b,$$

$$E\left(M_{m,k} - M\right)^2 \sim \left(M - \theta_{m\ell}\right)^2 \Psi(\ell, 2/\alpha)a'\Lambda a. \tag{4.4.20}$$

(Note that (4.4.19) is called the consistency condition and

$$a'b = \sum_{i=0}^{k} a_i \Gamma(1/\alpha + i + 1)/\Gamma(i+1) = 0 \tag{4.4.21}$$

is called the unbiasedness requirement.)

From (4.4.20) follows that a natural criterion for the optimal selection of parameters a is the quantity $a'\Lambda a$ which is to be minimized on the set of vectors a satisfying either restriction (4.4.19) or (4.4.19) together with (4.4.21). These optimization problems are similar to those which will be treated in Section 7.1, for the case of independent sampling. For instance, the vector $a=a^*$ determining the optimal consistent linear estimator is determined by (4.2.18).

Confidence intervals for M can be constructed by consistent estimates of the type (4.4.16) applying the asymptotic inequality

$$\Pr\left\{\left|M_{m,k} - M\right| \geq \varepsilon\right\} \lesssim \left(M - \theta_{m\ell}\right)^2 \Psi(\ell, 2/\alpha)a'\Lambda a/\varepsilon^2, \qquad m \to \infty,$$

which follows from (4.4.20) and the Chebyshev inequality. Another way of constructing confidence intervals for M is based on the following statement.

Lemma 4.4.1. Let the conditions of Theorem 4.4.1 be fulfilled. Then the sequence of statistics

$$\kappa_m = \left(M - \eta_{m,\ell}\right)/\left(\eta_{m,\ell} - \eta_{m,\ell-1}\right)$$

converges in distribution to a random variable χ_1 with the c.d.f. $F_1(u)=(u/(1+u))^{\alpha}$, $u>0$, for $m\to\infty$.

Proof. For the order statistics $\eta_{m,\ell-i}$, $i=0,1,\ldots,\ell-1$ an analogy of the Rényi representation

$$\eta_{m,\ell-i} = F_{*m}^{-1}\left(\exp\left\{-\left(\zeta_0/\ell + \zeta_1/(\ell-1) + \ldots + \zeta_i/(\ell-i)\right)\right\}\right)$$

is valid. Here $\zeta_0, \ldots, \zeta_i$ are independent random variables distributed exponentially with the density e^{-u}, $u>0$. Combining this with (4.4.6), we have

$$M-\eta_{m,\ell-i}\sim\left(M-\theta_{m\ell}\right)\left[\ell\left(1-\exp\left\{-\sum_{j=0}^{i}\zeta_j/(\ell-j)\right\}\right)\right]^{1/\alpha},\ m\to\infty.$$

(4.4.22)

For brevity write $w=(1+1/u)^{\alpha}$ and introduce

$$D_0=\left\{(x_0,y_0):\ x_0\ge 0,\ y_0\ge 0,\ \frac{1-\exp\left\{-\left(x_0/\ell+y_0/(\ell-1)\right)\right\}}{1-\exp\left\{-x_0/\ell\right\}}>w\right\},$$

$$D_1=\left\{(x_1,y_1):\ x_1\ge 0,y_1\ge 0,\ \left(1-\exp\left(-x_1-y_1\right)\right)/\left(1-e^{-x_1}\right)>w\right\},$$

$$D_2=\left\{(x_2,y_2):\ 0<x_2\le 1,\ 0<y_2\le 1,\ x_2\left(w-y_2\right)>w-1\right\},$$

$$D_3=\{(x,y):\ 0<x\le 1,\ 0<y\le 1,\ y<wx+1-w\}.$$

Taking into account that $w>1$, applying the asymptotic equation (4.4.22) for $i=0,1$ and changing variables

$$x_1=x_0/\ell,\qquad y_1=y_0/(\ell-1),$$
$$x_2=e^{-x_1},\qquad y_2=e^{-y_1},$$
$$x=x_2,\qquad y=x_2y_2,$$

we obtain for $m\to\infty$ the relations

$$\Pr\{\kappa_m<u\}\sim\Pr\left\{\frac{\left(1-\exp\left(-\zeta_0/\ell\right)\right)^{1/\alpha}}{\left(1-e^{-\zeta_0/\ell}\right)^{1/\alpha}-\left(1-\exp\left(-\zeta_0/\ell-\zeta_1/(\ell-1)\right)\right)^{1/\alpha}}<u\right\}=$$

$$=\int_{D_0}e^{-x_0-y_0}dx_0dy_0=\int_{D_1}\ell(\ell-1)\exp\left\{-\ell x_1-(\ell-1)y_1\right\}dx_1dy_1=$$

$$=\ell(\ell-1)\int_{D_2}x_2^{\ell-1}y_2^{\ell-2}dx_2dy_2=\ell(\ell-1)\int_{D_3}y^{\ell-2}dxdy=$$

$$= \ell(\ell-1) \int_{1-1/w}^{1} \left(\int_{0}^{wx+1-w} y^{\ell-2} dy \right) dx =$$

$$= \ell \int_{1-1/w}^{1} (wx+1-w)^{\ell-1} dx = 1/w = (u/(u+1))^{\alpha}.$$

The lemma is proved.

It follows from Lemma 4.4.1 that the way of constructing a confidence interval for M by the two maximal order statistics is identical to that used in the case of an independent sample. E.g., a one-sided confidence interval of confidence level 1-γ can be chosen as

$$\left[\eta_{m,\ell}, \eta_{m,\ell} + \left(\eta_{m,\ell} - \eta_{m,\ell-1}\right) / \left((1-\gamma)^{-1/\alpha} - 1\right)\right].$$

It follows from (4.4.7) that the mean length of this confidence interval for $m \to \infty$ asymptotically equals

$$\left(M - \theta_{m\ell}\right) \Psi(\ell, 1/\alpha) \Gamma(1 + 1/\alpha) / \left[\alpha\left((1-\gamma)^{-1/\alpha} - 1\right)\right]. \tag{4.4.23}$$

Note that this quantity is $1/\Psi(\ell, 1/\alpha)$ times smaller, than the asymptotic mean length of the analogous interval constructed through an independent sample. It also follows from Lemma 4.4.1 that the statistical hypothesis testing procedures for M using two maximal order statistics coincide for stratified and independent samples. (We remark that an attempt failed to generalize Lemma 4.4.1 for the case k>1: therefore the corresponding confidence intervals and statistical hypothesis tests for k>1 were not constructed for the case of stratified sampling.)

4.4.3 *Dominance of stratified over independent sampling*

We shall show first that stratified sampling dominates the independent one, when the quality criterion is the record value (current optimum estimate) of the objective function.

Let $\mathcal{F}=C(\mathcal{X})$ be the set of continuous functions on $\mathcal{X}$ and P be an absolutely continuous with respect to the Lebesgue measure probability measure on $(\mathcal{X}, \mathcal{B})$. Define

$$D = \mathcal{X}^N, \qquad \Xi = (x_1, \ldots, x_N), \qquad K[f] = \max_{x_i \in \Xi} f(x_i).$$

Let $N=m\ell$, where m and ℓ are natural numbers and a vector Ξ be chosen in D at random in accordance with a distribution $Q(d\Xi)$. We shall call an ordered pair $\pi=(K[f],Q)$ a global random search procedure for optimizing $f \in \mathcal{F}$ on $\mathcal{X}$.

According to the general concepts of domination, we shall say that a procedure π dominates a procedure π_o in $\mathcal{F}$ if $\Phi_f(\pi) \geq \Phi_f(\pi_o)$ for all $f \in \mathcal{F}$ and a function $g \in \mathcal{F}$ exists such that $\Phi_g(\pi) > \Phi_g(\pi_o)$. Here $\Phi_f(\pi)$ is a quality criterion, chosen below as

$$\Phi_f(\pi) = \Pr\{K[f] \geq t\}, \qquad t \in (\min f, \max f). \tag{4.4.24}$$

Note that this is a multiple criterion; the strict inequality

$$\Phi_f(\pi) > \Phi_f(\pi_0)$$

implies that in the probabilistic sense a record $K[f]$ of a function f is closer to $M = \max f$, when the procedure π is used than for π_o.

Denote by $\pi_m = (K_m[.], Q_m)$ a random search procedure based on stratified sampling with m stratifications $\mathcal{X}_1, \ldots, \mathcal{X}_m$ and 1 random points in each strata. The procedure π_1 corresponds to independent sampling from $\mathcal{X}$. Compare the procedures π_m and π_1, applying the criterion $\Phi_f(\pi)$.

Theorem 4.4.2. The procedure π_m for $m>1$ dominates the procedure π_1 in $C(\mathcal{X})$, that is

$$\Pr\{K_m[f] \geq t\} \geq \Pr\{K_1[f] \geq t\} \tag{4.4.25}$$

for all $f \in C(\mathcal{X})$, $t \in (-\infty, \infty)$ and there exists a function $g \in C(\mathcal{X})$ such that

$$\Pr\{K_m[g] \geq t\} > \Pr\{K_1[g] \geq t\}$$

for all $t \in (\min g, \max g)$.

Proof. Denote by $A_t = f^{-1}\{(-\infty, z)\}$ the inverse image of the set $(-\infty, t)$ for mapping f and let

$$\eta_{(i)} = \max_{x_j \in \Xi \cap \mathcal{X}_i} f(x_j)$$

be the record of f in $\mathcal{X}_i$. We have

$$\Pr\{K_1[f] \geq t\} = 1 - \Pr\{K_1[f] < t\} = 1 - \Pr\{f(x_1) < t, \ldots, f(x_N) < t\} =$$

$$= 1 - [\Pr\{f(x_1) < t\}]^N = 1 - (\Pr\{x_1 \in A_t\})^N = 1 - (P(A_t))^N,$$

$$\Pr\{K_m[f] \geq t\} = 1 - \Pr\{K_m[f] < t\} =$$

$$= 1 - \Pr\{\eta_{(1)} < t, \ldots, \eta_{(m)} < t\} = 1 - \left(\prod_{i=1}^{m} \beta_i\right)^{\ell}$$

where

$$\beta_i = P_i(A_t) = mP(A_t \cap X_i) \qquad \text{for } i = 1, \ldots, m$$

and P_i are defined as in Section 4.4.1.

Since $\{X_1, \ldots, X_m\}$ is a complete set of events, therefore

$$P(A_t) = \sum_{i=1}^{m} P(A_t \cap X_i)$$

Thus

$$\Pr\{K_1[f] \geq t\} = 1 - \left(\sum_{i=1}^{m} P(A_t \cap X_i)\right)^{m\ell} = 1 - \left(\frac{1}{m} \sum_{i=1}^{m} \beta_i\right)^{m\ell}.$$

The inequality (4.4.25) follows now from the classical inequality between arithmetic and geometric averages

$$\frac{1}{m} \sum_{i=1}^{m} \beta_i \geq \left(\prod_{i=1}^{m} \beta_i\right)^{1/m}. \qquad (4.4.26)$$

We know that (4.4.26) is valid for arbitrary nonnegative numbers $\beta_1, \ldots, \beta_m$ and becomes equality only for case of $\beta_1 = \ldots = \beta_m$: we shall show that there exists such a function $f \in C(X)$ for which not all $\beta_1, \ldots, \beta_m$ are equal (consequently, the inequalities (4.4.25) and (4.4.26) are strict). Choose a function f that is not equal to a constant, but identically equals $\min f$ on the set X_1. For such a function f, for each $t \in (\min f, \max f)$ one has

$$P(A_t) = \sum_{i=1}^{m} P(A_t \cap \mathcal{X}_i) = \frac{1}{m}\sum_{i=1}^{m} \beta_i < 1, \qquad \beta_1 = P_1(A_t) = 1.$$

Therefore some of the quantities β_i differ and the inequality (4.4.25) is strict for each $t \in (\min f, \max f)$. The proof is completed.

It should be noted that the above result does not use asymptotic extreme value theory and is rather general, i.e. valid under nonrestrictive assumptions. Moreover, it follows from the proof of the theorem that the set of continuous functions may be replaced by other, more narrow classes of functions, e.g. by

$$\mathcal{L}ip(\mathcal{X}, \mathcal{L}) \quad \text{or} \quad C^p(\mathcal{X}), \quad p \geq 1.$$

The following results on the domination of stratified over independent sampling are based on Theorem 4.4.1, together with the next statement.

Lemma 4.4.2. The function $\Psi(\ell, v)$ determined by formula (4.4.9) for each $v>0$ is strictly increasing in ℓ for $\ell \geq 1$, $\Psi(1,v)<1$ and

$$\lim_{\ell \to \infty} \Psi(\ell, v) = 1.$$

Proof. We have

$$\Psi(1,v) = 1/\Gamma(2+v) < 1$$

and by the Stirling formula

$$\lim_{\ell \to \infty} \Psi(\ell, v) = \lim_{\ell \to \infty} \frac{\ell^v \Gamma(\ell+1)}{\Gamma(\ell+1+v)} =$$

$$= \lim_{\ell \to \infty} \frac{\ell^v \sqrt{2\pi\ell}\, e^{-\ell} \ell^{\ell}}{\sqrt{2\pi(\ell+v)}\, e^{-\ell-v} (\ell+v)^{-\ell-v}} = 1.$$

Let us show finally that the function $\Psi(\ell, v)$ is a strictly increasing in ℓ, i.e. for all $v>0$, $\ell=1,2,...$ the inequality

$$\Psi(\ell+1, v)/\Psi(\ell, v) > 1$$

holds. Indeed

$$\Psi(\ell+1,v)/\Psi(\ell,v)=(1+1/\ell)^{v}/(1+v/(\ell+1)),$$

$$\psi(v)=\log\frac{\Psi(\ell+1,v)}{\Psi(\ell,v)}=v\log(1+1/\ell)-\log(1+v/(\ell+1)),$$

$$\psi'(v)=\log(1+1/\ell)-1/(\ell+1+v)>0.$$

Therefore the function $\psi(v)$ is monotone increasing for $v>0$ and $\psi(0)=0$, whence $\psi(v)>0$ for each $v>0$. Now it follows that

$$\psi(v)=\log\frac{\Psi(\ell+1,v)}{\Psi(\ell,v)}>0,\qquad \frac{\Psi(\ell+1,v)}{\Psi(\ell,v)}>1.$$

The lemma is proved.

Based on Theorem 4.4.1 and Lemma 4.4.2, we shall show below that for $m\to\infty$ the random search procedure π_m using a stratified sample dominates the corresponding procedure π_1 that uses an independent sample, with respect to criterion $\Phi_f(\pi)$ constructed on the basis of k+1 ($0\leq k\leq\ell-1$) records of f.

Set

$$K[f]=\left(K^{(0)}[f],\ldots,K^{(k)}[f]\right)$$

where $K^{(i)}[f]=\eta_{N-i}$, $\eta_1\leq\ldots\leq\eta_N$ are the order statistics corresponding to the sample $Y=\{y_i=f(x_i),\ x_i\in\Xi\}$.

For comparing global random search strategies we shall choose the vector criterion

$$\Phi_f(\pi)=\left(\Phi_f^{(0)}(\pi),\ldots,\Phi_f^{(k)}(\pi)\right) \tag{4.4.27}$$

where $0\leq k\leq\ell-1$,

$$\Phi_f^{(i)}(\pi)=M-EK^{(i)}[f].$$

Proposition 4.4.1. If the conditions of Theorem 4.4.1 hold, then the procedure π_m dominates the procedure π_1 with R-probability one.

The proof consists of using Lemma 7.1.2 (proved in Chapter 7) and (4.4.7) from which it follows that with R-probability one, the asymptotic relation

$$\left(M-EK_m^{(i)}[f]\right)/\left(M-EK_1^{(i)}[f]\right)\sim\Psi(\ell,1/\alpha),\qquad m\to\infty, \tag{4.4.28}$$

holds and applying Lemma 4.4.2 according to which the right hand side of (4.4.28) is less than 1.

From Proposition 4.4.1 follows that while applying stratified sampling the records of the objective function are closer to the value $M=\max f$ and, thus, using them more accurate statistical inference for M can be constructed. As it follows from (4.4.28) the best with respect to the criterion (4.4.27) is the stratified sampling with $l=1$, i.e. with a maximal degree of stratification. This finding suggests that *decreasing randomness* , in general, improves the efficiency of a global random search procedure.

Other quality criteria for global random search procedures π are possible: let us give three further such criteria and obtain for them results, analogous to Proposition 4.4.1.

As for $K[f]$, we choose now a consistent linear estimate of the parameter M, constructed on the base of the k+1 $(0\leq k\leq \ell-1)$ maximal order statistics $\eta_N,\ldots,\eta_{N-k}$ with fixed coefficients $a_0,\ldots,a_k$ (satisfying the condition (4.4.19), i.e.

$$K[f]=\sum_{i=0}^{k} a_i\eta_{N-i}. \qquad (4.4.29)$$

If we select the bias of (4.4.29) as $\Phi_f(\pi)$, i.e.

$$\Phi_f(\pi)=M-EK[f] \qquad (4.4.30)$$

then by virtue of Lemma 7.1.5 and (4.4.17), under the fulfilment of the conditions of Theorem 4.4.1 and the unbiasedness condition (4.4.19), with R-probability 1, the asymptotic relation

$$(M-EK_m[f])/(M-EK_1[f])\sim\Psi(\ell,1/\alpha),\qquad m\to\infty, \qquad (4.4.31)$$

holds, analogously to (4.4.28): its consequences are identical to those of Proposition 4.4.1, replacing the criterion (4.4.27) by (4.4.30).

Now let $K[f]$ be defined again by (4.4.29) and the criterion Φ be the mean square error of the estimator (4.4.29), i.e.

$$\Phi_f(\pi)=E(K[f]-M)^2$$

Then, by virtue of (4.2.17) and (4.4.20), under the conditions of Theorem 4.4.1, there holds

$$\Phi_f(\pi_m)/\Phi_f(\pi_1)\sim\Psi(\ell,2/\alpha),\qquad m\to\infty,$$

with R-probability one. Due to Lemma 4.4.2, $\Phi(\ell,2/\alpha)<1$ and thus the procedure π_m dominates the procedure π_1, according to this criterion.

Consider, finally, as $K[f]$ the confidence interval of confidence level $1-\gamma$

$$\left[\eta_N, \eta_N + (\eta_N - \eta_{N-1}) / \left((1-\gamma)^{-1/\alpha} - 1\right)\right]$$

for M, and its mean length as $\Phi_f(\pi)$. By virtue of Lemma 7.1.2 and (4.4.23), if the conditions of Theorem 4.4.1 are met, then with R-probability 1 the relation

$$\Phi_f(\pi_m)/\Phi f(\pi_1) \sim \Psi(\ell, 1/\alpha), \qquad m \to \infty,$$

is valid. Its consequences are identical to those formulated in Proposition 4.4.1 (with the indicated replacement of criterion $\Phi_f(\pi)$).

Some results of this subsection (namely, Theorem 4.4.2 and relation (4.4.31)) were formulated in Ermakov et al. (1988) which contains also the following formulation of the result concerning the admissibility of stratified sampling.

Proposition 4.4.2. Let $K[f]$, $\Phi_f(\pi)$ and $\mathcal{F}$ be the same as in Theorem 4.4.2, $\mathcal{M}$ be the set of all probability measures on $\mathcal{D}=\mathcal{X}^N$, m=N, l=1. Then the procedure $\pi_N=(K[f], Q_N)$ corresponding to stratified sampling with maximal stratification number is admissible for the functional class $\mathcal{F}$ in the set $\Pi=\{\pi=(K[f],Q), Q\in\mathcal{M}\}$ of all global random search procedures: in other words, there is no procedure $\pi\in\Pi$ such that π dominates π_N.

The proposition states that the global random search methods using stratified sampling can not be improved for all functions $f\in \mathcal{F}$ simultaneously.

This result is similar to the one of the admissibility of Monte Carlo estimates of integrals, see Ermakov (1975). For brevity,we shall not give a complete proof of the above proposition but present only its main ideas.

The proof of Proposition 4.4.2 uses the proof of Theorem 4.4.2 and contains two stages. The first one shows that the distribution $Q(d\Xi)$ of the procedure π that dominates, perhaps, the procedure π_N has the uniform marginal distribution of components of a vector Ξ and this distribution may be chosen symmetric. From the existence supposition of f from $\mathcal{F}$ such that for a symmetric $Q(d\Xi)$ with uniform marginal distribution of components, the probability $\Pr\{K[f]\geq t\}$ is strictly larger than

$$\Pr\{K_m[f] \geq t\} = 1 - \prod_{i=1}^{m}\left(mP(A_t \cap \mathcal{X}_i)\right)^{\ell}$$

for $\ell=1$, m=N and some t, the existence of $g\in\mathcal{F}$ such that for each $t\in(\min g, \max g)$ an inverse strict inequality holds, is deduced at the second stage. For details, see Ermakov et al. (1988).

4.5 Statistical inference in random multistart

Random multistart is the global optimization method consisting in several local searches starting at random initial points. It is the most well-known representative of the multistart technique described in Section 2.2.2. Random multistart is inefficient in its pure form, since it may waste much effort for repeated ascents (resp. descents). But some of its modifications, using cluster analysis procedures for preventing repeated ascents are rather efficient: they were already discussed in Section 2.2.2. This section follows mainly Zielinski (1981) and presents statistical inferences in random multistart methods that can be useful for controlling it as well as its modifications.

4.5.1 Problem statement

Let f be a continuous function on $\mathcal{X} \subset \mathbb{R}^n$, ℓ be an unknown number of local maximizers $x_1^*,...,x_\ell^*$ of f which are supposed to be isolated, P be a probability measure on $(\mathcal{X},\mathcal{B})$ (e.g. $P(dx)=\mu_n(dx)/\mu_n(\mathcal{X})$), and $\mathcal{A}$ be a local ascent algorithm. We shall write $\mathcal{A}(x)=x_i^*$ for $x \in \mathcal{X}$, if - whenever starting at the initial point x - the algorithm $\mathcal{A}$ leads to the local maximizer x_i^*.

Put $\theta_i=P(\mathcal{X}_i^*)$ for $i=1,...,\ell$ where

$$\mathcal{X}_i^* = \{x \in \mathcal{X}: \mathcal{A}(x) = x_i^*\}$$

is the region of attraction of x_i^*. The value θ_i will be referred to as the share of the i-th local maximizer x_i^* (with respect to the measure P). It is evident that

$$\theta_i > 0 \quad \text{for} \quad i = 1,\ldots,\ell \quad \text{and} \quad \sum_{i=1}^{\ell} \theta_i = 1.$$

A random multistart method is constructed as follows. An independent sample $\Xi=\{x_1,...,x_N\}$ from the distribution P is generated and the local optimization algorithm $\mathcal{A}$ is sequentially applied at each $x_j \in \Xi$. Let N_i be a number of points x_j belonging to $\mathcal{X}_i^*$ (i.e. N_i is a number of ascents to x_i^* among $\mathcal{A}(x_j)$, $j=1,...,N$). By definition

$$N_i \geq 0 \quad i = 1,\ldots,\ell, \qquad \sum_{i=1}^{\ell} N_i = N,$$

and the random vector $(N_1,...,N_\ell)$ follows multinomial distribution

$$\Pr\{N_1 = n_1,\ldots,N_\ell = n_\ell\} = \binom{N}{n_1,\ldots,n_\ell} \theta_1^{n_1} \ldots \theta_\ell^{n_\ell}$$

where

$$\sum_{i=1}^{\ell} n_i = N, \qquad \binom{N}{n_1,\ldots,n_\ell} = \frac{N!}{n_1!\ldots n_\ell!}, \qquad n_i \geq 0 \qquad (i = 1,\ldots,\ell).$$

The problem is to draw statistical inference on the number of local maximizers ℓ, the parameter vector $\theta=(\theta_1,\ldots,\theta_\ell)$, and the number N_* of trials that guarantees with a given probability that all local maximizers are found.

A main difficulty consists in that ℓ is usually not known. If an upper bound for ℓ is known, then standard statistical methods can be applied; the opposite case is more difficult and the Bayesian approach is applied.

4.5.2 *Bounded number of local maximizers*

Let U be a known upper bound for ℓ and $N \geq U$. Then $(N_1/N,\ldots,N_\ell/N)$ is the standard minimum variance unbiased estimate of θ where N_i/N is the estimator of the share of the i-th local maximizer x_i^*. Of course, for all N and $\ell>1$ it may happen that $N_i=0$ but $\theta_i>0$. So, the above estimator nondegenerately estimates only the share θ_i such that $N_i>0$.

Let W be the number of N_i's which are strictly positive. Then for a given ℓ and $\theta=(\theta_1,\ldots,\theta_\ell)$ we have

$$\Pr\{W = w|\theta\} = \sum_{\substack{n_1+\ldots+n_w=N \\ n_i>0}} \; \sum_{1\leq i_1<\ldots<i_w\leq\ell} \binom{N}{n_1,\ldots,n_w} \theta_{i_1}^{n_1}\ldots\theta_{i_w}^{n_w}.$$

For instance, the probability that the local search will lead to the single local maximizer is equal to

$$\Pr\{W = 1|\theta\} = \sum_{i=1}^{\ell} \theta_i^N,$$

furthermore, the probability that all local maxima will be found equals

$$\Pr\{W = \ell|\theta\} = \sum_{\substack{n_1+\ldots+n_\ell=N \\ n_i>0}} \binom{N}{n_1,\ldots,n_\ell} \theta_1^{n_1}\ldots\theta_\ell^{n_\ell}. \qquad (4.5.1)$$

The probability (4.5.1) is small if (at least) one of the θ_i's is a small number, even for large N. On the other hand, for any ℓ and θ we can find N_* such that for a given $q \in (0,1)$

we will have $\Pr\{W=\ell\,|\,\theta\}\geq q$ for all $N\geq N_*$. The problem of finding $N_*= N_*(q,\theta)$ means to find the (minimal) number of points in Ξ such that the probability that each local maximizer will be found is at least q.

Set

$$\delta = \min\{\theta_1,\ldots,\theta_\ell\} \leq 1/\ell$$

and note that

$$\Pr\{W=\ell|\theta\} \geq \sum_{n_1+\ldots+n_\ell=N} \binom{N}{n_1,\ldots,n_\ell}\delta^N =$$

$$= (\delta\ell)^N \Pr\{W=\ell|(\ell^{-1},\ldots,\ell^{-1})\}.$$

Hence the problem of finding $N_*(q,\theta)$ is reduced to that for $N_*(q,\theta_*)$ where $\theta_*=(\ell^{-1},\ldots,\ell^{-1})$. The latter is easy to solve for large N as

$$\Pr\{W=\ell|\theta_*\} = \ell^{-N} \sum_{n_1+\ldots+n_\ell=N} \binom{N}{n_1,\ldots,n_\ell} =$$

$$= \sum_{i=0}^{\ell}(-1)^i C_\ell^i (1-i/\ell)^N \sim \exp\{-\ell\exp\{-N/\ell\}\}, \qquad N\to\infty.$$

Solving the inequality $\exp(-\ell\exp(-N/\ell))\geq q$ with respect to N we find that

$$N_*(q,\theta_*) = [-\ell\log((-\log q)/\ell)]^+$$

For instance, for $q=0.9$ and $\ell=2$, 5, 10, 100 and 1000 the values of $N_*(q,\theta_*)$ are equal to 6, 19, 46, 686 and 9159, respectively. (Of course, $N_*(q,\theta_*)$ is greater than ℓ.)

4.5.3 Bayesian approach

Let there be given the prior probabilities α_j ($j=1,2,\ldots$) of events that the number ℓ of local maximizers of f equals j together with conditional prior measures $\lambda_j(d\theta^j)$ for the parameter vector $\theta^j=(\theta_1,\ldots,\theta_j)$ under the condition $\ell=j$. We shall assume that the measures $\lambda_j(d\theta^j)$ are uniform on the simplices

$$\Theta_j = \left\{ \theta^j = (\theta_1, \ldots, \theta_j) : \ \theta_i > 0, \ \sum_{i=1}^{j} \theta_i = 1 \right\}.$$

Thus, the parameter set Θ, on which the unknown parameter vector $\theta=(\theta_1,\ldots,\theta_\ell)$ can take its values, has the form

$$\Theta = \bigcup_{j=1}^{\infty} \Theta_j$$

and the prior measure $\lambda(d\theta)$ on Θ for θ equals

$$\lambda(d\theta) = \sum_{j=1}^{\infty} \alpha_j \lambda_j \left(d\theta^j \right). \tag{4.5.2}$$

It is natural to assume that λ is a probability measure.

Let $d=d(N_1,\ldots N_W)$ be an estimate of ℓ. The estimate

$$d_* = \arg\min_d \int_\Theta E_\theta \, 1_{\left[N_1,\ldots,N_W :\, d \neq \ell \right]} \lambda(d\theta)$$

is called optimal Bayesian estimate of ℓ; after some calculations it can be simplified to

$$d_* = \arg\max_{j \geq W} \alpha_j Q(j,W,N) \tag{4.5.3}$$

where

$$Q(j,W,N) = C_j^W \Gamma(j)/\Gamma(N+j).$$

Applying a quadratic loss function, the optimal Bayesian estimate for the total P-measure of the domains of attractions of the hidden ℓ-W local maximizers (i.e. of the sum of the θ_i's corresponding to the hidden maximizers) is given by

$$\sum_{j=W}^{\infty} \frac{j-W}{N+j} \alpha_j Q(j,W,N) \Big/ \sum_{j=W}^{\infty} \alpha_j Q(j,W,N).$$

The optimal Bayesian procedure for testing the hypothesis H_o: ℓ=W under the alternative H_1: ℓ>W is constructed in a similar way. According to it, H_o is accepted if

$$c_{01} \sum_{j=W+1}^{\infty} \alpha_j Q(j,W,N) \leq c_{10} \alpha_W \Gamma(W)/\Gamma(N+W),$$

otherwise H_o is rejected. Here c_{o1} is the loss arising after accepting H_o in the case of H_1's validity and c_{10} is the analogous loss due to accepting the hypothesis H_1 that is false.

The above statistical procedures were numerically investigated in Betró and Zielinski (1987). In some works (see Zielinski (1981), Boender and Zielinski (1985), Boender and Rinnooy Kan (1987)) the procedures were thoroughly investigated and modified also for the case of equal prior probabilities α_j (i.e. for the case $\alpha_j=1$, $j=1,2,...$). They are not presented here, since the equal prior probabilities assumption contradicts the finiteness of the measure (4.5.2) and is somewhat peculiar in the global optimization context (for instance, according to it, the prior probabilities of f having 2 or 10^{10} local maximizers coincide). Instead of these, we shall present below the following result.

Proposition 4.5.1. (Snyman and Fatti (1987)). Let the condition

$$\theta^* = \max_{1\le j\le \ell} \theta_j$$

hold, where θ^* is the share of the global maximizer, and a prior distribution for θ^* be beta distribution. Then the inequality

$$\Pr\left\{M_N^* = M\right\} \ge 1 - \Gamma(N)\Gamma(2N - r + 1)/[\Gamma(2N)\Gamma(N - r + 1)]$$

holds for the probability of the event that the record value M_N^* in the above described random multistart method equals $M=\max f$: here r is the number of sample points $x_i \in \Xi$ falling into the region of attraction of the maximizer with the function value M_N^*.

4.6 An approach based on distributions neutral to the right

This section follows mainly Betró (1983, 1984) and describes statistical inferences for quantiles of the c.d.f. (4.2.2). i.e.

$$F(t) = F_f(t) = \int_{f(x)<t} P(dx) \qquad (4.6.1)$$

under the supposition that this c.d.f. belongs to a subclass of the class of cumulative distribution functions neutral to the right, with a view on applications to global random search theory: essentially, the basis of the approach is a specific parametrization of the c.d.f. (4.6.1).

4.6.1 *Random distributions neutral to the right and their properties*

A random c.d.f. F(t) is said to be neutral to the right, if for each m>1 and

$$-\infty < t_1 < \ldots < t_m < \infty \qquad (4.6.2)$$

there exist nonnegative independent random variables $q_1,\ldots,q_m$ such that the random vector $(F(t_1),\ldots,F(t_m))$ is distributed as

$$\left(q_1, 1-(1-q_1)(1-q_2), \ldots, 1-\prod_{i=1}^{m}(1-q_i)\right).$$

Some properties of c.d.f.'s neutral to the right, are presented below without proofs.

P1. If F(t) is neutral to the right, then the normalized increments

$$F(t_1), (F(t_2)-F(t_1))/(1-F(t_1)), \ldots, (F(t_m)-F(t_{m-1}))/(1-F(t_{m-1}))$$

are mutually independent for each collection (4.6.2) such that $F(t_{m-1})<1$.

P2. A wide c.d.f. class can be approximated by c.d.f.'s neutral to the right.

P3. The posterior c.d.f. of a c.d.f. neutral to the right is neutral to the right.

P4. Most c.d.f.'s F(t) neutral to the right are such that the posterior c.d.f. $F(t\,|\,Y)$, given a sample $Y=\{y_1,\ldots,y_N\}$ from F, depends not only on the number N_A of x's that fall into A but also on where they fall within or outside A.

P5. A random c.d.f. F(t) is neutral to the right, if and only if it has the same probability distribution as $1-\exp\{-\xi(t)\}$ for some a.s. nondecreasing, right-continuous, independent-increment random process $\xi(t)$ with

$$\lim_{t\to-\infty} \xi(t) = 0 \quad \text{a.s.} \qquad \lim_{t\to\infty} \xi(t) = \infty \quad \text{a.s.}$$

Within the class of c.d.f.'s neutral to the right, apart from rather trivial cases, only the so-called Dirichlet processes do not hold the property P4. For this reason, the posterior distributions of these processes are easy to handle and are more widely used in applications (for instance, in some discrete problems generated by global random search, see Betró and Vercellis (1986), Betró and Schoen (1987)). But the Dirichlet processes are too simple and thus unable to approximate the c.d.f. (4.6.1) accurately enough. To this end, another subclass $\mathcal{T}$ of the c.d.f.'s neutral to the right will be considered.

A neutral to the right c.d.f. F(t) is an element of $\mathcal{T}$ if the corresponding random process

$$\xi(t) = -\log(1 - F(t)) \tag{4.6.3}$$

is a gamma process, i.e. $\xi(t)$ has gamma-distributed independent increments.

For a gamma process $\xi(t)$ the moment generation function

$$M_{\xi(t)}(v) = E\{\exp\{-v\xi(t)\}\}, \qquad v > 0, \tag{4.6.4}$$

has the form

$$M_{\xi(t)}(v) = \exp\left\{\gamma(t)\int_0^\infty (e^{-vs} - 1)e^{-\lambda s}s^{-1}ds\right\} = \left(\frac{\lambda}{\lambda + v}\right)^{\gamma(t)} \tag{4.6.5}$$

where λ is a positive number and $\gamma(t)$ is a nondecreasing function satisfying

$$\lim_{t\to-\infty} \gamma(t) = 0, \qquad \lim_{t\to\infty} \gamma(t) = \infty.$$

The expression (4.6.3) and (4.6.4) imply the representation

$$E\{[1 - F(t)]^m\} = M_{\xi(t)}(m)$$

for the moments of a neutral to the right c.d.f. F(t). For $F \in \mathcal{T}$, together with (4.6.5), this gives the expression

$$E\{[1 - F(t)]^m\} = [\lambda/(\lambda + m)]^{\gamma(t)}, \qquad m = 1,2,\ldots. \tag{4.6.6}$$

Denoting the prior c.d.f. of F(t) by $\beta(t)=EF(t)$, for $F \in \mathcal{T}$ we obtain

$$1 - \beta(t) = (\lambda/(\lambda + 1))^{\gamma(t)}$$

from (4.6.6) with m=1. This yields the representation

$$\gamma(t) = (\log(1 - \beta(t)))/\log(\lambda/(\lambda + 1)) \tag{4.6.7}$$

for $\gamma(t)$.

In order to see the role of parameter λ consider the variance of 1-F(t) which is represented as

$$\mathrm{var}(1-F(t)) = (1-\beta(t))^{g(\lambda)} - (1-\beta(t))^2 \qquad (4.6.8)$$

by (4.6.6) for m=1,2 and (4.6.7): here

$$g(\lambda) = (\log(\lambda/(\lambda+2)))/\log(\lambda/(\lambda+1)).$$

This is an increasing function, $g(\lambda)\to 1$ for $\lambda\to 0$, and $g(\lambda)\to 2$ for $\lambda\to\infty$. Thus, larger values of λ corresponds to smaller values of the variance (4.6.8), and smaller values of λ correspond to larger values of (4.6.8), i.e. λ measures the prior *strenght of belief* about $\beta(t)$.

The following proposition formulates an important feature of $\mathcal{T}$, viz., for each $F\in\mathcal{T}$ the characteristic function of the posterior distribution of $\xi(t)=-\log(1-F(t))$, given a sample from F, can be represented in an analytical form.

Proposition 4.6.1. Let $F\in\mathcal{T}$, $\eta_1,\ldots,\eta_N$ be the order statistics corresponding to an independent sample $Y=\{y_1,\ldots,y_N\}$ from F. Set $m_j=N+\lambda-j$, $\eta_0=-\infty$,

$$n_0(t) = \sum_{i:\,y_i<t} 1,\quad t_0=\eta_{n_0(t)},\quad m_0(t) = N+\lambda-n_0(t).$$

If the moment generation function for the process (4.6.3) has the form (4.6.5) and $\gamma(t)$ is continuous at the points $y_1,\ldots,y_N$, then the moment generation function corresponding to the posterior distribution is equal to

$$M_{\xi(t)}(v|Y) = \prod_{j=1}^{n_0(t)} \frac{\log\left[1+1/(m_j+v)\right]}{\log\left[1+1/m_j\right]}\times$$

$$\times\left[1+v/(m_j+1)\right]^{\gamma(\eta_{j-1})-\gamma(\eta_j)}\left[1+v/m_0(t)\right]^{\gamma(t_0)-\gamma(t)} \qquad (4.6.9)$$

Since for each t the characteristic function of a random variable $\xi(t)$ is

$$\Phi_{\xi(t)}(v) = M_{\xi(t)}(-iv), \qquad (4.6.10)$$

thus (4.6.9) presents the analytical form for the characteristic function of the posterior distribution of a gamma process $\xi(t)=-\log(1-F(t))$. This way one can obtain the posterior distribution of F by numerical evaluation of a Fourier integral. Once a posterior distribution of F(t), given the sample, is known, testing some statistical hypothesis about F can be performed in the framework of Bayesian statistics. We shall now describe how the statistical hypothesis about quantiles of F are tested.

4.6.2 *Bayesian testing about quantiles of random distributions*

It is shown here that in a natural setup the problem of testing a statistical hypothesis about random c.d.f. quantile is reduced to a single computation of the posterior probability.

Let F(t) be a random c.d.f., t_p be the p-th quantile of F, $Y=\{y_1,\ldots,y_N\}$ be an independent sample from F, t_* is a given constant. The problem is to test the hypothesis H_o: $t_p \le t_*$ which can be rewritten as H_o: $F(t_*) \ge p$. Let d(Y) be a decision function assuming two values d_o and d_1, corresponding to acceptance and rejection of H_o, and the losses connected with d_o and d_1 are

$$L(F,d_0) = c_0 1_{(t_*,\infty)}(t_p), \qquad L(F,d_1) = c_1 1_{(-\infty,t_*]}(t_p)$$

where c_o and c_1 are given positive values. The posterior mean values of $L(F,d_i)$ are

$$E\{L(F,d_0)\,|\,Y\} = c_0 \Pr\{t_p > t_*\,|\,Y\},$$

$$E\{L(F,d_1)\,|\,Y\} = c_1 \Pr\{t_p \le t_*\,|\,Y\}.$$

Thus the optimal decision function is

$$d^*(Y) = \begin{cases} d_0 & \text{if} \quad \Pr\{F(t_*) \ge p\,|Y\} \ge c_0/(c_0+c_1) \\ d_1 & \text{otherwise} . \end{cases} \qquad (4.6.11)$$

To construct d^* one needs to evaluate the posterior c.d.f. at $t=t_*$.

4.6.3 *Application of distributions neutral to the right to construct global random search algorithms*

It is shown below how the result cited can be applied for controlling the precision of some simple global random search algorithms (like Algorithm 3.1.1).

Let P be the uniform measure on $\mathcal{X}$, $\Xi=\{x_1,\ldots,x_N\}$ be an independent sample from P, f be a bounded measurable function, M=vrai sup f, f_* be the record, and the maximization problem of f is stated in terms of obtaining a point from $A(\varepsilon)=\{x: M-f(x) \le \varepsilon\}$ where ε is a given value. Note that

$$Y = \{y_i = f(x_i),\ i = 1,\ldots,N\} \qquad (4.6.12)$$

is an independent sample from c.d.f. (4.6.1) and

$$f_* \geq \max_{1 \leq i \leq N} f(x_i) = \eta_N$$

The statistical problem is to test the hypothesis H_0: $f_* \in A(\varepsilon)$ that can be rewritten in the forms H_0: $F(f_*) \geq 1-\varepsilon$ and H_0: $f_* \geq f_{1-\varepsilon}$ where $f_{1-\varepsilon}$ is the (1-ε)-quantile of the c.d.f. F(t) defined by (4.6.1).

Under the assumption that F is a random c.d.f., the decision rule (4.6.11) can be used to test the hypothesis H_0 (it is natural to choose e.g. $c_0=c_1=1$). In order to apply the results of Section 4.6.1, we suppose that $F \in \mathcal{T}$ and consider the choice problem for parameter λ and function γ(t) that determine the gamma process ξ(t)= - log (1-F(t)).

Using (4.6.7) we obtain for each $t \in \mathbb{R}$, $\delta \in (0,1)$

$$\Pr\{F(t) \geq 1-\delta\} = 1 - \Gamma(-\lambda \log \delta, -(\log(1-\beta(t)))/\log(1+1/\lambda)) \tag{4.6.13}$$

where β(t)=EF(t) is the prior c.d.f. corresponding to F(t) and

$$\Gamma(u,v) = \int_0^u t^{v-1} e^{-t} dt / \Gamma(v).$$

Before drawing statistical inference, we demand the following prior information: let for some pair $\alpha, \delta \in (0,1)$ a value $f_{\delta,\alpha}$ be given such that

$$\Pr\{f_{\delta,\alpha} \geq f_{1-\delta}\} = \alpha \tag{4.6.14}$$

It is natural to use e.g. α=0.5, δ=ε.

Equalities (4.6.13) and (4.6.14) imply the equation below for λ, given β(t):

$$1-\alpha = \Gamma\left(-\lambda \log \delta, -\left(\log\left(1-\beta\left(f_{\delta,\alpha}\right)\right)\right)/\log(1+1/\lambda)\right). \tag{4.6.15}$$

It may happen that (4.6.15) is unsolvable. In this case either (4.6.14) or β should be modified. The properties of the incomplete gamma function imply that the equation (4.6.15) has solution if $1-\beta(f_{\delta,\alpha}) \geq \delta(1-\alpha)$.

As for β(t), it is natural to set it in a parametric form with subsequent estimation of the parameters on the basis of the empirical c.d.f.

$$\hat{F}_N(t) = \sum_{i:y_i<t} 1/N.$$

For instance, if the parameters of β are the mean μ and variance σ^2, i.e.

$$\beta(t) = \beta_*((t-\mu)/\sigma) \tag{4.6.16}$$

where $\beta*$ is a given function, then

$$\hat{\mu} = N^{-1}\sum_{i=1}^{N} y_i, \qquad \hat{\sigma} = \left[\frac{1}{N-1}\sum_{i=1}^{N}(y_i - \hat{\mu})^2\right]^{1/2} \tag{4.6.17}$$

are natural estimates for μ and σ.

It should be noted that it is inappropriate to take $\hat{F}_N(t)$ as $\beta(t)$, since in this case $F(t)=1$ for all $t \geq \eta_N$ and hence $F(f*)=1$, i.e. $f* \geq f_{1-\varepsilon}$ and it is meaningless to test the hypothesis H_o.

Let us formulate a typical global random search algorithm based on statistical hypothesis testing for the quantiles of c.d.f. (4.6.1) under the assumption $F \in \mathcal{T}$.

Algorithm 4.6.1.

1. Assume that the prior c.d.f. $\beta(t)$ for $F(t)$ has the form of (4.6.16).
2. For some α, $\delta \in (0,1)$, take $f_{\delta,\alpha}$ satisfying (4.6.14).
3. Obtain a sample $\Xi=\{x_1,...,x_N\}$ by sampling a distribution P.
4. Evaluate $y_i=f(x_i)$ for $i=1,...,N$.
5. Estimate μ and σ by (4.6.17)
6. Find λ by solving (4.6.15). If the equation (4.6.15) has no solution, then increase σ until it becomes solvable.
7. Determine $\gamma(t)$ from (4.6.7).
8. Obtain $f*$ (for instance, set $f* = \eta_N = \max\{y_1,...,y_N\}$).
9. Estimate the value $\Pr\{F(f*) \geq 1-\varepsilon \,|\, Y\}$ by numerical evaluation of a Fourier integral from the characteristic function determined by (4.6.9) and (4.6.10).
10. Test the hypothesis H_o: $f* \in A(\varepsilon)$ by (4.6.11).
11. If H_o is accepted, then terminate the algorithm, otherwise sample N_o times the distribution P and return to Step 4 substituting $N+N_o$ for N (N_o is some fixed natural number).

Let us comment on the essence and possibility of applications of Algorithm 4.6.1.

The essence of Algorithm 4.6.1 is in setting a particular parametrization of the c.d.f. (4.6.1): it consists of using (4.6.5), (4.6.7), and a parametrization for $\beta(t)$. Only by a thorough theoretical and numerical study can be seen the quality of parametrization for specific classes of multiextremal problems.

The complexity of Algorithm 4.6.1 and its modifications is rather high, since several times one needs to solve the complicated equation (4.6.15) and compute a value of c.d.f. via the characteristic function. Hence, if f is easy to evaluate, it is unprofitable to use the approach presented.

A natural way of applying the above described testing procedures of statistical hypothesis, concerning quantiles of c.d.f.'s neutral to the right, consists of using them in branch and probability bound methods for evaluating a prospectiveness criterion. To construct the prospectiveness criterion, one can apply Steps 1-10 of Algorithm 4.6.1, substituting Z for X and the record value of f in Z for f_*. Acceptance of the hypothesis H_0: $f_* \geq f_{1-\varepsilon}$ for $\varepsilon \approx 0$ corresponds to the decision that Z does not contain a global maximizer of f, i.e. Z is unpromising for further search.

CHAPTER 5. METHODS OF GENERATIONS

The methods studied in this chapter consist of sequential multiple sampling of probability distributions, asymptotically concentrated in a vicinity of the global optimizer. These methods form the essence of numerous heuristic global random search algorithms and can be regarded as a generalization of simulation annealing type methods in the sense that groups of points are transformed to groups, rather than points to points.

The methods of generations are rather simple to realize, but are not very efficient for solving *easy* global optimization problems. Nevertheless, numerical results demonstrate that they can be applied even for solving very complicated problems (the author used them for solving some location problems in which the number of variables exceeded 100).

Many methods of generations are suitable for the case when the objective function is subject to random noise: this is the case generally considered in this chapter. For convenience, we shall suppose here (as well as in the preceding chapter) that the maximization problem of f is considered. Besides, the condition $\mathcal{X}\subset\mathbb{R}^n$ is relaxed in this chapter, and the feasible region $\mathcal{X}$ is supposed to be a compact metric space of an arbitrary kind.

The theoretical study of the methods of generations is the main aim of the present chapter which is divided into four sections. Section 5.1 describes some approaches to algorithm construction and formulates the basic model. Section 5.2 investigates the convergence and the rate of convergence of the sequences of probability measures generated by the basic model. Section 5.3 studies homogeneous variants of the methods: its results are closely connected with the theory of Monte Carlo methods. Finally, Section 5.4 modifies the main methods of generations in a sequential fashion and investigates the convergence of these modifications.

5.1 *Description of algorithms and formulation of the basic probabilistic model*

5.1.1 *Algorithms*

Beginning in the late 1960's, many authors suggested heuristic global random search algorithms based on the next three ideas: (i) new points at which to evaluate f are determined mostly not far from some best previous ones, (ii) the number of new points in the vicinity of a previously obtained point must depend on the function value at this point, (iii) it is natural to decrease the search *span* while approaching to a global optimizer. Such algorithms are described e.g. in McMurty and Fu (1966), Rastrigin (1968), Bekey and Ung (1974), Ermakov and Mitioglova (1977), Ermakov and Zhigljavsky (1983), Zhigljavsky (1981, 1987), Price (1983, 1987), Masri et al. (1980), Pronzato et al. (1984) and in many other works.

A general algorithm relying upon the above ideas is as follows.

Algorithm 5.1.1.

1. Sample some number N_1 times a distribution P_1, obtain points $x_1^{(1)},...,x_{N_1}^{(1)}$; set $k=1$.

2. From the points $x_j^{(i)}$ ($j=1,...,N_i$; $i=1,...,k$) choose ℓ points $x_{1*}^{(k)},...,x_{\ell *}^{(k)}$ having the greatest values of f.

3. Determine the natural numbers

$$n_{kj} \quad \left(j=1,\dots,\ell \qquad \sum_{j=1}^{\ell} n_{kj} = N_{k+1} \right)$$

applying some rule.

4. Sample n_{kj} times the distributions $Q_k(x_{j*}^{(k)},dx)$ for all $j=1,2,...,\ell$ and thus obtain the points

$$x_1^{(k+1)},\dots,x_{N_{k+1}}^{(k+1)}. \tag{5.1.1}$$

5. Substitute k+1 for k and go to Step 2.

Algorithm 5.1.1 becomes a special case of Algorithm 3.1.5 (i.e. the general scheme of global random search algorithms) if n_{kj} of Algorithm 5.1.1 is used in the latter as N_k.

The lack of search *span* decrease is equivalent to defining the transition probabilities $Q_k(z,.)$ so as to meet $Q_{k+1}(z,B(z,\varepsilon))\geq Q_k(z,B(z,\varepsilon))$ for all $k\geq 1$, $z\in \mathcal{X}$, $\varepsilon>0$. The particular choice of the rate of this decrease depends on the prior information about f, on magnitudes of N_k and on the requirements to the accuracy of extremum determination. As it was established in Section 3.2, if one does not want to *miss* the global optimizer, then the *span* must decrease slowly. For the sake of simplicity, the sampling algorithm for $Q_k(z,.)$ is often defined as follows: a uniform distribution on $\mathcal{X}$ is sampled with a small probability $p_k\geq 0$ and a uniform distribution on a ball or a cube (their volume being dependent on k) with centre in z_∞ is sampled with probability $1-p_k$. In this case the condition

$$\sum_{k=1}^{\infty} p_k = \infty$$

is sufficient for the convergence of the algorithm.

Although various versions of Algorithm 5.1.1 can be succesfully used in practical applications, for theoretical research it seems to be inconvenient because (i) it is not clear how to choose ℓ (the choice l=1 leads to the most well-known version of Algorithm 5.1.1) and (ii) much depends on the choice of numbers n_{kj} which is not yet formalized. Let us introduce randomization into the choice rule of n_{kj} with a view of overcoming these disadvantages.

Algorithm 5.1.2.

1. Sample N_1 times a distribution P_1, obtain $x_1^{(1)},...,x_{N_1}^{(1)}$; set k=1.

2. Construct an auxiliary nonnegative function f_k using the results of evaluating f at the points $x_j^{(i)}$ ($j=1,...,N_i$, $i=1,...,k$)

3. Sample the distribution

$$P_{k+1}(dx) = \sum_{j=1}^{N_k} p_j^{(k)} Q_k\left(x_j^{(k)}, dx\right) \tag{5.1.2}$$

where

$$p_k^{(k)} = f_k\left(x_j^{(k)}\right) / \sum_{i=1}^{N_k} f_k\left(x_i^{(k)}\right)$$

and thus obtain the points (5.1.1) of the next iteration.

4. Go to Step 2 substituting k+1 for k.

The distribution (5.1.2) is sampled by the superposition method: at first the discrete distribution

$$\varepsilon_k = \begin{Bmatrix} x_1^{(k)},\dots,x_{N_k}^{(k)} \\ \\ p_1^{(k)},\dots,p_{N_k}^{(k)} \end{Bmatrix} \tag{5.1.3}$$

is sampled and then the distribution $Q_k(x_j^{(k)},.)$, if $x_j^{(k)}$ is the realization obtained.

Since the functions f_k are arbitrary, they may be chosen in such a way that the mean values En_{kj} of numbers n_{kj} of Algorithm 5.1.2 correspond to the numbers n_{kj} of Algorithm 5.1.1. Allowing for this fact and for the procedure of determining quasi-random points with distribution (5.1.3), one can see that Algorithm 5.1.1 is a special case of Algorithm 5.1.2.

In theoretical studies of Algorithm 5.1.2 (more precisely, of its generalization Algorithm 5.1.4) it will be assumed that the discrete distribution (5.1.3) is sampled in a standard way, i.e. independent realizations of a random variable are generated with this distribution. In practical calculations, it is more advantageous to generate quasi-random points from this distribution by means of the following procedure that is well known in the regression design theory (see Ermakov and Zhigljavsky (1987)) as the construction procedure of exact design from approximate ones. (Note that this is also a simple variant of the main part extraction method used for reduction of Monte Carlo estimate variance at calculation of integrals, Ermakov (1975)). Set $r_j^{(k)}$, the greatest integer, in $Np_j^{(k)}$,

$$N_{(k)} = \sum_{j=1}^{N_k} r_j^{(k)}, \qquad N^{(k)} = N_k - N_{(k)}, \qquad \alpha_j^{(k)} = p_j^{(k)} - r_j^{(k)}.$$

Then

$$\varepsilon_k = \left(N_{(k)}/N_k\right)\varepsilon_k^{(1)} + \left(N^{(k)}/N_k\right)\varepsilon_k^{(2)}$$

where

$$\varepsilon_k^{(1)} = \begin{Bmatrix} x_1^{(k)} & ,\ldots, & x_{N_k}^{(k)} \\ r_1^{(k)}/N_{(k)}, & \ldots, & r_{N_k}^{(k)}/N_{(k)} \end{Bmatrix}, \qquad \varepsilon_k^{(2)} = \begin{Bmatrix} x_1^{(k)} & ,\ldots, & x_{N_k}^{(k)} \\ \alpha_1^{(k)}/N^{(k)}, & \ldots, & \alpha_{N_k}^{(k)}/N^{(k)} \end{Bmatrix}.$$

Instead of N_k-fold sampling of (5.1.3), one can sample $\varepsilon_k^{(2)}$ $N^{(k)}$ times and choose $r_j^{(k)}$ times the points $x_j^{(k)}$ for $j=1,\ldots,N_k$. The above procedure reduces the indeterminacy in the selection of points $x_j^{(k)}$ in whose vicinity the next iteration points are chosen according to $Q_k(x_j^{(k)},.)$. In the case of using this procedure, these points include some *best* points of the preceding iteration with probability one. Besides the procedure is independent of the method of determining $p_j^{(k)}$ (and is, therefore, usable in the case when evaluations of f are subject to random noise).

The quality of the variants of Algorithm 5.1.2 greatly depends on the choice of f_k that should reflect the properties of f (e.g, be on the average greater, where f is great or smaller, where f is small). Their construction should be based on some technique of objective function estimation or on a technique of extracting and using information about the objective function during the search. Various estimates

$\hat{f}_k$ of f

can be used as f_k (in this case

$$\hat{f}_k(x)\mu(dx)/\int \hat{f}_k(z)\mu(dz)$$

may be used, instead of (5.1.2), where μ is a finite measure on $(\mathcal{X},\mathcal{B})$) as well as prospectiveness criteria (see Section 4.3) and other functions related to f (e.g. solutions of (5.3.5)). A simple but practically important way of choosing f_k is $f_k = f$. The resulting algorithm is readily generalized to the situation when a random noise is present. This generalization is given below under the supposition that the result of evaluating f at a point $x \in \mathcal{X}$ at iteration k is a random variable $y_k(x)=f(x)+\xi_k(x)$ taking values on the set of nonnegative numbers.

Algorithm 5.1.3.

1. Choose a distribution P_1, set k=1.

2. Sample N_k times the distribution P_k, obtain points $x_1^{(k)},\ldots,x_{N_k}^{(k)}$. Evaluate f at these points. If

$$\sum_{j=1}^{N_k} y_k\left(x_j^{(k)}\right) = 0,$$

repeat the sampling.

3. Take the distribution P_{k+1} in the form (5.1.2) where

$$p_j^{(k)} = y_k\left(x_j^{(k)}\right) / \sum_{i=1}^{N_k} y_k\left(x_i^{(k)}\right). \tag{5.1.4}$$

4. Go to Step 2 substituting k+1 for k.

The heuristic meaning of Algorithm 5.1.3 will be discussed in the next section : its essence is the fact that for great N_k the unconditional distributions of the random elements $x_j^{(k)}$ are close to the distributions

$$f^k(x)\mu(dx)/\int f^k(z)\mu(dz)$$

and, therefore, weakly converge for $k\to\infty$ to the distribution concentrated at the global maximizer of f.

Stopping rules for these or similar algorithms may be constructed through the recommendations of Chapter 4 (termination takes place when reaching the accuracy desired); the simplest rule (the prescribed number of iterations is executed) can be chosen as well.

Algorithms 5.1.1 - 5.1.3 (as well as the Algorithm 5.1.4 presented below) will be called methods of generations. This name originates from the fact that these algorithms are analogous to or direct generalizations of Algorithms 5.3.3., 5.3.4, for which *methods of generations* is the standard terminology in the theory of the Monte Carlo methods.

5.1.2 The basic probabilistic model

The algorithm presented below is a generalization of Algorithm 5.1.3 to the case, when f_k are evaluated with random noise: it is a mathematical model of Algorithms 5.1.1 through 5.1.3 and their modifications.

Algorithm 5.1.4.

1. Choose a distribution P_1 on $(\mathcal{X},\mathcal{B})$ and set k=1.
2. Sample N_k times P_k and obtain points $x_1^{(k)},\ldots,x_{N_k}^{(k)}$.
3. Evaluate the random variables $y_k(x_j^{(k)})$ at the points $x_j^{(k)}$, where $y_k(x)=f_k(x)+\xi_k(x)\geq 0$ with probability one. If

$$\sum_{j=1}^{N_k} y_k\left(x_j^{(k)}\right) = 0,$$

return to Step 2 (repeat sampling).

4. Take the distribution P_{k+1} in the form (5.1.2) where $p_j^{(k)}$ are defined by (5.1.4).
5. Go to Step 2, substituting k+1 for k.

Measures $P_{k+1}(dx)$, k=1,2,... , defined through (5.1.2), are the distributions of random points $x_j^{(k+1)}$ conditioned on the results of preceding evaluations of f. We shall study in this chapter their unconditional (average) distributions which will be denoted by $P(k+1,N_k;dx)$. Obviously, the unconditional distribution of $x_j^{(1)}$ is $P_1(dx)=P(k,N_o;dx)$.

5.2 Convergence of probability measure sequences generated by the basic model

5.2.1 Assumptions

For the sake of convenience, the assumptions used in this chapter and their explanation are formulated separately. Assume that

(a) $\xi_k(x)$ for any $x \in \mathcal{X}$ and $k=1,2,\dots$ are random variables with a zero-expectation distribution $F_k(x,d\xi)$ concentrated on a finite interval $[-d,d]$; the random variables $\xi_{k1}(x_1)$, $x_{k2}(x_2),\dots$ are mutually independent for any $k_1,k_2,\dots$ and $x_1,x_2,\dots$ from $\mathcal{X}$;

(b) $y_k(x)=f_k(x)+\xi_k(x)\geq c_1>0$ with probability one for all $x \in \mathcal{X}$, $k=1,2,\dots$;

(c) $0<c_1\leq f_k(x)\leq M_k=\sup f_k(x)\leq C<\infty$ for all $x \in \mathcal{X}$, $k=1,2,\dots$;

(d) the sequence of functions $f_k(x)$ converges to $f(x)$ for $k\to\infty$ uniformly in x;

(e) $Q_k(z,dx)=q_k(z,x)\mu(dx)$,

$$\sup_{z,x\in\mathcal{X}} q_k(z,x) \leq L_k < \infty$$

for all $k=1,2,\dots$ where μ is a probability measure on $(\mathcal{X},\mathcal{B})$;

(f) the random elements $\chi_1,\dots,\chi_N$ with a distribution $R(dx_1,\dots dx_N)$ defined on $\mathcal{B}_N=\sigma(\mathcal{X}\times\dots\times\mathcal{X})$ are symmetrically dependent;

(g) the probability distribution $P_M(dx_1,\dots dx_N)$ on $\mathcal{B}_M$ is expressed in terms of the distribution $R_N(dx_1,\dots dx_N)$ through

$$P_M(dx_1,\dots,dx_M) = \int_{Z^N} \Pi(d\Theta_N)\prod_{j=1}^{M} a(\Theta_N)\sum_{i=1}^{N}\Lambda(z_i,\xi_i,dx_j) \qquad (5.2.1)$$

where

$$\Theta_N = \{z_1,\dots,z_N,\xi_1,\dots,\xi_N\}, \qquad Z=\mathcal{X}\times[-d,d],$$

$$\Pi(d\Theta_N) = R_N(dz_1,\dots,dz_N)F(z_1,d\xi_1)\dots F(z_N,d\xi_N),$$

$$a(\Theta_N) = 1/\sum_{j=1}^{N}\left(f(z_j)+\xi_j\right), \qquad \Lambda(z,\xi,dx) = (f(z)+\xi)Q(z,dx);$$

(h) the global maximizer x^* of f is unique, and there exists $\varepsilon>0$ such that f is continuous in the set $B(x^*,\varepsilon)=B(\varepsilon)$;

(i) μ is a probability measure on $(\mathcal{X},\mathcal{B})$ such that $\mu(B(\varepsilon))>0$ for any $\varepsilon>0$;

(j) there exists $\varepsilon_o>0$ such that the sets $A(\varepsilon)=\{x\in\mathcal{X}: f(x^*)-f(x)\le\varepsilon\}$ are connected for any ε, $0<\varepsilon\le\varepsilon_o$;

(k) for any $x\in\mathcal{X}$ and $k\to\infty$ the sequence of probability measures $Q_k(x,dz)$ weakly converges to $\varepsilon_x(dz)$ which is the probability measure concentrated at the point x;

(l) for any $x\in\mathcal{X}$ and $k\to\infty$, the sequence of probability measures $R(k,N_k,x;dz)$ weakly converges to $\varepsilon_x(dz)$;

(m) for any $\varepsilon>0$ there are $\delta>0$ and a natural k_o such that $P_k(B(\varepsilon))\ge\delta$ for all $k\ge k_o$;

(n) for any $\varepsilon>0$ there are $\delta>0$ and a natural k_o such that $P(k,N_{k-1};B(\varepsilon))\ge\delta$ for all $k\ge k_o$;

(o) the functions f_k (k=1,2,...) are evaluated without random noise;

(p) the transition probabilities $Q_k(x,.)$ are defined by

$$Q_k(x,A)=\int 1_{\left[z\in A,\, f_k(x)\le f_k(z)\right]}T_k(x,dz)+1_A(x)\int 1_{\left[f_k(z)<f_k(x)\right]}T_k(x,dz) \tag{5.2.2}$$

where $T_k(x,dz)$ are transition probabilities, weakly converging to $\varepsilon_x(dz)$ for $k\to\infty$ and for all $x\in\mathcal{X}$;

(q) $P_1(B(x,\varepsilon))>0$ for all $\varepsilon>0$, $x\in\mathcal{X}$;

(r) the transition probabilities $Q_k(x,dz)$ are defined by

$$Q_k(x,dz)=c_k(x)\varphi((z-x)/\beta_k)\mu_n(dz) \tag{5.2.3}$$

where φ is a continuous symmetrical finite density in $\mathbb{R}^n$,

$$\beta_k>0,\qquad \sum_{k=1}^{\infty}\beta_k<\infty,\qquad c_k(x)=1/\int\varphi((z-x)/\beta_k)\mu_n(dz);$$

(s) $f_k(x)=f(x)$, $\xi_k(x)=\xi(x)$, $Q_k(x,dz)=Q(x,dz)$ for each k=1,2,...; and

(t) $f_k(x)=f(x)$ for k=1,2,...

Let us comment now on the assumptions formulated.

Condition (a) requires that the evaluation noises be independent and concentrated on a finite interval. The independence requirement does not seem to be of basic nature, although this issue has not been investigated. The requirement of evaluation noise finiteness, on the contrary, is necessary: if for example, the noise in a point distant from x* is positive and very large, then all the evaluations will take place in the vicinity of this point with high probability and therefore the search process will leave the global extremum vicinity even if it was already there.

Superficially, the condition (b) seems to be restrictive, but the fact that one can perform transformations on observed values enables one to set up optimization problems in a form where (b) is fulfilled. Indeed, if (b) is not met for a regression function f_k, then one can determine $a_k \geq 0$ in such a way that the probability of the event $\{\sup |\xi_k(x)| \leq a_k\}$ is equal or almost equal to 1 and, instead of $y_k(x)$, compute

$$\tilde{y}_k(x) = \tilde{f}_k(x) + \tilde{\xi}_k(x) \geq c_1$$

where

$$\tilde{f}_k(x) = \begin{cases} f_k(x) - y_k(x_0) + 2a_k & \text{if } y_k(x) - y_k(x_0) + 2a_k \geq c_1 \\ c_1 & \text{otherwise} \end{cases}$$

$$\tilde{\xi}_k(x) = \begin{cases} \xi_k(x) & \text{if } y_k(x) - y_k(x_0) + 2a_k \geq c_1 \\ 0 & \text{otherwise} \end{cases}$$

and x_0 is an arbitrary point from $\mathcal{X}$. In this case a function, that is made arbitrarily close to $\max\{c_1, f_k(x)+\text{const}\}$ by appropriate choice of a_k, is the regression function, not $f_k(x)$.

The assumption (c) whose major part is a corollary of (b) will be used for convenience in some formulations.

The assumptions (h), (i) and (j) are natural and non-restrictive. The uniqueness requirement concerning x* imposed in order to simplify the formulations. This requirement may be relaxed: considering the results presented below one can see that, in fact, one deals with distribution convergence to some distribution concentrated on the set

$$\bigcap_{\varepsilon>0} A(\varepsilon) \supset \{\arg\max f(x)\}$$

rather than with convergence to $\varepsilon_{x^*}(dx)$ and, therefore, if the unique maximizer requirement is dropped, then convergence can be understood in this sense.

Conditions (e), (k) and (l) formulate necessary requirements on the parameters of Algorithm 5.1.4 that may always be satisfied.

Assumptions (f), (g) and (s) are nor requirements but only auxiliary tools for formulating Lemma 5.2.1. For Theorem 5.2.1, a similar role is played by (m) and (n) that can be also regarded as conditions imposed on the parameters of Algorithm 5.1.4. They are not constructive, however, and therefore easily verifiable conditions sufficient for validity of (m) or (n) are of interest. The conditions (p), (q), (r) serve these aims for two widely used forms of transition probabilities. The choice of a realization y_k of the random element with the distribution $Q_k(x,dy_k)$ as defined through (5.2.2) implies that first one has to determine a realization ζ_k of the random element with distribution $T_k(x,d\zeta_k)$ and then set

$$y_k = \begin{cases} \zeta_k & \text{if } f_k(\zeta_k) \geq f_k(x) \\ x & \text{otherwise .} \end{cases}$$

The transition probabilities $Q_k(x,.)$ may be determined by (5.2.2) if the functions f_k ($k=1,2,...$) are evaluated without random noise. In presence of random noise the choice through (5.2.3) is a natural way of determining transition probabilities for $\mathcal{X}\subset\mathcal{R}^n$. To obtain a realization y_k of the random vector with the distribution $Q_k(x,dy_k)$ as defined through (5.2.3), one must obtain a realization ζ_k of the random vector distributed with the density φ, to check the occurence $x+\beta_k\zeta_k\in\mathcal{X}$ (and, otherwise, to determine a new realization ζ_k) and to assume $y_k=x+\beta_k\zeta_k$. Note also that the transition probabilities $T_k(x,.)$ of (5.2.2) in the case of $\mathcal{X}\subset\mathcal{R}^n$ can be naturally chosen using (5.2.3).

Let us finally comment on condition (q) presenting both requirements to $\mathcal{X}$ and P_1. This condition is best understandable in the case, when $\mathcal{X}$ is a subset of $\mathcal{R}^n$ of non-zero Lebesgue measure because then (q) means that the P_1-measure of any non-empty ball in R^n with the centre in $\mathcal{X}$ is larger than zero and that $\mathcal{X}$ has no *appendices* , i.e. parts for which there exist non-empty balls in $\mathcal{R}^n$ having centres in these parts and with zero Lebesgue measure of the intersection of these balls and $\mathcal{X}$. The same simple interpretation of (q) is valid in other practically important cases, where $\mathcal{X}$ is a part of an m-surface in $\mathcal{R}^n$, is discrete set, or is a subset of a simply structured functional space (such as L^2 or $C([0,1])$).

5.2.2 *Auxiliary statements*

The two Lemmas below are more important than the stages of proof in Theorem 5.2.1. Lemma 5.2.1 will be used in Sections 5.3 and 5.4, and the statement of Lemma 5.2.2 contains very weak conditions sufficient for weak convergence of the distribution sequence (5.2.9) to $\varepsilon^*(.)=\varepsilon_{x^*}(.)$, that are of independent interest.

Lemma 5.2.1. Let the assumptions (a), (b), (c), (e), (f), (g) and (s) be fulfilled. Then

1. the random variables with the distribution $P_M(dx_1,...,dx_M)$ are symmetrically dependent;

2. the marginal distributions $\tilde{P}_M(dx)=P_M(dx,\mathcal{X},...,\mathcal{X})$ is representable as

$$\tilde{P}_M(dx)=\left[\int\tilde{R}_N(dz)f(z)\right]^{-1}\int\tilde{R}_N(dz)f(z)Q(z,dx)+\Delta_N(dx) \qquad (5.2.4)$$

where $\tilde{R}_N(dz)=R_N(dz,\mathcal{X},...,\mathcal{X})$, the signed measures Δ_N converge to zero in variation for $N\to\infty$ with the rate $N^{-1/2}$, i.e. $\mathrm{var}(\Delta_N)=O(N^{-1/2})$, $N\to\infty$.

Proof. The first statement follows from (f) and (g) and from the definition of symmetrical dependence. Let us represent the marginal distribution $\tilde{P}_M(dx)$ as follows:

$$\tilde{P}_M(dx) = \int_{Z^N} \Pi(d\Theta_N) a(\Theta_N) \sum_{i=1}^{N} \Lambda(z_i, \xi_i, dx) =$$

$$= \sum_{i=1}^{N} \int_{Z^N} \Pi(d\Theta_N) a(\Theta_N) \Lambda(z_i, \xi_i, dx) =$$

$$= \int_{Z^N} \Pi(d\Theta_N)[Na(\Theta_N)] \Lambda(z_i, \xi_i, dx).$$

The resulting relation is represented in the form of (5.2.4) with

$$\Delta_N(dx) = \int_{Z^N} \Pi(d\Theta_N) \Lambda(y_1, \xi_1, dx) \left\{ Na(\Theta_N) - \left[\int \mathfrak{f}(z) \tilde{R}_N(dz)\right]^{-1} \right\}. \tag{5.2.5}$$

We shall show that $\Delta_N \to 0$ in variation for $N \to \infty$. With regard to (e), this convergence is equivalent to the fact that

$$\int |v_N(x)| \mu(dx) \to 0.$$

where

$$v_N(x) = \int_{Z_N} \Pi(\Theta_N)(\mathfrak{f}(z_1) + \xi_1) q(z_1, x) \left\{ Na(\Theta_N) - \left[\int \mathfrak{f}(z) \tilde{R}_N(dz)\right]^{-1} \right\}$$

In order to prove this, we prove that for any $\delta > 0$ and $x \in \mathbf{X}$ there exists $N_* = N_*(\delta, x)$ such that for $N \geq N_*$ there holds

$$|v_N(x)| < \delta. \tag{5.2.6}$$

The symmetrical dependence of the random variables $y(\chi_i) = \mathfrak{f}(\chi_i) + \xi(\chi_i)$ $(i=1,...,N)$ follows from the symmetrical dependence of random elements $\chi_1,...,\chi_N$ and condition (a). In virtue of the above and Loeve (1963), Sec. 29.4, the random variables

$$\beta_N = N^{-1} \sum_{i=1}^{N} y(\chi_i)$$

converge in mean for $N \to \infty$ to some random variable β independent of all β_i, $y(\chi_i)$ $i=1,2,...,$ and

$$E\beta = Ey(\chi_i) = \int f(z)\tilde{R}_N(dz).$$

This can be formulated as follows: for any $\delta_1>0$ there exists $N_*\geq 1$ such that $E|\beta_N-\beta|<\delta_1$ for $N\geq N_*$. Denote $\psi=(f(\chi_1)+\xi(\chi_1))q(\chi_1,x)$. Exploiting the independence of β from β_N and ψ, the condition (a), (b), (c) and (e) for the case (s), and also the fact that

$$\text{vrai sup}\,\psi \leq (\sup f(x) + d)L = \tilde{L}$$

(the constant L is from the condition (e)), we obtain

$$|v_N(x)| = |E(\psi/\beta_N) - E\psi/E\beta| = |E(\psi\beta/\beta_N) - E\psi|/E\beta \leq$$

$$\leq \inf (E\beta)^{-1}|E(\psi\beta/\beta_N - \psi)| \leq c_1^{-1}E(\psi|\beta - \beta_N|/\beta_N) \leq$$

$$\leq c_1^{-1}\text{vrai sup}\,\psi(\text{ vrai sup}\,\beta_N)^{-1}E|\beta - \beta_N| \leq c_1^{-2}\tilde{L}E|\beta - \beta_N|.$$

Thus, if one takes $\delta= \delta_1 c_1^2/\tilde{L}$, then (5.2.6) will be met for $N\geq N_*$. Moreover, it follows from the last chain of inequalities that var $(\Delta_N)\leq c_1^{-2}\tilde{L}E|\beta-\beta_N|$. From the central limit theorem for symmetrically dependent random variables, see Blum et al. (1958), and the inequality

$$E|\beta - \beta_N| \leq N^{-1/2} + \text{vrai sup}|\beta - \beta_N| \Pr\{|\beta - \beta_N| \geq N^{-1/2}\}$$

which is a special case of the inequality given in Loeve (1963), Sec. 9.3, it follows that $E|\beta-\beta_N|=O(N^{-1/2})$, $N\to\infty$. Consequently, var $(\Delta_N)=O(N^{-1/2})$, $N\to\infty$. The lemma is proved.

By substituting f_k, N_k, N_{k+1}, $P(k,N_{k-1};.)$, $P(k+1,N_k;.)$, $P(k+1,N_k;dx)=P(k+1,N_k;dx,\mathcal{X},...,\mathcal{X})$, $\Delta(k,N_k,.)$, respectively, for f, N, M, $R_N(.)$, $P_M(.)$, $\tilde{P}_M(dx)$, and $\Delta_N(.)$ and applying Lemma 5.2.1, we obtain the following assertion.

Corollary 5.2.1. Let (a), (b), (c) and (e) be met. Then for any k=1,2,... and N_k=1,2,... the following equality holds for the unconditional distribution of random elements $x_j^{(k)}$:

$$P(k + 1,N_k;dx) =$$

$$= [\int P(k,N_{k-1};dz)f_k(z)]^{-1} \int P(k,N_{k-1};dz)f_k(z)R(k,N_k,z;dx) \qquad (5.2.7)$$

where

$$R(k,N_k,z;dx) = Q_k(z,dx) + \Delta(k,N_k;dx),$$

for any k=1,2,... the signed measures $\Delta(k,N_k;.)$ converge in variation to zero for $N\to\infty$ with the rate of order $N_k^{-1/2}$.

The next corollary follows from the above.

Corollary 5.2.2. Let (a), (b), (c) and (e) be met. Then for any k=1,2,... the sequence of distributions $P(k+1,N_k;.)$ converges in variation for $N_k\to\infty$ to the limit distributions $P_k(.)$ and

$$P_{k+1}(dx) = \left[\int P_k(dz) f_k(z)\right]^{-1} \int P_k(dz) f_k(z) Q_k(z,dx). \tag{5.2.8}$$

Lemma 5.2.2. Let the conditions (c), (d), (h), i and (j) be met. Then the sequence of distributions

$$f^m(x)\mu(dx)/\int f^m(z)\mu(dz) \tag{5.2.9}$$

weakly converges to $\varepsilon^*(dx) = \varepsilon_{x^*}(dx)$ for $m\to\infty$.

Proof. Set $B_i = B(\varepsilon_i)$,

$$\mathfrak{D}_i = \overline{\mathfrak{X}\setminus B_i} = \{x \in \mathfrak{X}: \|x - x^*\| \geq \varepsilon\}, \qquad i = 1,\ldots,4, \qquad K_1 = \sup_{x\in\mathfrak{D}_1} f(x).$$

Choose an arbitrary value $\varepsilon_1 > 0$. It follows from (h) that for any $\varepsilon_1 > 0$ there exists ε_2, $0 < \varepsilon_2 < \varepsilon_1$, such that

$$K_2 = \inf_{x\in B_2} f(x) > K_1.$$

For any m>0, we have

$$\int_{B_1} \left(\frac{f(x)}{K_1}\right)^m \mu(dx) > \int_{B_2} \left(\frac{f(x)}{K_1}\right)^m \mu(dx) \geq \int_{B_2} \left(\frac{K_2}{K_1}\right)^m \mu(dx)$$

By passing to the limit (for $m\to\infty$) in the inequality

$$\int_{D_1} \mu(dx) / \int_{B_2} \left(\frac{K_2}{K_1}\right)^m \mu(dx) \geq \int_{D_1} \left(\frac{f(x)}{K_1}\right)^m \mu(dx) / \int_{B_1} \left(\frac{f(x)}{K_1}\right)^m \mu(dx) \geq 0$$

we obtain

$$\int_{D_1} f^m(x)\mu(dx) / \int_{B_1} f^m(x)\mu(dx) \to 0,$$

whence

$$c_m \int_{D_1} f^m(x)\mu(dx) \to 0, \qquad c_m \int_{B_1} f^m(x)\mu(dx) \to 1 \tag{5.2.10}$$

where $m \to \infty$, and

$$c_m = 1/\int f^m(x)\mu(dx).$$

Choose now an arbitrary function $\psi(x)$ that is continuous on $\mathcal{X}$. By the definition of weak convergence, it suffices to demonstrate that

$$\lim_{m \to \infty} c_m \int f^m(x)\psi(x)\mu(dx) = \psi(x^*). \tag{5.2.11}$$

For any $\delta>0$ there exists $\varepsilon_3>0$ such that $|\psi(x)-\psi(x^*)|<\delta$ for all $x \in B_3$. Setting $\varepsilon_4=\min\{\varepsilon_1,\varepsilon_3\}$ we have

$$\left|c_m \int f^m(x)\psi(x)\mu(dx) - \psi(x^*)\right| \leq$$

$$\leq c_m \int_{B_4} f^m(x)|\psi(x) - \psi(x^*)|\mu(dx) + c_m \int_{D_4} f^m(x)|\psi(x) - \psi(x^*)|\mu(dx) \leq$$

$$\leq \delta c_m \int_{B_4} f^m(x)\mu(dx) + 2\max|\psi(x)|\, c_m \int_{D_4} f^m(x)\mu(dx)$$

whence the validity of (5.2.11) follows in virtue of (5.2.10). The lemma is proved.

5.2.3 Convergence of the sequences (5.2.7) and (5.2.8) to $\varepsilon^*(dx)$

Below sufficient conditions are determined for weak convergence of the distribution sequences (5.2.7) and (5.2.8) to $\varepsilon^*(dx)$ for $k\to\infty$.

Theorem 5.2.1. Let the conditions (c) through (e) and (h) through (j) be satisfied as well as (k) and (m) or (l) and (n). Then the distribution sequence determined through (5.2.8) (or, respectively, through (5.2.7)) weakly converges to $\varepsilon^*(dx)$ for $k\to\infty$.

Proof. We consider only (5.2.8), because for (5.2.7) the proof is essentially the same, but the formulas are more tedious. Choose from (5.2.8) a weakly convergent subsequence $P_{k_i}(dx)$ (this is possible in virtue of Prokhorov's theorem, see Billingsley (1968)) and denote the limit by $\mu(dx)$ where μ stands for a probability measure on $(\mathcal{X},\mathcal{B})$. It follows from (5.2.8) that the subsequence $P_{k_i+1}(dx)$ weakly converges to the distribution $Q_1(dx)=c_1 f(x)\mu(dx)$, where c_1 is the normalization constant, and, similarly, $P_{k_i+m}(dx)$ converges to $Q_m(dx)$ of the form of (5.2.9). By means of the diagonalization one can show that there exists a subsequence $P_{k_j}(dx)$ that converges weakly to $\varepsilon^*(dx)$.

In virtue of Theorem 2.2. of Billingsley (1968), the set of all finite intersections of open balls with centers from countable and everywhere dense in $\mathcal{X}$ set and with rational radii is the countable set defining convergence. Extract from the above set a subset $\mathfrak{I}$ consisting of sets of Q_1-continuity. Enumerate the elements of $\mathfrak{I}$,

$$\mathfrak{I}=\{A_j\}_{j=1}^{\infty}$$

Fix a monotone sequence of numbers

$$\{\varepsilon_m\}_{m=1}^{\infty}, \qquad \varepsilon_m>0, \qquad \varepsilon_m\to 0 \text{ as } m\to\infty.$$

Since $P_{k_i+m}(A)\to Q_m(A)$ for any $A\in\mathfrak{I}$ as $i\to\infty$, there exists a subsequence

$$R_{1,m}(dx)=P_{k_{i_m}+m}(dx)$$

for which the inequality $|R_{1,m}(A_1)-Q_m(A_1)|<\varepsilon_m$ is valid for any $m=1,2,...$ Extract in a similar manner from the sequence

$$\{R_{j-1,m}(dx)\}_{m=1}^{\infty} \qquad (j=2,3,\ldots.)$$

a subsequence

$$\{R_{j,m}(dx)\}_{m=1}^{\infty} \quad \text{for which} \quad \left|R_{j,m}(A_j)-Q_m(A_j)\right|<\varepsilon_m.$$

For any $A_i \in \mathfrak{I}$, $i \leq m$, the diagonal subsequence $\{R_{m,m}(dx)\}$ has the property $|R_{m,m}(A_i) - Q_m(A_i)| < \varepsilon_m$. This subsequence weakly converges to $\varepsilon^*(dx)$: indeed, for all $A_i \in \mathfrak{I}$,

$$\left|R_{m,m}(A_i) - \varepsilon^*(A_i)\right| \leq \left|R_{m,m}(A_i) - Q_m(A_i)\right| + \left|Q_m(A_i) - \varepsilon^*(A_i)\right|.$$

Here the first term for $m \geq i$ does not exceed ε_m and therefore approaches zero for $m \to \infty$; the second term approaches zero in virtue of Lemma 5.2.2 where (m) plays the role of (i).

Thus there is a subsequence $P_{k_j}(dx)$ converging to $\varepsilon^*(dx)$. It follows from (5.2.8) that $P_{k_j+1}(dx)$ converges to the same limit and thus any subsequence of $P_k(dx)$ converges to this limit. The same holds for the sequence itself. The theorem is proved.

Let us note that all the previously used conditions (with the exception of (m) and (n)) are evident and natural. It is desirable to derive conditions that imply (m) and (n). The corollaries of Theorem 5.2.1 presented below formulate the sufficient conditions for distribution convergence to $\varepsilon^*(dx)$ for two theoretically most important ways of choosing the transition probabilities $Q_k(z,dx)$.

Corollary 5.2.3. Let the conditions (c), (d), (e), (h), (i), (j), (o), (p), (q), (t) and also (k) for the transition probabilities $T_k(x,dz)$ of (5.2.2) be satisfied. Then the sequence of distributions determined by (5.2.8) weakly converges to $\varepsilon^*(dx)$ for $k \to \infty$.

Proof. It follows from (q) and (h) that $P_1(A(\delta))>0$ for any $\delta>0$, and from (5.2.2) that $P_k(A(\delta)) \geq \ldots \geq P_1(A(\delta))$ for any $\delta>0$, $k=2,3,\ldots$ and therefore (m) is met. All conditions of Theorem 5.2.1 concerning the sequence (5.2.8) are satisfied: the corollary is proved.

Corollary 5.2.4. Let the conditions (t), (e), (h), (i), (j), (q) and (r) be satisfied. Then the sequence of distributions determined through (5.2.8) weakly converges to $\varepsilon^*(dx)$ for $k \to \infty$.

Proof. Under our assumptions the distributions (5.2.8) have continuous densities with respect to the Lebesgue measure. Denote them by $p_k(x)$, $k=1,2,\ldots$. It follows from (5.2.3) that $p_k(x)>0$ for any $k \geq 1$ and those $x \in \mathcal{X}$ for which $f(x) \neq 0$. Let us show that (m) is satisfied. Fix $\delta>0$. It follows from (5.2.8) and the finiteness of φ that for any k and $\varepsilon>0$ the following inequality holds

$$P_{k+1}(A(\varepsilon + \varepsilon_k)) \geq P_k(A(\varepsilon)) \tag{5.2.12}$$

where $\varepsilon_k \geq 0$ are defined in terms of β_k and the sizes of the support of density φ,

$$\sum_1^{\infty} \varepsilon_k = \text{const} \sum_1^{\infty} \beta_k < \infty.$$

Choose k_0 so as to make

$$\sum_{k_0}^{\infty} \varepsilon_k < \delta/2 \qquad \text{and let} \qquad \delta_1 = P_{k_0}(A(\delta/2)).$$

For any $k \geq k_0$ we have

$$P_k(A(\delta)) \geq P_{k_0}\left(A\left(\delta/2 + \sum_{k_0}^{k} \varepsilon_i\right)\right) \tag{5.2.13}$$

thus (m) is satisfied: the corollary is proved.

Like Theorem 5.2.1, Corollaries 5.2.3 and 5.2.4 may be reformulated so as to assert convergence of (5.2.7) to $\varepsilon^*(dx)$. Let us reformulate Corollary 5.2.4 which is more non-trivial of the two.

Corollary 5.2.5. Let the conditions formulated in Corollaries 5.2.1 and 5.2.4 be satisfied. Then there exists a sequence of natural numbers N_k ($N_k \to \infty$ for $k \to \infty$) such that the sequence of distributions $P(k+1, N_k; dx)$ determined by (5.2.7) weakly converges to $\varepsilon^*(dx)$ for $k \to \infty$.

Proof. Repeat the proof of Corollary 5.2.4 changing only (5.2.12) and (5.2.13). Let us require that N_k be so large that for any k the inequality

$$P(k+1, N_k; A(\varepsilon + \varepsilon_k)) \geq (1-\delta_k) P(k, N_{k-1}; A(\varepsilon))$$

is satisfied instead of (2.12), where

$$0 < \delta_k < 1 \qquad (k = 1, 2, \ldots), \qquad \sum_{1}^{\infty} \delta_k < \infty. \tag{5.2.14}$$

This is possible in virtue of Lemma 5.2.1. Instead of (5.2.13), we have

$$P(k+1, N_k; A(\delta)) \geq \prod_{i=k_0}^{k} (1-\delta_i) P\left(k_0, N_{k_0 - 1}; A\left(\delta/2 + \sum_{i=k_0}^{k} \delta_i\right)\right) \geq$$
$$\geq \delta_1 \prod_{i=k_0}^{\infty} (1-\delta_i).$$

To complete the proof, it remains to exploit the fact that if (5.2.14) is satisfied then

$$\prod_{k=1}^{\infty} (1-\delta_k) > 0.$$

The corollary is proved.

5.3 *Methods of generations for eigen-measure functional estimation of linear integral operators*

5.3.1 *Eigen-measures of linear integral operators*

Let us introduce some notations that will be used throughout this section.

Let $\mathcal{X}$ be a compact metric space; $\mathcal{B}$ be the σ-algebra of Borel-subsets of $\mathcal{X}$; $\mathcal{M}$ be the space of finite signed measures, i.e. regular (countable) additive functions on $\mathcal{B}$ of bounded variation; $\mathcal{M}_+$ be the set of finite measures on $\mathcal{B}$ ($\mathcal{M}_+$ is a cone in the space $\mathcal{M}$); $\mathcal{M}^+$ be the set of probability measures on $\mathcal{B}(\mathcal{M}^+ \subset \mathcal{M}_+)$; $C_+(\mathcal{X})$ be the set of continuous non-negative functions on $\mathcal{X}$ ($C_+(\mathcal{X})$ is a cone in $C(\mathcal{X})$, the space of continuous functions on $\mathcal{X}$); $C^+(\mathcal{X})$ be the set of continuous positive functions on $\mathcal{X}$ ($C^+(\mathcal{X})$ is the interior of the cone $C_+(\mathcal{X})$); a function $K: \mathcal{X} \times \mathcal{B} \to \mathcal{R}$ be such that $K(.,A) \in C_+(\mathcal{X})$ for each $A \in \mathcal{B}$ and $K(x,.) \in \mathcal{M}_+$ for each $x \in \mathcal{X}$. The analytical form of K may be unknown, but it is required that for any $x \in \mathcal{X}$ a method be known for evaluating realizations of a non-negative random variable $y(x)$ such that

$$Ey(x) = f(x) = K(x, \mathcal{X}), \qquad \text{var}\; y(x) \le \sigma^2 < \infty,$$

and of sampling the probability measure $Q(x,dz)=K(x,dz)/f(x)$ for all $x \in \{x \in \mathcal{X}: f(x) \neq 0\}$.

Denote by $\mathcal{K}$ the linear integral operator from $\mathcal{M}$ to $\mathcal{M}$ by

$$\mathcal{K}\nu(.) = \int \nu(dx) K(x,.). \tag{5.3.1}$$

The conjugate operator $\mathcal{L} = \mathcal{K}^*: C(\mathcal{X}) \to C(\mathcal{X})$ is defined as follows

$$\mathcal{L}h(.) = \int h(x) K(.,dx). \tag{5.3.2}$$

As it is known from the general theory of linear operators (see Dunford and Schwartz (1958)), any bounded linear operator mapping from a Banach space into $C(\mathcal{X})$ is representable as (5.3.2) and $\|\mathcal{L}\| = \|\mathcal{K}\| = \sup f(x)$. Moreover, the operators $\mathcal{K}$ and $\mathcal{L}$ are completely continuous in virtue of the compactness of $\mathcal{X}$ and continuity of $K(.,A)$ for all $A \in \mathcal{B}$.

As is known from the theory of linear operators in a space with a cone, a completely continuous and strictly positive operator $\mathcal{L}$ has eigen-value λ that is maximal in modulus, positive, simple and at least one eigen-element belonging to the cone corresponds to it; the conjugate operator $\mathcal{L}^*$ has the same properties.

In the present case, the operator $\mathcal{L}$ is determined by (5.3.2). It is strictly positive provided that for any non-zero function $h \in C_+(\mathcal{X})$ there exists $m=m(h)$ such that $\mathcal{L}^m h(.) \in C^+(\mathcal{X})$ where $\mathcal{L}^m$ is the operator with kernel

$$\int \dots \int K(\cdot,dx_1)K(x_1,dx_2)\dots K(x_{m-2},dx_{m-1})K(x_{m-1},\cdot).$$

Thus, if the operator $\mathcal{L}=\mathcal{K}^*$ is strictly positive (which is assumed to be the case), the maximal in modulus eigen-value λ of $\mathcal{K}$ is simple and positive; a unique eigen-measure P in $\mathcal{M}^+$ defined by

$$\lambda P(dx) = \int P(dz)K(z,dx) \tag{5.3.3}$$

corresponds to it and λ is expressed in terms of this measure as

$$\lambda = \int f(x)P(dx). \tag{5.3.4}$$

It is evident from (5.3.3) and (5.3.4) that if $\lambda \neq 0$, then the necessary and sufficient condition that P is a unique in $\mathcal{M}^+$ eigen-measure of $\mathcal{K}$ is as follows: P is a unique in $\mathcal{M}^+$ solution of the integral equation

$$P(dx) = \left[\int f(z)P(dz)\right]^{-1} \int P(dz)K(z,dx). \tag{5.3.5}$$

Assume that for any $x_1,x_2,\dots$ from $\mathcal{X}$ an algorithm is known for evaluation of the random variables $\zeta(x_1),\zeta(x_2),\dots$ that are mutually independent and for any $x\in\mathcal{X}$ are such that $E\zeta(x)=h(x)$, var $\zeta(x)\leq\sigma_1^2<\infty$ where h is some function from $C(\mathcal{X})$.

In the following algorithms will be contructed and studied for estimation of the functional

$$\mathfrak{I} = (h,P) = \int h(x)P(dx) \tag{5.3.6}$$

of the probabilistic eigen-measure of the operator $\mathcal{K}$. In virtue of (5.3.4), this problem includes the estimation of the maximal eigen-value of the integral operator (5.3.1) known as estimation of branching process critical parameter or the problem of critical system calculation (see Mikhailov (1966), Khairullin (1980), Kashtanov (1987)); the so-called *method of generations with constant number of particles* was developed for solving this problem. It finds wide applications in practical calculations and for its study a special technique has been developed. This method is studied below by the apparatus of Section 5.2.

Together with its modifications, it will go under the name *generation method.*

The connection between the problem under consideration and that of searching of the global extremum of f is two-fold: in addition to the interrelation between the methodology and technique of the algorithm investigation mentioned above, it turns out that the extremal problems arise from the problems of estimating functionals of eigen-measures as limit problems. This is discussed in the next section.

5.3.2 Closeness of eigen-measures to ε(dx)*

In this section we demonstrate that, in a number of fairly general situations, the problem of determining the global maximizer of f can be regarded as the limit case of determining the eigen-measures P of integral operators (5.3.1) with kernels $K_\beta(x,dz)= f(x)Q_\beta(x,dz)$ where the Markovian transition probabilities $Q_\beta(x,dz)$ weakly converge to $\varepsilon_x(dz)$ for $\beta\to 0$.

In order to relieve the presentation of unnecessary *details*, assume that $\mathcal{X}=\mathbb{R}^n$, $\mu=\mu_n$ and that $Q_\beta(x,dz)$ are chosen by (5.2.3) with $\beta_k=\beta$, i.e.

$$Q_\beta(x,dz)=\beta^{-n}\varphi((z-x)/\beta)\mu_n(dz). \tag{5.3.7}$$

Lemma 5.3.1. Let the transition probability $Q= Q_\beta$ have the form (5.3.7), where φ is a continuously differentiable distribution density on $\mathbb{R}^n$,

$$\int \|x\|\varphi(x)\mu_n(dx)<\infty$$

f be positive, satisfy the Lipschitz condition with a constant L, attain the global maximum at the unique point x^*, and $f(x)\to 0$ for $\|x\|\to\infty$. Then for any $\varepsilon>0$ and $\delta>0$, there exists $\beta>0$ such that $P(B(\delta))\geq 1-\varepsilon$ where P is the probabilistic solution of (5.3.5).

Proof. Multiply (5.3.5) by f, integrate it with respect to $\mathcal{X}$ and let β approach zero. Exploiting the Lipschitz property of f and the inequality from Kantorovich and Akilov (1977), p. 318, obtain that the variance of the random variable $f(\xi_\beta)$ (where ξ_β is a random vector with distribution $P=P_\beta$) tends to zero for $\beta\to 0$ and, therefore $f(\xi_\beta)$ converges in probability to some constant M. To complete the proof, we shall show that $M= f(x^*)$. Assume the contrary; then there exist c, $\varepsilon_1>0$ such that $P(D_\beta)>0$ and $\mu_n(D_\beta)\geq c$ for all $\beta>0$, where

$$D_\beta=\{x\in\mathcal{X}: f_1(x)>1+\varepsilon_1\},\qquad f_1(x)=f(x)/\int f(z)P(dz).$$

From (5.3.5) we obtain

$$P(D_\beta)=\int P(dz)f_1(z)Q(z,D_\beta)\geq$$

$$\geq\int_{D_\beta} P(dz)f(z)Q(z,D_\beta)\geq(1+\varepsilon_1)P(D_\beta)(1-\delta_\beta)$$

with $\delta_\beta \to 0$ for $\beta \to 0$ which follows from (5.3.7): but this contradicts to the assumption. The lemma is proved.

Heuristically, the statement of Lemma 5.3.1 can be illustrated by the following reasoning. In the case studied, P(dx) has a density p(x) that may be obtained as the limit (for $k \to \infty$) of recurrent approximations

$$p_{k+1}(x) = s_{k+1} \int p_k(z) f(z) \varphi_\beta(x-z) \mu_n(dz) \tag{5.3.8}$$

where

$$\varphi_\beta(x) = \beta^{-n} \varphi(x/\beta), \qquad s_{k+1} = 1 / \int p_k(z) f(z) \mu_n(dz).$$

(5.3.8) implies that p_{k+1} is a kernel estimator of the density $s_{k+1} p_k f$, where the parameter β is called *window width* . One can anticipate that for a small β the asymptotic behaviour of densities (5.3.8) should not differ very much from that of distribution densities (5.2.9) which converge to $\varepsilon^*(dx)$, in virtue of Lemma 5.2.2.

Numerical calculations have revealed the fact that in problems resembling realistic ones (for not *too bad* functions f) the eigen-measures $P=P_\beta$ explicitly tend, for small β, to concentrate mostly within a small vicinity of the global maximizer x^* (or the maximizers). Moreover, the tendency mentioned manifests itself already for not *very small* β (say, of the order of 0.2 to 0.3, under the unity covariance matrix of the distribution with the density φ).

The following example illustrates to some extent the issue on closeness of P(dx) and $\varepsilon^*(dx)$.

Example. Let $\mathcal{X}=\mathbb{R}$, $f=N(a,\sigma^2)$, i.e. f is the density of the normal distribution with mean a and variance σ^2, Q(x,dz) be chosen via (5.3.7), where $\varphi=N(0,\beta^2)$. Now, one can readily see from (5.3.5) that the normal distribution density with mean a and variance

$$\beta\left(\beta + \sqrt{\beta^2 + 4\sigma^2}\right)/2$$

is the density of P(dx). A similar result holds for the mutidimensional case, since the coordinates may be considered independently, following an orthogonal transformation.

Now let us establish the possibility of using the generation method of the next subsection for searching the global maximum of f: the essence being that all search points in these algorithms have asymptotically the distribution P(dx) that can be brought *near* to a distribution concentrated at the set $\mathcal{X}^*=\{\arg\max f\}$ of global maximizers by an appropriate choice of the transition probability Q(x,dz). If this is the case, then the majority of points determined by the generation methods are in the limit sufficiently close to $\mathcal{X}^*$ and this is highly desirable for (random) optimization algorithms. (After carrying out a sufficient number of evaluations of f in the vicinity of a point $x^* \in \mathcal{X}^*$, its position can be determined more exactly by constructing a regression model, say, polynomial second-order one). This property is of special importance, when f is evaluated with a random noise. Another positive aspect of the generation methods described below as global optimization algorithms is their easy comparability in terms of closeness of the distributions $P(dx)$ and $\varepsilon^*(dx)$. It is noteworthy that the algorithms of independent random sampling of points in $\mathcal{X}$ (Algorithms 3.1.1 and 3.1.2) can be also classified as belonging to the generation methods described below if one assumes that $Q(x,dz)=P_1(dz)$. For these algorithms $P(dx)= P_1(dx)$ and the points generated by them, therefore, do not tend to concentrate in the vicinity of $\mathcal{X}^*$ and, from the viewpoint of asymptotic behaviour, they are inferior to those generation methods whose distribution $P(dx)$ is concentrated near to $\mathcal{X}^*$. (This way, the situation here is quite similar to the situation concerning the simulated annealing method, see Section 3.3.2.)

5.3.3 Description of the generation methods

Let $P_1(dx)$ be some probability distribution on $(\mathcal{X},\mathcal{B})$, that usually is taken to be uniform, and

$$Q(z,dx) = \begin{cases} K(z,dx)/f(z) & \text{if } f(z) = K(z,\mathcal{X}) \neq 0 \\ P_1(dx) & \text{if } f(z) = 0. \end{cases}$$

It is assumed in the description of Algorithms 5.3.1 through 5.3.3. that $\mathcal{X} \subset \mathbb{R}^n$, $P_1(dx)$ is the uniform distribution on $\mathcal{X}$ and the random variable $\xi(x)$ takes values on the set $\{0,1,\ldots\}$.

The most straightforward algorithm used for a long time for estimating λ is based on the N-fold sampling of the general branching process (see Harris (1963)) defined by $K(z,dx)$ and consists in the following.

Algorithm 5.3.1.

1. Sample $N_1=N$ times $P_1(dx)$, obtain $x_1^{(1)},\ldots,x_{N_1}^{(1)}$ and set $k=1$.
2. Set $i=1$, $N_{k+1}=0$.
3. Sample the random variable $\xi(x_i^{(k)})$, obtain a realization $r_i^{(k)}$.
4. Sample $r_i^{(k)}$ times $Q(x_i^{(k)},dx)$, obtain

$$x^{(k+1)}_{N_{k+1}+1},\ldots,x^{(k+1)}_{N_{k+1}+r_i^{(k)}}.$$

5. Put $N_{k+1}= N_{k+1}+ r_i^{(k)}$.
6. If $i< N_k$, put $i=i+1$ and go to Step 3.
7. If $k\leq I$, put $k=k+1$ and go to Step 2, otherwise the calculations are stopped.

In this method and the subsequent algorithms the number of iterations is defined by a number I.

Since the random vectors $x_i^{(k)}$ asymptotically (for $N_1 \to \infty$, $k \to \infty$) follow the distribution P(dx) (see Harris (1963)), Algorithm 5.3.1 may be applied to the estimation of the functional (5.3.6). The estimator is constructed in a standard fashion (used in Monte-Carlo methods):

$$\lambda \sim \left[\sum_{k=I_0}^{I} N_k\right]^{-1} \sum_{k=I_0}^{I} \sum_{i=1}^{N_k} \zeta\left(x_i^{(k)}\right) \tag{5.3.9}$$

where $1 \le I_0 \le I$. In particular, for $I_0=I$, $I \to \infty$, one obtains the well-known estimator for λ: $\lambda \sim N_{I+1}/N_I$.

From the computational point of view, Algorithm 5.3.1 is inconvenient in the sense that for $\lambda<1$ the process rapidly degenerates (all the particles, i.e. points $x_i^{(k)}$, die or leave $\mathcal{X}$), and for $\lambda>1$ the number of particles grows with k in geometric progression so that their storage soon becomes impossible. Modifications have been made of the algorithm with the purpose of overcoming the latter inconvenience: they are called *generation methods with constant number of particles* and are described below.

Algorithm 5.3.2.

If the amount of descendants (i.e. points $x_i^{(k+1)}$) at the k-th step of Algorithm 5.3.1 is $N_{k+1}>N_1$, then $N=N_1$ *particles* (points) of the next generation are randomly chosen from them. If $N_{k+1}<N$, *particles* of the preceding generation are added in the same manner until their number in the new generation becomes N.

Algorithm 5.3.3.

The new generation is formed in Algorithm 5.3.1 by N-fold random choice with return of N_{k+1} descendants of the *particles* of the previous generation. If $N_{k+1}=0$, the sampling is repeated until $N_{k+1}>0$.

With the use of the above algorithms, the distributions of the random vectors $x_i^{(k)}$ tend to P(dx) (for $k \to \infty$, $N \to \infty$): therefore, with $N_k=N$, the estimate (5.3.9) of (5.3.6) is asymptotically accurate.

Obviously, the efficiency of Algorithm 5.3.2 still depends on λ: this algorithm is not Markovian and, thus is difficult to study. Algorithm 5.3.3 whose rate of convergence can be investigated by a special technique (see Khairullin (1980)) is more attractive. Let us write Algorithm 5.3.3 in a slightly more general and convenient form. To this end, note that at the k-th iteration of Algorithm 5.3.3, the random choice with return from the set

$$\left\{x_1^{(k+1)}, \ldots, x_{r_1^{(k)}}^{(k+1)}, \ldots, x_{N_{k+1}}^{(k+1)}\right\}$$

is performed N times which is equivalent to N-fold sampling of the discrete distribution concentrated on the set

$$\left\{x_1^{(k)},\ldots,x_N^{(k)}\right\} \tag{5.3.10}$$

with probabilities

$$p_i^{(k)}=r_i^{(k)}/\sum_{j=1}^{N} r_j^{(k)} \tag{5.3.11}$$

and subsequent sampling of the Markovian transition probability Q. This interpretation of Algorithm 5.3.3 eliminates the integrality requirement concerning the random variables $\xi(x)$, while their non-negativeness is still prescribed.

Algorithm 5.3.4.

1. Choose a probability distribution $P_1(dx)$ on $(\mathcal{X},\mathcal{B})$, set k=1.
2. Sample N times the distribution $P_k(dx)$ to obtain the points (5.3.10).
3. Sample the random variables $\xi(x_i^{(k)})$ and obtain their realizations $r_i^{(k)}$ (i=1,2,...,N). If

$$\sum_{i=1}^{N} r_i^{(k)} = 0,$$

repeat the sampling until this sum differs from zero.
4. Set

$$P_{k+1}(dx) = \sum_{i=1}^{N} p_i^{(k)} Q\left(x_i^{(k)},dx\right)$$

where $p_i^{(k)}$ are defined through (5.3.11).
5. Put k=k+1; if $k \leq I$, return to Step 2; otherwise stop.

Although Algorithms 5.3.3 and 5.3.4 coincide in the probabilistic sense, their interpretations in terms of *particles* may differ, see Ermakov and Zhigljavsky (1985) (this work describes also some other approaches to the estimation problem of (5.3.6)).

5.3.4 Convergence and rate of convergence of the generation methods

This section deals with the generation method as formulated in the form of Algorithm 5.3.4. The analysis, like that of Section 5.2, relies upon Lemma 5.2.1, because Algorithm 5.3.4 is a special case of Algorithm 5.1.4 (in which $N_k=N$, $f_k=f$, $\xi_k=\xi$,

$Q_k(z,dx)=Q(z,dx)$) and Lemma 5.2.1 defines some fundamental properties of the latter algorithm.

First let us prove an auxiliary assertion.

Lemma 5.3.2. Let the operator $\mathcal{L}=\mathcal{K}^*$, defined by (5.3.2), be strictly positive, λ be the maximal eigen-value of $\mathcal{K}^*$, and P(dx) be the probabilistic eigen-measure corresponding to this eigen-value. The operator $\mathcal{U}$ acting from $\mathcal{M}$ to $\mathcal{M}$ according to

$$\mathcal{U}v(.) = v(.) + \lambda^{-1}P(.)\int f(z)v(dz) - \lambda^{-1}\int v(dz)f(z)Q(z,.) \qquad (5.3.12)$$

has, then, a continuous inverse operator.

Proof. In virtue of Corollary from Kantorovich and Akilov (1977), p. 454, it suffices to demonstrate that the equation $\mathcal{U}v=0$ has no non-trivial solution belonging to $\mathcal{M}$. As follows from Fredholm's alternative result, this is equivalent to $\mathcal{U}^*u=0$, i.e.

$$u(z) + \lambda^{-1}f(y)\int u(x)P(dx) - \lambda^{-1}f(z)\int u(x)Q(z,dx) = 0, \qquad (5.3.13)$$

has no non-trivial solutions belonging to $C(\mathcal{X})$. In order to show this, multiply (5.3.13) by P(dz) and then integrate with respect to $\mathcal{X}$. If u satisfies (5.3.13), then it satisfies $\mathcal{K}^*u=\lambda u$ and $\int u(z)P(dz)=0$: these relations together can be satisfied only by a function that is identically equal to zero since, in virtue of the property mentioned in Section 5.3.1, the non-zero eigen-function of $\mathcal{K}^*$, corresponding to the eigen-value λ, is either strictly positive or strictly negative. That proves the lemma.

Theorem 5.3.1. Let the conditions (a), (b), (c), (e) and (s) of Section 5.2 be satisfied; assume that $Q(z,dx)\geq c_2\mu(dx)$ for μ-almost all $z\in\mathcal{X}$ where $c_2>0$ and the probability measure μ is the same as in the condition (e) of Section 5.2. Then

1) for any N=1,2,... the random elements $a_k=(x_1^{(k)},\dots,x_N^{(k)})$, k=1,2,... , (as defined in Algorithm 5.3.4) constitute a homogeneous Markov chain with stationary distribution $R_N(dx_1,\dots,dx_N)$, the random elements with this distribution being symmetrically dependent;

2) for any $\varepsilon>0$ there exists $N_*\geq 1$ such that for $N\geq N_*$ the marginal distribution

$$R_{(N)}(dx) = R_N(dx,\mathcal{X},\dots,\mathcal{X})$$

differs in variation from P(dx) at most by ε.

Proof. Consider Algorithm 5.3.4 as that for sampling a homogeneous Markov chain in $D=\mathcal{X}^N$. Denote the elements of D by

$$a = (x_1, \dots, x_N), \qquad a_k = \left(x_1^{(k)}, \dots, x_N^{(k)}\right), \qquad b = (z_1, \dots, z_N).$$

The initial distribution of the chain is

$$Q_1(da) = P_1(dx_1) \dots P_1(dx_N).$$

The transition probability is

$$Q(a, db) = \int_{-d}^{d} \dots \int_{-d}^{d} F(x_1, d\xi_1) \dots F(x_N, d\xi_N) \times$$

$$\times \prod_{j=1}^{N} \sum_{i=1}^{N} (f(x_i) + \xi_i) Q(x_i, dz_j) / \sum_{\ell=1}^{N} (f(x_\ell) + \xi_\ell). \tag{5.3.14}$$

Note that this transition probability is Markovian, because the Markovian assumption concerning the transition probability $Q(z,dx)$.

Let us prove that the recurrently defined distributions

$$Q_{k+1}(db) = \int_D Q_k(da) Q(a, db)$$

converge to a limit in variation for $k \to \infty$. Indeed, this follows from (5.3.14) and the conditions of theorem, as

$$Q(a, db) \ge \prod_{j=1}^{N} \sum_{i=1}^{N} \left[c_1 / (c_1 + (N-1)(\max f + d))\right] Q(x_i, dz_j) \ge$$

$$\ge \prod_{j=1}^{N} N c_1 c_2 \left[c_1 + (N-1)(\max f + d)\right]^{-1} \mu(dz_j) = c_3 \tilde{\mu}(db)$$

where

$$\tilde{\mu}(db) = \mu(dz_1) \dots \mu(dz_N)$$

is the probability measure on $(\mathfrak{D}, \mathfrak{B}_N)$, and

$$c_3 = \left[N c_1 c_2 / (c_1 + (N-1)(\max f + d))\right]^N$$

(obviously, $0 < c_3 < 1$). Now, it follows from the above said and Neveu (1964), Supplement to Section V.3, that

$$\Delta_0 = \sup_{a,b \in \mathfrak{D}} \sup_{B \in \mathfrak{B}_N} \{Q(a,B) - Q(b,B)\} \le 1 - c_3.$$

In virtue of the exponential convergence criterion (see Loeve (1963), Section 27.3) the distributions $Q_k(da)$ converge in variation for $k \to \infty$ to a distribution $R_N(da)$ which is the unique positive solution of

$$R_N(da) = \int_D R_N(db) Q(b,da), \tag{5.3.15}$$

moreover, the following relations are valid:

$$\sup_{B \in \mathfrak{B}_N} |Q_k(B) - R_N(B)| \le \Delta_0^{k-1} \le (1 - c_3)^{k-1}. \tag{5.3.16}$$

Using the notation of assumption (g) of Section 5.2, rewrite (5.3.15) as

$$R_N(dx_1, \dots, dx_N) = \int_{Z^N} \Pi(d\Theta_N) \prod_{j=1}^{N} a(\Theta_N) \sum_{i=1}^{N} \Lambda(z_i, \xi_i, dx_j).$$

Using Lemma 5.2.1, we obtain that the random elements a_k with distribution $Q_k(da_k)$ are symmetrically dependent for all k=1,2,... . Let us show that the random elements with distribution $R_N(da)$ are symmetrically dependent as well. Assume for any $B=(B_1,\dots,B_N) \in \mathfrak{B}_N$ that

$$\mathfrak{F}(B) = \left\{ A = \left(B_{i_1}, B_{i_2}, \dots, B_{i_N}\right) \in \mathfrak{B}_N \right.$$

where $(i_1, i_2, \dots, i_N)$ is an arbitrary permutation of (1,2,...,N)}. Choose any two sets $B \in \mathfrak{B}_N$, $A \in \mathfrak{F}(B)$. In virtue of the fact that $Q_k(B)=Q_k(A)$, and for all k=1,2,... (5.3.16) is satisfied, we obtain that

$$|R_N(B) - R_N(A)| \le |R_N(B) - Q_k(B)| + |R_N(A) - Q_k(A)| \le 2\Delta_0^{k-1}.$$

The left hand side of the inequality is not influenced by k, and the right hand side tends to zero for $k \to \infty$. Therefore, $R_N(B)=R_N(A)$ for any $B \in \mathfrak{B}_N$, $A \in \mathfrak{F}(B)$, that is equivalent to the symmetrical dependence of random elements with probability distribution $R_N(da)$.

Now, let us make use again of Lemma 5.2.1 with M=N, $P_N=R_N$: it follows that $R_{(N)}(dx)$ is representable as

$$R_{(N)}(dx) = \left[\int f(z) R_{(N)}(dz)\right]^{-1} \int R_{(N)} dz\, f(z) Q(z,dx) + \Delta_N(dx) \tag{5.3.17}$$

where $\Delta_N \to 0$ in variation for $N\to\infty$ with a rate of the order $N^{-1/2}$.

Finally, let us consider the operator T mapping $\mathcal{M}\times\mathcal{M}$ into $\mathcal{M}$ by

$$T(\Delta, R)(dx) = R(dx) - \left[\int f(z)R(dz)\right]^{-1} \int R(dz)f(z)Q(z,dx) - \Delta(dx).$$

T is Frechet-differentiable with respect to the second argument at the point (0,P), the derivative being $T_R'(0,P)=\mathcal{U}$ where the operator $\mathcal{U}$ is defined by (5.3.12). In virtue of Lemma 5.3.2, the inverse operator $\mathcal{U}^{-1}$ exists and is continuous and, therefore, one can apply the implicit function theorem to (5.3.17). This completes the proof.

If the conditions formulated in Theorem 5.3.1 are met, then one can estimate the convergence rate of P(k,N,dx) to P(dx). Indeed, using (5.3.16) we obtain for all k=1,2,...

$$\sup_{A\in\mathcal{B}} \left| P(k,N;A) - R_{(N)}(A) \right| \le \sup_{B\in\mathcal{B}_N} \left| Q_k(B) - R_N(B) \right| \le (1-c_3)^{k-1}$$

i.e. the distribution $Q_k(dx)$ converge with the rate of geometric progression for $k\to\infty$. On the other hand, it follows from Lemma 5.2.1 and the implicit function theorem (see e.g. Kantorovich and Akilov (1977) §4 of Ch.17) that $\mathrm{var}(R_{(N)}-P)\le cN^{-1/2}$ where c>0 is a constant.

Thus, if the conditions of Theorem 5.3.1 are satisfied, then the distributions P(k,N,dx) of the random elements $x_j^{(k)}$ (j=1,...,N) are close (in variation) to P(dx) for sufficiently large N and k and, therefore, the estimator (5.3.9) is applicable to (5.3.6). In the case of $I_o=I$ the estimate of mean square error is readily derived:

$$E\left(\mathfrak{I} - \frac{1}{N}\sum_{i=1}^{N}\zeta\left(x_i^{(I)}\right)\right)^2 \le \left(\mathfrak{I} - \int h(x)P(I,N,dx)\right)^2 + \frac{1}{N}\left(\sigma_1^2 + \mathrm{var}\, h\left(x_1^{(I)}\right)\right).$$

The first term on the right side of the inequality (the systematic component) can be estimated by means of the above estimates; the order of the second term (the random component) is N^{-1}, $N\to\infty$.

Two facts should be mentioned that follow from the above results and may prove useful, together with the discussion of Section 5.3.2 concerning the generation method as a global optimization algorithm.

First, if f_o is close (in the norm of space $C(\mathcal{X})$) to f, then the solutions of (5.3.5) corresponding to these functions will be close. This follows from the implicit function theorem used in the proof of Theorem 5.3.1, Lemma 5.3.2 and the fact that Fréchet-derivative with respect to the second argument of V acting from $C(\mathcal{X})\times\mathcal{M}$ to $\mathcal{M}$ according to

$$V(f_0, R)(dx) = R(dx) - \left[\int f_0(z)R(dz)\right]^{-1} \int R(dz) f_0(z) Q(z,dx)$$

is $V_R'(f,P)=\mathcal{U}$ in the point (f,P) where $\mathcal{U}$ is defined by (5.3.9). This fact justifies the use of Algorithm 5.3.4 for optimization of estimating function f, if the evaluations of f itself are too expensive.

Assume now that the optimization problem is stated in terms of estimating the point x^* on the basis of a fixed (but sufficiently large) number N_o of evaluations of f (possibly, with a random noise). If one applies Algorithm 5.3.4 and chooses Q (under the assumption $\mathcal{X} \subset \mathbb{R}^n$) according to (5.3.7), then the following may be recommended for choosing the algorithm parameters β, N, and I: first, β is chosen so that

$$\delta^2(\beta) = \mathrm{var}(P - \varepsilon^*) \qquad \text{or} \qquad \sigma^2(\beta) = \int (f(\zeta) - f(x^*))^2 P(d\zeta)$$

be small; then using the convergence rate with respect to N and prior information about f (approximate number of local extrema, Lipschitz constant etc.) N is chosen; finally, I is taken to be the integer part of N_o/N.The closeness of the distribution of random vectors $x_i^{(I)}$ obtained at the last step of Algorithm 5.3.4 to the distribution $\varepsilon^*(dx)$ can be estimated using the estimates of the convergence rate and $\delta^2(\beta)$, whose value can be estimated by means of the results of Section 5.3.2. Indeed, one has

$$\mathrm{var}(P(I,N,.) - \varepsilon^*) \le \mathrm{var}\left(P(I,N,.) - R_{(N)}\right) + \mathrm{var}\left(R_{(N)} - P\right) + \mathrm{var}(P - \varepsilon^*) \le$$

$$\le (1 - c_3)^{I-1} + cN^{-1/2} + \delta^2(\beta) \qquad (5.3.18)$$

On the basis of this inequality, one can formulate the problem of optimal choice of β, N and I, as determination of the minimum of the right part of (5.3.18) under the constraint $NI \le N_o$. The numerical solution of this problem encounters significant computational difficulties, due to the lack of or incomplete knowledge of the constants involved in the estimate.

The investigation of the generation method, as described in this section, is based on the apparatus for analysing global search algorithms. That is why the results obtained are of fairly general character (in the sense of the techniques used), but need somewhat specific assumptions that are natural in constructing global random search algorithms. Let us remark that Mikhailov (1966), Khairullin (1980) and Kashtanov (1987) studied the convergence rate of the generation method as presented in the form of Algorithms 5.3.3 and 5.3.4. The convergence rate estimate with respect to N was proved to have the form $O(N^{-1})$, $N \to \infty$, under assumptions that slightly differ from the above ones and are, generally speaking, more natural for this problem. The approach described is, nevertheless, still sensible because (i) it enables one to detect a number of qualitative features of eigen-measure behaviour and (ii) it may be used, in virtue of its generality, for the investigation of algorithms differing from the generation method (e.g. for sequential algorithms described in the following section).

5.4 Sequential analogues of the methods of generations

The algorithms of this section are modifications of those described in Sections 5.2, 5.3: the basic difference is in the possibility of using the points obtained at earlier iterations - rather than only those obtained at the preceding one - for determination of the subsequent points.

5.4.1 Functionals of eigen-measures

The two algorithms described below are modifications of Algorithm 5.3.4 and can be used like that algorithm for estimating functionals of the form (5.3.6), for sampling random elements whose distribution is the eigen-measure of (5.3.1), or for estimating the maximum of a function.

Algorithm 5.4.1.

1. Sample N_1 times a probability distribution $P_1(dx)$, obtain $x_1,...,x_{N_1}$. Evaluate $y(x_i)$ at these points where $y(x)=f(x)+\xi(x)\geq 0$. If

$$\sum_{i=1}^{N_1} y(x_i) = 0,$$

then the sampling procedure is repeated.

2. Set $k=N_1$.
3. Set

$$P_{k+1}(dx) = \sum_{i=1}^{k} p_i Q(x_i, dx) \qquad \text{where} \qquad p_i = y(x_i)/\sum_{j=1}^{k} y(x_j).$$

4. Obtain a point x_{k+1} by sampling the distribution $P_{k+1}(dx)$, evaluate $y(x_{k+1})$.
5. If $k\leq I$, then the algorithm is terminated. Otherwise return to Step 3, substituting $k+1$ for k.

Similarly to the study of Algorithm 5.3.4, let us consider the asymptotic behavior of the unconditional distributions $P(k,dx)$ corresponding to $P_k(dx)$. Note that $P(k,dx)=P_1(dx)$ for $k\leq N_1$.

Using the symbols of the assumption (g) of Section 5.2, the distributions $P(k+1,dx)$ for $k\geq N_1$ can be represented as

$$P(k+1,dx) = \int_{Z^k} \Pi(d\Theta_k) a(\Theta_k) \sum_{i=1}^{k} \Lambda(z_i, \xi_i, dx) \tag{5.4.1}$$

where $R_k(dx_1,...,dx_k)$ is the joint probability distribution of the random elements $x_1,...,x_k$ and

$$R_k(\mathcal{X},\ldots,\mathcal{X},dx_i,\mathcal{X},\ldots,\mathcal{X}) = P(i,dx) \qquad \text{for } i = 1,..,k.$$

Theorem 5.4.1. Let the conditions (a), (b), (c), (e) and (s) of Section 5.2 be satisfied and the operator $\mathcal{K}^*$ be strictly positive. Then the distributions P(k,dx) defined by (5.4.1) weakly converge for $k\to\infty$ and $N_1\to\infty$ to the eigen-measure of $\mathcal{K}$, P(dx) corresponds to the maximal eigen-value λ.

Proof. The distributions P(k,dx) converge in variation for $k\to\infty$ to some probability distribution S(dx): indeed, the following takes place for any $m\geq 1$, $k\geq N_1+m$:

$$P_{k+m}(dx) = (1-p_{k+m-1})P_{k+m-1}(dx) + p_{k+m-1}Q(x_{k+m-1},dx) =$$

$$=\ldots= \prod_{i=k}^{k+m-1}(1-p_i)P_k(dx) + \prod_{i=k+1}^{k+m-1}(1-p_i)p_kQ(x_k,dx) +\ldots+ p_{k+m-1}Q(x_{k+m-1},dx),$$

$$\text{var}(P(k+m,.)-P(k,.)) \leq \text{vrai sup var}(P_{k+m}-P_k) \leq$$

$$\leq (m-1)\text{vrai sup}\left[\max_{k\leq i\leq k+m-1} p_i\right] \leq (m-1)(\max f + d)/(c_1k) \to 0$$

for $k\to\infty$, i.e. the sequence of distributions {P(k,.)} is fundamental in variation. Let us show that the limit S(dx) of the sequence coincides with P(dx). For any $A\in\mathcal{B}$ we obtain

$$S(A) = \lim_{k\to\infty} P(k+1,A) = \lim_{k\to\infty}\frac{1}{k}\sum_{i=1}^{k}\int_{Z^k}\Pi(\Theta_k)[ka(\Theta_k)]\Lambda(z_i,\xi_i,A).$$

The assertion will be proved, if one can show that for any i=1,2,...,k $\delta_{i,k}\to 0$ is valid (in variation, for $k\to\infty$), where $\delta_{i,k}\in\mathcal{M}$ is defined by

$$\delta_{i,k}(dx) = \int_{Z^k}\Pi(d\Theta_k)\Lambda(z_i,\xi_i,dx)\left[ka(\Theta_k) - 1/\int f(z)S(dz)\right]$$

But this fact is proved by almost literally repeating the second part of the proof of Lemma 5.2.1, considering the fact that, for uniformly bounded sequences of random variables, convergence in probability is equivalent to convergence in mean. The theorem is proved.

The following modification of Algorithm 5.4.1 is more convenient for practical purposes.

Algorithm 5.4.2.

In Algorithm 5.4.1 assume that for k>N (where N is a fixed number)

$$p_i = y(x_i) / \sum_{j=k-N}^{k} y(x_j);$$

in all other details, repeat the operations of Algorithm 5.4.1.

Unlike Algorithm 5.4.1, in the present algorithm a constant number of points (N) is used at each k-th step for k≥N. Of course, the number N should be chosen in Algorithm 5.4.2 much greater, than in Algorithms 5.3.2, 5.3.3 and 5.3.4.

It is worth to note that in Algorithms 5.4.1 and 5.4.2 the distributions $P_{k+1}(dx)$ are expressed in terms of $P_k(dx)$ in a recurrent manner: this facilitates the construction of sampling algorithms for these distributions. We also remark that although further (both numerical and theoretical) studies of Algorithms 5.4.1 and 5.4.2 are required, the first experimental results are encouraging.

5.4.2 Sequential maximization algorithms

Everything said in Section 5.3.2 about the capabilities of Algorithm 5.3.4 as that of optimization, completely applies to Algorithm 5.4.2. Moreover, the latter algorithm may be improved when the objective function f is evaluated without random noise.

Algorithm 5.4.3.

Perform the same operations as in Algorithm 5.4.2, but for k>N disregard the point where f has the least value of all the points included into the set of N points rather than the point x_i, i=k-N.

The question of convergence in this case, for any set of parameters N_1, N is solved in a simple manner: if f is continuous in the vicinity of at least one of its global maximizers and if $Q(x,B(z,\varepsilon)) \geq \delta(\varepsilon) > 0$ for all $x,z \in \mathcal{X}$ and $\varepsilon > 0$, then the sequence $\{f(x_k)\}$ converges in probability to $f(x^*)$ for $k \to \infty$.

As opposed to the parameters $N_1$and N of Algorithm 5.4.2, their counterparts in Algorithm 5.4.3 may be chosen in an arbitrary manner. If it is *a priori* improbable that f reaches a local maximum far from the global one, having a value near to max f, then even N_1=1 and N=1 become acceptable. The resulting algorithm becomes Markovian (see Section 3.3), converges under the above conditions, and the limiting distribution of points x_k is concentrated on the set of global maximizers.

Algorithm 5.4.3 is rather similar to the well-known controlled random search procedure of Price: its essence is as follows (for more details, see Price (1983, 1987)).

At the k-th iteration (k≥N>n), n+1 distinct points $z_1, ..., z_{n+1}$ are chosen from the collection Z_k consisting of N points in store; these points define a simplex in $\mathbb{R}^n$. Here z_1 has the greatest objective function value evaluated so far, and the other n points are randomly drawn from the remaining N-1 points of Z_N. A point x_{k+1} is determined as the image point of the simplex's pole z_{n+1} with respect to the centroid $\bar{z}$ of the other n points, i.e. $x_{k+1}=2\bar{z}-z_{n+1}$. If $x_{k+1}\notin \mathcal{X}$, then the operation is repeated. If

$$f(x_{k+1}) \geq \max_{x\in Z_k} f(x)$$

then let $Z_{k+1}=Z_k$, otherwise x_{k+1} is included into Z_{k+1} instead of the point from Z_k with the least objective function value. From the abovesaid it follows that Algorithm 5.4.3 and Price's algorithm differ only in the way of choosing the next trial points.

5.4.3 Narrowing the search area

As it is demonstrated by numerical calculations, one can somewhat improve convergence of Algorithm 5.2.4 by gradually narrowing the search area, i.e. by tending $Q_k(z,dx)$ to $\varepsilon_z(dx)$ for $k\to\infty$. For sequential algorithms a similar problem is still to be analysed. It is evident only that the narrowing operation should be made slowly enough, to avoid missing the global maximizer. For such a narrowing, one can prove an assertion similar to that of Theorem 5.2.1 on convergence of the sequence of unconditional distributions generated by the algorithm to $\varepsilon^*(dx)$. Since its rigorous proof would require too much space, only the key points are given below.

Let the transition probabilities Q(z,dx) in Algorithm 5.4.1 depend on the iteration count: $Q=Q_k$, for $k\to\infty$ the transition probabilities $Q_k(z,dx)$ be uniformly convergent with respect to z in variation to some transition probability $\tilde{Q}(z,dx)$. Then an analogue of Theorem 5.3.1 is valid - only a term will appear in the estimate var(P(k+m,.)-P(k,.)) that tends to zero. Analysing the equation for the limit measure that has the form (5.3.5) with $K(z,dx)=f(x)\varepsilon_z(dx)$, one obtains that the limit distribution is concentrated on the set $\{x\in\mathcal{X}: f(x)=c\}$, where c is some constant. Finally, in order to show that c=max f, one needs to impose on the transition probabilities $Q_k(z,dx)$ conditions, similar to those of Section 3.2, that ensure that

$$\lim_{k\to\infty} \inf \|x_k - x^*\| = 0$$

with probability one.

CHAPTER 6. RANDOM SEARCH ALGORITHMS FOR SOLVING SPECIFIC PROBLEMS

One may encounter various difficulties while applying the above developed technique to specific problem-classes. In this chapter we shall discuss the ways of overcoming some difficulties arising in constrained, infinite dimensional, discrete and multicriterial optimization problems.

6.1 Distribution sampling in random search algorithms for solving constrained optimization problems

The application of random search methods to constrained optimization with equality-type constraints inevitably involves sampling of some probability distributions on the surface defined by the binding constraints. Below, algorithms are constructed for distribution sampling on general and on some particular manifolds in $\mathfrak{R}^k$, relying upon well-known facts of mathematical analysis, multi-dimensional geometry and probability theory.

6.1.1 Basic concepts

Let $\mathcal{X}$ be a Borel subset of $\mathfrak{R}^n$, $n \geq 1$: ($\mathcal{X}$ is the parameter space), $\mathfrak{B}$ be the σ-algebra of Borel subsets of $\mathcal{X}$, $\mu_n(\mathcal{X})>0$, Φ be a continuously differentiable mapping of $\mathcal{X}$ into $\mathfrak{R}^k$ ($k \geq n$), $\mathfrak{B}_Y$ be the σ-algebra of Borel subsets of the set $Y=\Phi(\mathcal{X})$, $x=(x_1,...,x_n)$, $y=(y_1,...,y_k)$, $\Phi=(\varphi_1,...,\varphi_k)$. With this notation, $y=\Phi(x)$ means that

$$\begin{cases} y_1 = \varphi_1(x_1,\dots,x_n) \\ \dots\dots\dots\dots\dots\dots\dots \\ y_k = \varphi_k(x_1,\dots,x_n). \end{cases}$$

For any $x \in \mathcal{X}$, set

$$d_{ij}(x) = \sum_{\ell=1}^{k} \frac{\partial \varphi_\ell(x)}{\partial x_i} \frac{\partial \varphi_\ell(x)}{\partial x_j} \qquad (i,j = 1,\dots,n),$$

$$D(x) = \left(\det \left\| d_{ij}(x) \right\|_{i,j=1}^{n}\right)^{1/2}$$

where the determinant under the root sign is always non-negative, in virtue of non-negative definiteness of the matrix $\| d_{ij}(x) \|$. The following relation

$$\int_B f(s)ds = \int_{\Phi^{-1}(B)} f(\Phi(x))D(x)\mu_n(dx)$$

is known to be valid (see Schwartz (1967), §10 of Ch.4) for any $\mathfrak{B}_Y$-measurable function f defined on Y and any set B from the set $\{B: B=\Phi(A), A\in\mathfrak{B}\}$ where ds is the surface measure on $Y=\Phi(\mathfrak{X})$. Hence, for any measurable non-negative function f defined on Y and satisfying the condition

$$\int f(\Phi(x))D(x)\mu_n(dx) = 1,$$

the probability measure

$$P(dx) = f(\Phi(x))D(x)\mu_n(dx)$$

induces the distribution $f(s)ds$ on the manifold $Y=\Phi(\mathfrak{X})$. In the important particular case of

$$c = \int D(x)\mu_n(dx) < \infty$$

the distribution

$$P(dx) = c^{-1}D(x)\mu_n(dx)$$

induces a uniform distribution $c^{-1}ds$ on the manifold.

It follows from the abovesaid that distribution sampling on Y is reduced to that in the parameter space $\mathfrak{X}$: indeed, in order to obtain a realization ξ of a random vector in $\mathbb{R}^k$ with the distribution $f(s)ds$ on $(Y,\mathfrak{B}_Y)$, it suffices to obtain a realization ζ of a random vector in $\mathbb{R}^n$ with the distribution $f(\Phi(x))D(x)\mu_n(dx)$ on $(\mathfrak{X},\mathfrak{B})$ and to take $\xi=\Phi(\zeta)$.

6.1.2 *Properties of D(x)*

Below some of the properties of D(x) are listed that will be used in the construction of distribution sampling algorithms on particular surfaces.

In the sequal Φ_i (i=1,2) will be understood as the mapping

$$\Phi_i = \left(\varphi_1^{(i)},\dots,\varphi_k^{(i)}\right)$$

of $\mathfrak{X}_i\subset\mathbb{R}^n$ on $Y_i\subset\mathbb{R}^k$ of the (smooth) function of class C^1,

$$D_i(x) = \left[\det\left\|d_{\ell j}^{(i)}(x)\right\|_{\ell,j=1}^{n}\right]^{1/2}$$

where

$$d_{\ell j}^{(i)}(x) = \sum_{t=1}^{k} \frac{\partial \varphi_t^{(i)}(x)}{\partial x_\ell} \frac{\partial \varphi_t^{(i)}(x)}{\partial x_j}.$$

Lemma 6.1.1.

1) Let $Y_1=Y_2=Y$, H be C^1-diffeomorphism of $\mathcal{X}_1$ on $\mathcal{X}_2$ such that $\Phi_1=\Phi_2 \circ H$, f be a $\mathcal{B}_Y$-measurable function, $f \geq 0$,

$$\int_Y f(s)ds = 1.$$

Then the distributions $f(\Phi_i))D_i(x)\mu_n(dx)$ on $\mathcal{X}_i$ (i=1,2) induce the same distribution $f(s)ds$ on Y.

2) Let $\mathcal{X}_1=\mathcal{X}_2=\mathcal{X}$, $\Phi_1=c\Phi_2+b$ where c is a constant and b is a constant vector. Then $D_1(x)=|c|^n D_2(x)$ for all $x \in \mathcal{X}$.

3) If Φ is linear with respect to each coordinate, then D(x)=const.

4) Let $\mathcal{X}_1=\mathcal{X}_2=\mathcal{X}$, $\Phi_1(x)=B\Phi_2(x)$, where B is an orthogonal (k×k)-matrix (i.e. $BB'=I_k$). Then $D_1(x)=D_2(x)$ for all $x \in \mathcal{X}$.

5) If n=k, then $D(x)=|\partial\Phi/\partial x|$ is the Jacobian of the transformation Φ.

6) If k=n+1, $\varphi_j(x)=x_j$ (j=1,2,...,n), then

$$D(x) = \left[1 + \sum_{i=1}^{n} \left(\partial\varphi_{n+1}(x)/\partial x_i\right)^2\right]^{1/2}.$$

Proof. The first assertion readily follows from Theorem 106 of Schwartz (1967). The second, third and sixth statements are verified by direct calculation of the determinant. The fifth follows from the fact that

$$\left\| d_{ij}(x) \right\|_{i,j=1}^{n} = \frac{\partial \Phi}{\partial x} \frac{\partial \Phi}{\partial x}.$$

Let us now prove the fourth assertion; we have

$$\varphi_i^{(1)}(x) = \sum_{j=1}^{k} b_{ij} \varphi_j^{(2)}(x) \qquad (i = 1, \ldots, k)$$

where

$$\sum_{t=1}^{k} b_{kj} b_{k\ell} = \delta_{j\ell} = \begin{cases} 0 & \text{if } j \neq \ell, \\ 1 & \text{if } j = \ell. \end{cases}$$

Let us show that

$$d_{i\ell}^{(1)}(x) = d_{i\ell}^{(2)}(x)$$

for all $x \in \mathcal{X}$ and $i, \ell = 1, \ldots, n$. Indeed,

$$d_{i\ell}^{(1)} = \sum_{m=1}^{k} \left[\sum_{j=1}^{k} b_{mj} \frac{\partial \varphi_j^{(2)}}{\partial x_i} \sum_{t=1}^{k} b_{mt} \frac{\partial \varphi_t^{(2)}}{\partial x_\ell} \right] =$$

$$= \sum_{j=1}^{k} \sum_{t=1}^{k} \frac{\partial \varphi_j^{(2)}}{\partial x_i} \frac{\partial \varphi_t^{(2)}}{\partial x_\ell} \sum_{m=1}^{k} b_{mj} b_{mt} = \sum_{j,t=1}^{k} \frac{\partial \varphi_j^{(2)}}{\partial x_i} \frac{\partial \varphi_t^{(2)}}{\partial x_\ell} \delta_{jt} = d_{i\ell}^{(2)}.$$

The lemma is proved.

The above results enable us in some cases to simplify the distribution sampling on manifolds. For instance, it follows from the fourth assertion that a uniform distribution on an n-surface may be defined to within an orthogonal transformation. Let us remark finally that the validity of statements 2 - 6 of Lemma 6.1.1 follow the diffeomorphism theorem, see Schwartz (1967).

6.1.3 General remarks on sampling

The relation between a distribution on the set $\mathcal{X}$ having a non-zero Lebesgue measure and on the manifold $Y=\Phi(\mathcal{X})$ enables the reduction of sampling on Y to that on $\mathcal{X}$ which is usually much simpler and can be solved by standard methods (such as the inversion formula, the acceptance-rejection method or other procedures described in Ermakov (1975) or in Devroye (1986)). Some methods of sampling complicated distributions on Y can be used directly on manifold Y without any changes (which, of course, corresponds to applying the method on Y). For convenience of references, let us briefly describe the acceptance-rejection method on Y.

Let two distributions be defined

$$P_i(ds) = \varphi_i(s)ds \qquad (i = 1,2), \qquad g(y) = \varphi_2(y)/(c\varphi_1(y)) \leq 1, \quad c > 0.$$

The acceptance-rejection method consists in sequential sampling of pairs of independent realizations $\{\xi_j, \alpha_j\}$ of a random vector with distribution $P_1(ds)$ and a random variable with the uniform distribution on [0,1], until the inequality $g(\xi_j) \leq \alpha_j$ is observed. The latter of ξ_j is a realization of a random vector with distribution $P_2(ds)$. The mean number of pairs produced by the method equals c, for generating a realization that follows P_2.

Below we describe how this general philosophy can be applied to distribution sampling on various surfaces of first and second orders. In doing so, we shall use the notation

$$Y^+ = \{y = (y_1, \ldots, y_k)' \in Y: y_k \geq 0\}.$$

Obviously, the construction of a sampling algorithm for sampling a part of a distribution on Y^+ and of a similar algorithm for $Y \backslash Y^+$ is equivalent to the construction of a sampling algorithm for the distribution itself.

6.1.4 Manifold defined by linear constraints

Let Y be a non-empty set of the form

$$Y = \{y \in \mathcal{R}^k: G_1 y \leq E_1, G_2 y = E_2\}$$

where G_i is an $m_i \times k$-matrix, E_i is an m_i-vector, and the simultaneous equations $G_2 y = E_2$ define an n-dimensional plane. By virtue of the fourth assertion of Lemma 6.1.1, one can regard the left part of G_2 as an unity matrix I_n. Assume that $x_j = y_j$ (j=1,2,...,n) and obtain from the equalities $G_2 y = E_2$ that $y = G_3 x$ where G_3 is an $k \times n$-matrix whose upper left part is I_n. Assume that

$$\mathcal{X} = \{x \in \mathcal{R}^n: \; G_1 G_3 x \leq E_1\} \quad \text{and} \quad \Phi: \Phi(x) = G_3 x.$$

The mapping Φ transforms the uniform distribution on $\mathcal{X}$ into the uniform distribution on Y, because D(x)=const., by vurtue of the third assertion of Lemma 6.1.1. To sample a uniform distribution on Y, it suffices, therefore, to sample the same on $\mathcal{X}$. A similar (but distribution-specific) assertion holds for non-uniform distributions.

6.1.5 Uniform distribution on an ellipsoid

As it is well known, any ellipsoid can be reduced by means of an orthogonal transformation and translation to the following form

$$Y = \left\{y \in \mathcal{R}^k: \; \sum_{i=1}^{k} a_i^2 y_i^2 = 1\right\} \quad \text{where} \quad a_i > 0 \quad (i = 1, \ldots, k).$$

By virtue of the second and fourth assertions of Lemma 6.1.1, the uniform distribution on Y is uniform on the original ellipsoid. Therefore, it suffices to sample only the first of these distributions.

Set $n=k-1$, $\Phi_1: \mathcal{X} \to Y$,

$$\mathcal{X} = \left\{ x \in \mathcal{R}^n: \ \sum_{i=1}^{n} a_i x_i^2 \le 1 \right\},$$

$$y = \Phi_1(x) = \left(x_1, \ldots, x_{k-1}, \left(1 - \sum_{i=1}^{n} a_i^2 x_i^2\right)^{1/2} / a_k \right)'. \tag{6.1.1}$$

The following distribution

$$P(dx) = c_1 D_1(x) \mu_n(dx), \qquad c_1 = 1/\int D_1(x) \mu_n(dx),$$

corresponds to the uniform distribution on Y^+, where $2/c_1$ is the volume of ellipsoid Y and

$$D_1(x) = \left[\left(1 + \sum_{i=1}^{n} \left(a_i^2/a_k^2 - 1\right) a_i^2 x_i^2 \right) \Big/ \left(1 \sum_{i=1}^{n} a_i^2 x_i^2\right) \right]^{1/2}.$$

The representation $\Phi_1 = \Phi_4 \circ \Phi_3 \circ \Phi_2$ is valid, where Φ_2 is the mapping of $\mathcal{X}$ into

$$B_{(n)} = \left\{ x \in \mathcal{R}^n: \ \sum_{i=1}^{n} x_i^2 \le 1 \right\}$$

as defined by

$$z = \Phi_2(x) = (a_1 x_1, \ldots, a_n x_n)',$$

$$\Phi_3: B_{(n)} \to S^+_{(k)} = \left\{ y \in \mathcal{R}^k: \ \sum_{i=1}^{k} y_i^2 = 1, \ \ y_k \ge 0 \right\},$$

$$y = \Phi_3(x) = \left(x_1, \ldots, x_n, \left(1 - \sum_{i=1}^{n} x_i^2\right)^{1/2} \right)',$$

$$\Phi_4: S^+_{(k)} \to Y^+, \qquad \Phi_4(y) = (y_1/a_1, \ldots, y_k/a_k)'.$$

Mapping by Φ_2, the distribution P(dx) on $\mathcal{X}$ becomes

$$c_2\left[1+\left(\sum_{i=1}^{n} a_i^2 x_i^2\right)\Big/\left(a_k^2\left(1-\sum_{i=1}^{n} x_i^2\right)\right)\right]^{1/2} \mu_n(dx), \qquad c_2=c_1/(a_1 \dots a_n),$$

on $B_{(n)}$ which induces on $S_{(k)}^+$ the distribution

$$F(ds)=c_3\left(\sum_{i=1}^{k} a_i^2 s_i^2\right)^{1/2} ds, \qquad c_3=c_2/a_k.$$

Sampling the above four distributions (on Y^+, $S_{(k)}^+$, $\mathcal{X}$ and $B_{(n)}$) is equivalent. Most naturally, $F(ds)$ on $S_{(k)}^+$ is to be sampled by means of the acceptance-rejection method, where $P_2(ds)=F(ds)$, $P_1(ds)=c_4 ds$, and $c_4=2\pi^{-k/2}\Gamma(k/2+1)$ with $2/c_4$ standing for the volume of unit sphere $S_{(k)}$. Sampling $P_1(ds)$ is well-known, see e.g. Zhigljavsky (1985), Devroye (1986).

Algorithms for sampling non-uniform distributions on an ellipsoid may be constructed in a similar manner.

6.1.6 *Sampling on a hyperboloid*

Let the distribution $F(ds)=f(s)ds$ ($f \geq 0$, $\int_{Y+} f(s)ds=1$) be defined on a hyperboloid which will be written without loss of generality as ($k\geq 3$, $1\leq m\leq k-1$)

$$Y=\left\{y \in \mathcal{R}^k: \sum_{i=1}^{k} b_i y_i^2=1, \quad b_i<0 \quad (i=1,\dots,m), \quad b_j>0 \quad (j=m+1,\dots,k)\right\}.$$

Set $n=k-1$,

$$\mathcal{X}=\left\{x \in \mathcal{R}^n: \sum_{i=1}^{n} b_i x_i^2 \leq 1\right\}, \qquad \Phi: \mathcal{X} \to Y^+,$$

$$y=\Phi(x)=\left(x_1,\dots,x_n,\left[\left(1-\sum_{i=1}^{n} b_i x_i^2\right)\Big/b_k\right]^{1/2}\right)'.$$

The distribution $P_1(dx)=f_1(x)\mu_n(dx)$ on $\mathcal{X}$ corresponds to $F(ds)$ on Y^+, where

$$f_1(x)=f(\Phi(x))D_1(x),$$

$$D_1(x) = \left[1 + \left(\sum_{i=1}^{n} b_i^2 x_i^2\right) \Big/ \left[b_k\left(1 - \sum_{i=1}^{n} b_i x_i^2\right)\right]\right]^{1/2}.$$

Assume further that

$$\mathcal{X}_1 = \left\{z \in \mathcal{R}^n:\ \sum_{i=m+1}^{n} z_i^2 \le 1\right\}$$

and define the mapping $\Phi_1: \mathcal{X} \to \mathcal{X}_1$ as follows:

$$z = \Phi_1(x), \qquad z_i = |b_i|^{1/2} x_i \qquad \text{(for } i = 1,\ldots,m),$$

$$z_j = b_j^{1/2} x_j \left(1 + \sum_{i=1}^{m} b_i x_i^2\right)^{-1/2} \qquad \text{(for } j = m+1,\ldots,n).$$

The Jacobian of the transformation Φ_2, inverse to Φ_1, is

$$D_2(z) = \left(\prod_{i=1}^{n} |b_i|^{-1/2}\right)\left(1 + \sum_{j=1}^{m} z_j^2\right)^{(n-m)/2}$$

and the transformation $\Phi_2 = \Phi_1^{-1}$ itself is defined as

$$x = \Phi_2(z), \qquad x_i = |b_i|^{-1/2} z_i \qquad \text{(for } i = 1,\ldots,m)$$

$$x_j = b_j^{-1/2} z_j \left(1 + \sum_{i=1}^{m} z_i^2\right)^{1/2} \qquad \text{(for } j = m+1,\ldots,n).$$

For mapping Φ_1, the distribution

$$P_2(dz) = f(\Phi(\Phi_2(z)))D_1(\Phi_2(z))D_2(z)\mu_n(dz)$$

on $\mathcal{X}_1$ corresponds to $P_1(dx)$ on $\mathcal{X}$.

Thus, in order to determine a realization ξ of a random vector with distribution $F(ds)$ on Y^+, one has to obtain a realization ζ of a random vector with distribution $P_2(dz)$ on $\mathcal{X}_1$, and to take $\xi = \Phi(\Phi_2(\zeta))$. Sampling distributions on a cylinder $\mathcal{X}_1$ does not encounter any serious difficulty.

6.1.7 *Sampling on a paraboloid*

Let a distribution $F(ds)=f(s)ds$ be defined on a paraboloid

$$Y=\left\{y\in\mathcal{R}^k:\ \sum_{i=1}^{k-1} b_i y_i^2=y_k\right\} \qquad \text{where} \quad b_i\neq 0 \quad \text{for} \quad i=1,\ldots,k-1.$$

Set $n=k-1$, $\mathcal{X}=\mathcal{R}^n$, $\Phi: \mathcal{X}\to Y$, and

$$y=\Phi(x)=\left(x_1,\ldots,x_n,\ \sum_{i=1}^{n} b_i x_i^2\right)'.$$

In this case

$$D(x)=\left(1+4\sum_{i=1}^{n} b_i^2 x_i^2\right)^{1/2}$$

and the distribution $f(\Phi(x))D(x)\mu_n(dx)$ on $\mathcal{R}^n$ corresponds to $F(ds)$ on Y.

6.1.8 *Sampling on a cone*

Let a distribution $F(ds)=f(s)ds$ be defined on a cone

$$Y=\left\{y\in\mathcal{R}^k:\ \sum_{i=1}^{k} b_i y_i^2=0\right\}$$

where $b_i>0$ (for $i=1,\ldots,m$), $b_j<0$ (for $j=m+1,\ldots,k-1$), $b_k=-1$, $k/2\leq m<k$.
Set $n=k-1$,

$$X=\left\{x\in\mathcal{R}^n:\ \sum_{i=1}^{n} b_i x_i^2\geq 0\right\}, \qquad \Phi: \mathcal{X}\to Y^+,$$

$$y=\Phi(x)=\left(x_1,\ldots,x_n,\left(\sum_{i=1}^{n} b_i x_i^2\right)^{1/2}\right)'.$$

In this case

$$D(x)=\left[1+\left(\sum_{i=1}^{n} b_i^2 x_i^2\right)\Big/\sum_{i=1}^{n} b_i x_i^2\right]^{1/2}$$

and the distribution

$$P(dx) = f(\Phi(x))D(x)\mu_n(dx)$$

on $\mathcal{X}$ corresponds to F(ds) on Y^+. If m=k-1, then $\mathcal{X}=\mathcal{R}^n$ and the sampling P(dx) presents no basic difficulties. Now let m<k-1. Assume that

$$\mathcal{X}_1 = \left\{ x \in \mathcal{R}^n: \sum_{i=m+1}^{n} x_i^2 \leq 1 \right\}$$

is an unbounded cylinder, $\Phi_1: \mathcal{X} \to \mathcal{X}_1$, $z=\Phi(x)$: $z_i=b_i^{1/2}x_i$ for i=1,...,m,

$$z_j = |b_j|^{1/2} x_j \left(\sum_{i=1}^{m} b_i x_i^2 \right)^{-1/2} \qquad \text{for} \quad j = m+1, \ldots, n.$$

The Jacobian of the mapping $\Phi_2 = \Phi_1^{-1}$ equals

$$D_2(z) = \left(\prod_{j=1}^{n} |b_j|^{-1/2} \right) \left(\sum_{j=1}^{m} z_j^2 \right)^{(n-m)/2}$$

and the mapping $\Phi_2: \mathcal{X}_1 \to \mathcal{X}$ is defined as follows:

$$x = \Phi_2(z): \quad x_i = b_i^{-1/2} z_i \qquad \text{for} \quad i = 1, \ldots, m,$$

$$x_j = |b_j|^{-1/2} z_j \left(\sum_{i=1}^{m} z_i^2 \right)^{1/2} \qquad \text{for} \quad j = m+1, \ldots, n.$$

For the mapping $\Phi \cdot \Phi_2$, the distribution

$$f(\Phi(\Phi_2(z)))D(\Phi_2(z))D_2(z)\mu_n(dz)$$

on $\mathcal{X}_1$ corresponds to F(ds) on Y^+.

Thus, sampling random vectors on non-degenerate second-order surfaces has been reduced by means of the results obtained at the beginning of this section to sampling on sufficiently simple sets. In the case of degenerate second-order surfaces, the above discussion needs only a slight modification.

6.2 Random search algorithm construction for optimization in functional spaces, in discrete and in multicriterial problems

Many of the above global optimization methods may be applied without essential modifications to problems in which the feasible set $\mathcal{X}$ is discrete or is a subset of a functional space. Nevertheless, these optimization problems have some distinctive features whose careful consideration could contribute to the efficiency of the methods used. For instance, in discrete extremal problems, the usual difficulty lies in selecting metrics on $\mathcal{X}$, while optimizing in functional spaces the basic difficulty is in selecting a method reducing the problem to a finite-dimensional one.

The aim of this section is to consider some aspects of the above-mentioned problem features as well as possible ways of using random search algorithms in multicriterial optimization problems.

6.2.1 *Optimization in functional spaces*

Optimization in functional spaces means that the set $\mathcal{X}$ belongs to a functional space, and $\mathfrak{F}$ is a set of functionals $f: \mathcal{X} \to \mathbb{R}$. For example, numerous problems of mechanics and control are reducible to such optimization. As usual, the consideration of problem specific features enables one to develop specific and fairly efficient solution methods: e.g. the carefully elaborated calculus of variations is usually employed for optimization of integral functionals dependent on an unknown function and its derivatives.

Attempts to apply general numerical methods to optimization in functional spaces usually do not meet with basic difficulties cf. e.g. Vasil'ev (1981). Formally, many of the random (including global) search methods also can be used for functional optimization, although this gives rise to some specific problems related to the distribution sampling in functional spaces.

Two major problems - *uniform* random choice of functions from $\mathcal{X}$ and choice of a function close to a given one - occur in distribution sampling which corresponds to sampling of stochastic processes or fields. The way of *uniform* choice process organization should be completely dependent on $\mathcal{X}$: if $\mathcal{X}$ is a subset of a space of the C[a,b] type, the Wiener measure may be chosen as the *uniform* measure in $\mathcal{X}$; if there is some prior information about smoothness of the functions in $\mathcal{X}$, one of the Gaussian measures whose trajectories have the desired smoothness should be chosen instead of the Wiener measure. Gaussian measures are preferable because they easily let themselves to theoretical study and there exist quite a few algorithms for their sampling.

Sampling a random function close to a given one is equivalent to sampling a random function close to zero: for solving this problem, one can also employ the methods of sampling Gaussian measures in a special manner, but the following two methods provide a more convenient way to parametrization of this problem under the assumption that $\mathcal{X}$ is a subset of one-dimensional functions. The first method is based on the fact that any Gaussian process is representable as

$$x(t,\omega) = \sum_{i=1}^{\infty} \lambda_i \xi_i(\omega) \varphi_i(t) \tag{6.2.1}$$

where λ_i and $\varphi_i(t)$ (i=1,2,...) are the eigen-values and corresponding orthonormalized eigen-functions of the correlation operator of the process, and $\xi_1,\xi_2,...$ are independent normally distributed random variables with zero mean and unit variance. Sampling is defined by a finite number defining the number of terms in the decomposition (6.2.1) or a distribution on the set of these numbers, by fixing a basic set $\{\varphi_i(t)\}$ or several sets among which one is randomly chosen each time, and by fixing small values of $\lambda_1,\lambda_2,...$ or defining a corresponding distribution on the set of these numbers. Now the desired realizations of a close-to-zero Gaussian process are obtained through sampling the random variables $\xi_1,\xi_2,...$ and all the specified distributions and substituting them into (6.2.1). The second method of sampling a close-to-zero random parametrically defined function consists in the preliminary reduction of a given class of random functions z(t) to the class of parametrically defined $z(t,\theta)$, $\theta\in\Theta$, functions with subsequent sampling of parameters $\theta\in\Theta$. As for the parametrization, it is natural to take it as non-linear, since this provides a great variety of forms and profiles of the curves $z(t,\theta)$ with a small number of unknown parameters.

For defining on the parameter set a distribution to be sampled, the only point to mention is that the quasi-uniform distributions (see Kolesnikova et al. (1987)) defined by the condition of equal probability that the cross-section $z(t_0,\theta)$ of $z(t,\theta)$ passes through any point of a given interval $[z_1,z_2]$:

$$\Pr_\theta\{z(t_0,\theta)\in[a,b]\}=(b-a)/(z_2-z_1) \qquad \text{for } z_1\le a\le b\le z_2,$$

seem to be promising. Whereas the quasi-uniform distributions enable one to define uniformity on the set of values of chosen random functions, the uniform distribution on the parameter set could result in a situation where mostly *unreasonable* functions $z(t,\theta)$ are chosen.

When optimizing in functional spaces, one usually passes to finite dimensional optimization. In doing so, the simplest and widely used way consists in substituting functions $x(t)\in\mathcal{X}$, $t\in T$, by their values on a point set $\{t_1,...,t_m\}$, $t_i\in T$, with subsequent (usually, piecewise-linear) approximation of x(t). Here the values $x(t_i)$, i=1,...,m, serve as optimization parameters. This optimization is hardly successful because (i) many points t_i should be defined in order to support the desired accuracy of the approximation, thus the finite dimensional optimization becomes multiparametric and, as a rule, multiextremal: (ii) it is very difficult to take into consideration *a priori* information on the smoothness of the functions from $\mathcal{X}$ - this may be done only by defining the correlations at random choice of unknown parameters of $x(t_i)$, i=1,...,m; and (iii) if the optimization parameters are regarded as independent (this usually is the case, meaning that prior information of smoothness and degree of multiextremality of the optimal function $x^*\in\mathcal{X}$ is completely ignored), the result may be meaningless.

Reduction of the set $\mathcal{X}$ to the set

$$\mathcal{X} \cap \left\{ x(t,\theta) = \sum_{i=1}^{m} \theta_i \varphi_i(t) \right\}$$

of functions, linear with respect to the unknown parameters $\theta=(\theta_1,...,\theta_m)$ to be optimized, is a more general and, heuristically, more attractive approach as compared with the above one. This technique was used, for instance, by Chen and Rubin (1986). Its drawbacks lie in the uncertainty of choosing $\{\varphi_i(t)\}$ and in the presence of all the disadvantages of the first approach (although to a smaller degree).

The transition to functions that are nonlinearly dependent on the parameters provides a third way of passing from infinite dimensional optimization to finite dimensionality. The case in which $T=[0,\infty)$ and the optimal function $x^*(t)\in\mathcal{X}$ is known to be unimodal, smooth, positive for $t>0$, and $x^*(0)=0$, $x^*(+\infty)=0$ indicates a successful transition to functions that are nonlinearly dependent on the parameters. In this case it is natural to narrow $\mathcal{X}$ to the set of parametrically defined functions

$$\mathcal{X} \cap \{x(t) = x(t,\theta)\},$$

where $x(t, \theta)$ are, for instance, functions of the following form

$$x(t,\theta) = \theta_1 t^{\theta_2} \exp\left\{-\theta_3 t^{\theta_4}\right\} \tag{6.2.2}$$

where $\theta_i>0$ $(i=1,2,3,4)$ are unknown parameters. Functions of the form of (6.2.2) are well known to approximate with a high accuracy any function with the above properties. Other methods of parametrization are in existence as well.

6.2.2 *Random search in multicriterial optimization problems*

In complicated practical problems, one often comes across a situation where points of $\mathcal{X}$ (called admissible solutions) are to be compared by multiple criteria rather than by a single one, i.e. the preformance of the decision variants is evaluated by vector functions $F=(f_1,..., f_m)'$, f_i: $\mathcal{X}\to\mathbb{R}$ $(i=1,...,m)$. A vector y from $Y=F(\mathcal{X})\subset\mathbb{R}^m$ will be called an estimate, and an estimate y_* from Y such that there is no $y\in Y$ for which $y\neq y_*$ and $y\leq y_*$ (i.e. the inequality $\leq$ holds for each component) will be referred to as an admissible estimate. The set of admissible estimates is called the Pareto set, and the corresponding set of feasible solutions is called the effective (Pareto-optimal) solution set.

Although a large portion of the literature dealing with multicriterial optimization is devoted to the analysis of the Pareto set $\mathbf{P}$ and similar subsets of Y, for practical purposes the description of the effective solution set $\varepsilon\subset\mathcal{X}$ for given criteria F is of primary importance. In practice, one can obtain this description only by forming a fairly representative sample of ε and then approximating it (e.g. by piecewise linear or piecewise quadratic approximation): considerations below are given to the generation of such a sample.

Some points of the set $\mathcal{E}$ may be obtained solving minimization problem of scalar trade-off criteria

$$f_{\lambda}(x) = \lambda F(x), \quad \lambda \in S_m = \left\{ \lambda = (\lambda_1, \ldots, \lambda_m) : \lambda_i \geq 0, \quad \sum_{i=1}^{m} \lambda_i = 1 \right\}$$

(consideration may be given also to other trade-off criteria sets); note that all the above points constitute the entire set $\mathcal{E}$, if all the scalar criteria f_i (i=1,...,m) are convex. Although the individual minimizers of $f_{\lambda}(x)$ are not sufficient for describing the whole set $\mathcal{E}$, they are usually sufficient for obtaining a representative sample from $\mathcal{E}$.

Theoretically, the minimization of $f_{\lambda}(x)$ under various parameters λ is the simplest way of determining points from $\mathcal{E}$. This way, however, may be inefficient, because of the difficulties related to the solution of the single-criterion problems: indeed, for non-convex (but, possibly, uniextremal) criteria f_i (i=1,...,m), the trade-off criteria $f_{\lambda}(x)$ are (already) multiextremal. Moreover, small variations of λ can result in abrupt changes of the global minimizer location. We indicate a number of approaches that are based on ideas of random search, readily yield themselves to algorithmization and programming and might prove useful in solving the problem of describing $\mathcal{E}$.

First, it is natural to take such values λ that are independent realizations of a random vector uniformly distributed on S_m.

Second, the search of minima of $f_{\lambda}(x)$ should be carried out for all functions in a simultaneous manner, all the points obtained being tested for feasibility and rejected, if found inadmissible.

Third, local random search is possible in various versions of the algorithm below if $\mathcal{E}$ is *a priori* known to be connected. At the first iteration, one or more points of $\mathcal{E}$ are determined as the minimizers of criteria $f_{\lambda}(x)$; having several points of $\mathcal{E}$, construct at the k-th iteration Λ_k - the linear hull of points - and then determine in a random fashion several new points in $\mathcal{X}$ near and far from Λ_k, compute F in these points, test the points'membership to $\mathcal{E}$, and pass to the next iteration.

Fourth, the following natural approach (see Sobol and Statnikov (1981)) may be used for numerical construction of $\mathcal{E}$: choose in $\mathcal{X}$ a grid Ξ_N with good uniform characteristics (e.g. a Π_{τ}-grid), compute F in the grid points, then construct (in a finite number of comparisons) the set of effective points on Ξ_N that is an approximation of $\mathcal{E}$, for large N. Substitution of the set $\mathcal{X}$ by a finite number of points selected in a special manner is the essence of this approach.

Fifth, using the results of Chapter 4 and 7, one can formulate random search algorithms, where the prospectiveness of uniformly chosen points and corresponding subsets of $\mathcal{X}$ is determined in a probabilistic, rather than deterministic manner - after determination of confidence intervals for the values of $f_\lambda(x)$ minima, under randomly chosen $\lambda \in S_m$.

The following algorithm is simple, but it reflects the principal features discussed. Begin with determining N independent realizations $x_1,\dots,x_N$ of the random vector uniformly distributed in $\mathcal{X}$; compute next the vector function F in these points and take ℓ independent values $\lambda_{(1)},\dots,\lambda_{(\ell)}$ of the random vector with uniform distribution on S_m. Perform for all $i=1,\dots,\ell$ the following procedure: choose that point x_i^* of the obtained ones where

$$f_{\lambda_{(i)}}(x)$$

is maximal; construct a ball $B(x_i^*,\varepsilon)$ with the centre in x_i^* and radius ε selected so that only a small portion of the points $x_1,\dots,x_N$ belong to the ball; using (4.2.25) construct the confidence interval of a fixed level $1-\gamma$ for

$$\min_{x\in\mathcal{X}\setminus B(x_i^*,\varepsilon)} f_{\lambda_{(i)}}(x)$$

using $x_j \in \mathcal{X}\setminus B(x_i^*,\varepsilon)$ $(j=1,\dots,N)$; if

$$f_{\lambda_{(i)}}(x_i^*)$$

falls into the confidence interval, then construct in a similar manner in the set $\mathcal{X}\setminus B(x_i^*,\varepsilon)$ the ball of the same radius until

$$f_{\lambda_{(i)}}(x_i^*)$$

falls into the last confidence interval. The union of all the balls constructed is regarded as an approximation to the set of efficient points. This union can be considered as the *truncated* search domain on which the same operation can be performed, but with smaller ε. For $N,\ell\to\infty$ and natural regularity (say, convexity) conditions on f_i $(i=1,\dots,m)$, one can directly prove that the probability of missing global minimizer of a randomly chosen $f_\lambda(x)$ tends to a value not exceeding γ.

6.2.3 *Discrete optimization*

Development and application of the general global search philosophy to discrete optimization requires a more thorough consideration of the specific features of the feasible set, than in the continuous case $\mathcal{X}\subset\mathbb{R}^n$. In particular, the choice of a suitable metric (or pseudo-metric) ρ on a discrete set $\mathcal{X}$ is usually not evident, not defined *a priori*. Metrics, then may be chosen in the course of the search being, by the following heuristic considerations. For optimization of a function f, the pseudo-metric

$$\rho_f(z,x) = |f(z) - f(x)|, \qquad x, y \in \mathcal{X},$$

generated by f, is the most suitable one. The Lipschitz-constant of f in ρ_f is 1; the ball

$$B(x,\varepsilon,\rho_f) = \{z \in \mathcal{X} : \rho_f(x,z) \le \varepsilon\}$$

contains the points of $\mathcal{X}$, where $f(z)$ differs from $f(x)$ by ε at most. The objective of minimizing the Lipschitz constant leads to the following way of choosing a metric, best fitted to a given function f, from a given set $\Pi=\{\rho_1,\ldots,\rho_\ell\}$ of metrics or pseudo-metrics: fix a number k_0; normalize the metrics ρ_i so that the number of points in all the balls $B(x,\varepsilon,\rho_i)$ of a fixed radius ε (say, $\varepsilon=1$) approximately equals k_0; estimate the Lipschitz constant of f by (2.2.33) for each $\rho_i \in \Pi$ and use for the optimization that metric for which the Lipschitz constant estimate is minimal.

The set Π might contain metrics that are standard for the sets under consideration and, if possible, pseudometrics ρ_g for functions g that are *close* in some sense to f (e.g. estimates of f).

Let us demonstrate how the methods of branch and probability bounds may be used for optimizing discrete functions. The only basic difficulty lies in the fact that the order statistics apparatus cannot be formally applied. Indeed, since there is a positive probability of getting on the bound (i.e. a global optimizer of f),there is no sense to consider conditions like (a) of Section 4.2, as the corresponding limit simply does not exist. Moreover, the order staistics of discrete distributions do not form a Markov chain, see Nagaraja (1982). But for a very great number m of points of $\mathcal{X}$ (only this case is of practical interest), one may assume that the probability of getting exactly to the bound is negligible and that the discrete c.d.f. $F(t)=\Pr\{x: f(x)<t\}$ is approximated, to a high accuracy, by a continuous c.d.f. for which one can apply extreme order statistics theory and, therefore, the apparatus of Sections 4.2 - 4.4 and of Chapter 7. This way, for very large m the accuracy of statistical inference made under the continuity assumption diminishes only insignificantly. Such an approach was applied in Dannenbring (1977), Golden and Alt (1979), Zanakis and Evans (1981) (but, of course, these works do not use the more up-to-date statistical apparatus described in Chapters 4, 7).

Now we shall turn to the study of relative efficiency of a discrete random search following Ustyuzhaninov (1983) in the formulation of the problem.

6.2.4 Relative efficiency of discrete random search

Below a problem-type is considered, for which an exact result on the relative efficiency of random search can be obtained.

Let $\mathcal{X}$ and Y be finite sets, $\mathcal{X}$ consists of m elements, $f: \mathcal{X}\to Y$ be an algorithmically given function. A non-empty point set $M(f)$ is associated with f, e.g. $M(f)$ is a set of its global minimizers. It is required to find a point $x\in M(f)$ through sequential evaluations of f. Function f is known to belong to a finite set $\mathcal{F}$ consisting of ℓ functions. Thus, a table with ℓ rows and m columns is given in such a way that a function $f\in\mathcal{F}$ corresponds to a row and a point $x\in\mathcal{X}$ to a column: the value $f(x)$ is the intersection of row f and column x.

Consider now the scheme of random search algorithms solving the given problem.

An algorithm involves two stages. The first stage contains not more than s(a) iterations. At each k-th iteration either the transition to the second stage is established (perhaps, at random) or a point of evaluating f is chosen, according to a (conditional) probability distribution

$$P_k(dz \mid x_1,\dots,x_{k-1}, f(x_1),\dots,f(x_{k-1}))$$

At the second stage a point x is chosen that is thought to belong to $M(f)$: herein a mistake is possible.

An algorithm is called deterministic, if all indicated probability distributions are degenerate (i.e. all decisions are deterministic), otherwise it is called probabilistic.

Define the problem π as a pair $\pi=(\mathcal{F},\mathcal{M})$ consisting of a class of functions $\mathcal{F}$ and a family of sets $\mathcal{M}=\{M(f), f\in\mathcal{F}\}$. A deterministic algorithm d is called applicable to a problem π if for any function $f\in\mathcal{F}$, it gives an element from $M(f)$ without mistake. Denote the class of all such algorithms as $D(\pi)$. The maximal number N(d) of evaluations of f needed using an algorithm d is called the problem hardness with respect to d. The problem complexity is defined as

$$N_D = \min_{d\in D(\pi)} N(d).$$

Let $p(M(f) \mid f,r)$ be the probability of that the application of an algorithm r to a function f yields the correct solution. We shall call an algorithm r p-solves a problem π if $p(M(f) \mid f,r)\geq p$ for any $f\in\mathcal{F}$. Denote by $R_p(\pi)$ the class of p-solving algorithms for a problem π and set

$$N_p = \inf_{r\in R_p(\pi)} N(r).$$

Since $D(\pi)\subset R_p(\pi)$, thus $N_p\leq N_D$.

According to Ustyuzhaninov (1983), we shall characterize the relative hardness of random search for problem π by the ratio $\gamma= N_D/(N_p+1)$ which is called the problem index. We shall call a problem π full, if for any unknown function $f\in\mathcal{F}$ and any collection $\Xi=\{x_1,...,x_k\}$ of points $\mathcal{X}$, with the property $\Xi\cap M(f)\neq\emptyset$, given Ξ, $f(x_i)$ $(i=1,...,k)$ and the fact that $\Xi\cap M(f)\neq\emptyset$ an element $x\in\Xi\cap M(f)$ can be indicated. E.g., a full problem is the search problem for a point in which the value of a multiextremal function is distinct from a known minimal value not more than ε.

Consider full problems $\pi=\pi_m$ in which classes $\mathcal{F}$ consist of m different functions and let $0<p<1$, $q=1-p$. Then Ustyuzhaninov (1983) showed that

$$0\leq\gamma\leq(\log(pm))/\log(1/q)+1/p \tag{6.2.3}$$

for the index γ of a full problem π. The inequality (6.2.3), however, is wrong. In order to show this, it suffices to let p tend to zero on the right hand side of (6.2.3):

$$\lim_{p\to 0}\left(\frac{\log pm}{-\log(1-p)}+\frac{1}{p}\right)=-\infty$$

This leads to the contradiction $0\leq-\infty$.

While proving (6.2.3), the following error was done: The estimate $N_D\leq\gamma/(1-\gamma)$, trivially following from the equality $N_D=\gamma(N_p+1)$ and the inequality $N_p\leq N_D$, is valid only for $\gamma<1$, but Ustyuzhaninov (1983) used this estimate and applied for $\gamma\leq 1/p$ (thus including cases where $\gamma\geq 1$).

Following the reasoning of the mentioned work, let us prove the correct inequality with respect to γ.

If $0\leq\gamma\leq 1-1/m$, then $\gamma\leq N_D\leq\gamma/(1-\gamma)$. The same holds for $0\leq\gamma<1$, since the values of γ can not fall into the interval $(1-1/m,1)$. Let $\gamma\geq 1$. Taking the valid inequality below from Ustyuzhaninov (1983)

$$N_D\leq mq^v/(1-v/\gamma),\qquad 0\leq v\leq[\gamma], \tag{6.2.4}$$

choose v to make its right side minimal. The function

$$\varphi(v)=mq^v/(1-v/\gamma)$$

attains its minimal value for $v_o=\gamma+1/\log q$. If $q>1/\ell\approx 0.368$, then

$$\gamma+1/\log q<[\gamma],$$

thus the minimum of φ is reached at the point v_o. In the inequality (6.2.4) set $v=[\gamma+1/\log q]$, express this quantity as $v=\gamma+1/\log q-\sigma$ (where $0\leq\sigma<1$), substitute it into (6.2.4) obtaining $N_D\leq mq^{\gamma}\gamma/\chi(\sigma)$, where

$$\chi(\sigma) = q^{\sigma-1/\log q}(\sigma - 1/\log q).$$

The function χ decreases on the interval [0,1), since $\chi'(0)=0$ and $\chi'(\sigma)<0$ for $\sigma>0$. Therefore $\chi(\sigma)>\chi(1)$ for $0\leq\sigma<1$, whence $N_D\leq mq^{\gamma}\gamma/\chi(1)$. Using this inequality and the relation $\gamma\leq N_D$, one obtains $\gamma\leq\kappa(q,m)$ where

$$\kappa(q, m) = 1 - (\log m)/\log q - (1 + \log((-\log q)/(1 - \log q)))/\log q. \tag{6.2.5}$$

If $v_o\leq 0$, i.e. $\gamma\leq -1/\log q$, then the inequality (6.2.4) is equivalent to $N_D\leq m$: this way $N_D\leq m-1$ for $m\leq q(1-1/\log q)$. If $q<1/\ell$, then the restrictions on m are absent. Summing up the above, the indices γ of full problems π_m satisfy the following conditions:

a) $\gamma\leq m-1$ for $q>1/e$, $m\leq q(1-1/\log q)$,

b) $-1/\log q\leq\gamma\leq\kappa(q,m)$ for $q>1/e$, $m>q(1-1/\log q)$,

c) $0\leq\gamma\leq\kappa(q,m)$ for $q\leq 1/e$

where $\kappa(q,m)$ is determined by (6.2.5).

As it follows from condition b), the maximal acceleration due to using random search is s+1/s-1, and it is attained for $q=\exp(s/(s-1))$, where s is the solution of the equation $ms=\exp(-s)$.

PART 3. AUXILIARY RESULTS

CHAPTER 7. STATISTICAL INFERENCE FOR THE BOUNDS OF RANDOM VARIABLES

This chapter describes and studies statistical procedures having a significant place in the theory of global random search (these procedures are included into some of the methods of Chapter 4). Most attention is paid to linear statistical procedures that are simple to realize.

Section 7.1 states the problems and considers the case, when the value of the tail index of the extreme value distribution is known. This is a typical situation in global random search theory, as follows by the results of Section 4.2.6.

Section 7.2 deals with the case, when the value of the tail index is unknown. Finally, Section 7.3 investigates the asymptotic normality and optimality of the best linear estimates.

7.1 Statistical inference when the tail index of the extreme value distribution is known

7.1.1 Motivation and problem statement

Let an independent sample $Y=\{y_1,...,y_N\}$ from the values of a random variable y with a continuous c.d.f. F(t) be given, y being concenrtrated on an interval [L,M] where $-\infty\leq L<M<\infty$. More precisely, let the upper bound

$$M = \text{vrai sup } y = \inf\{a: F(a) = 1\} \tag{7.1.1}$$

of y be a.s. finite and consider statistical inference for M throughout the chapter. Statistical inference for the lower bound L=vrai inf y under the supposition $L>-\infty$ are constructed similarly or can be elementarily obtained from results related to M and hence will not be considered here.

Various approaches can be used for constructing statistical inference. In particular the parametric approach is based on the supposition that an anlytical form of the c.d.f. F(t) is accepted (identified), but some of its parameters are unknown being estimated from the sample. This approach is of moderate interest in the context of global random search theory and is not considered below.

The *yearly maximum* approach, thoroughly described by Gumbel (1958) and studied in some works, the most valuable of which is Cohen (1986), involves the partition of the sample Y of size $N=\ell r$ into r equal subsamples, and the estimation of the extreme value distribution parameters as if the maximal elements of the subsamples have this distribution, is generally inefficient as well. After realizing this inefficiency, many works have been devoted to the problem: among them the work of Robson and Whitlock (1964) was the first and most of them became known in the 1980's. These works (including those of the present author) construct and study statistical inference about M, based on using some k+1 elements of the maximal order statistics

$$\eta_N \geq \eta_{N-1} \geq \ldots \geq \eta_{N-k} \qquad \text{(where } k = k(N) \geq 1, \ k/N \to 0 \text{ for } N \to \infty) \tag{7.1.2}$$

from the set $H=\{\eta_1,...,\eta_N\}$ of the order statistics derived from the sample Y, rather than using the whole sample Y. The following arguments may be put in favour of this approach: (i) according to a heuristic reasoning, the order statistics not belonging to (7.1.2) are *far enough* from M and so not carry much valuable information concerning M, (ii) the theoretical considerations presented below as well as the corresponding numerical results show that if k is sufficiently large, then further increase of k (under $N\to\infty$) may lead to either an insignificant improvement or even the deterioration of the statistical procedure precision (due to the inaccuracy of computations), (iii) using (7.1.2), the asymptotic theory of extreme order statistics can be applied to construct and investigate the decision procedures. We shall confine ourselves to this approach being a semiparametric one (at present not seeing any satisfactory alternative).

The following classical result from the theory of extreme order statistics (see, for instance, Galambos (1978)) is essential for the theory presented later.

Theorem 7.1.1. Assume that the conditions below are satisfied:

a) the upper bound (7.1.1) is finite and the function $V(v)=1-F(M-v^{-1})$, $v>0$, regularly varies at infinity with some exponent $-\alpha$, $0<\alpha<\infty$, i.e.

$$\lim_{v\to\infty} V(tv)/V(v) = t^{-\alpha}$$

holds for each $t>0$.

Then

$$\lim_{N\to\infty} F^N(M+(M-\theta_N)z) = \Psi_\alpha(z) \tag{7.1.3}$$

where θ_N is the (1-1/N)-quantile of the c.d.f. F, i.e.

$$F(\theta_N) = 1 - 1/N, \tag{7.1.4}$$

$$\Psi_\alpha(z) = \begin{cases} \exp\{-(-z)^\alpha\} & \text{for } z<0 \\ 1 & \text{for } z \geq 0 \end{cases} \tag{7.1.5}$$

The asymptotic relation (7.1.3) implies that (under $N\to\infty$) the sequence of random variables $(\eta_N-M)/(M-\theta_N)$ converges in distribution to the random variable with the c.d.f. (7.1.5) which is called the extreme value c.d.f. and together with $\Psi_\infty(z)=\exp\{-\exp(-z)\}$ is the only nondegenerate limit of the c.d.f. sequences for $(\eta_N+a_N)/b_N$ (where $\{a_N\}$ and $\{b_N\}$ are arbitrary numerical sequences).

The parameter α of the c.d.f. (7.1.5) is called the tail index (or the shape parameter) of the extreme value distribution. This section deals with the case where condition a) of Theorem 7.1.1 holds and the value of the parameter α is known and Section 7.2 will treat the case of unknown α.

7.1.2 Auxiliary statements

In this chapter we shall often use the following well-known results. The first is the so-called Rényi representation (see e.g. Galambos (1978))

$$\eta_{N-i} = F^{-1}\left(\exp\left(-\left(\zeta_0/N + \zeta_1/(N-1) + \ldots + \zeta_i/(N-i)\right)\right)\right) \tag{7.1.6}$$

where $\zeta_0, \zeta_1, \ldots, \zeta_i$ are mutually independent exponential random variables with the density e^{-x}, $x \geq 0$. The second is the asymptotic relation

$$M - F^{-1}(x) \sim (M - \theta_N)(-N \log x)^{1/\alpha}, \qquad N \to \infty, \tag{7.1.7}$$

due to Cook (1979), being a simple consequence of (7.1.3). Here the notation $a_N \sim b_N$ $(N \to \infty)$ means that the limit values

$$\lim_{N\to\infty} a_N \qquad \text{and} \qquad \lim_{N\to\infty} b_N$$

exist and are equal. (Here the convergence in distribution is considered, if $\{a_N\}$ and $\{b_N\}$ are sequences of random variables.)

The next statement immediately follows from (7.1.6) and (7.1.7).

Lemma 7.1.1. If a) holds then for $N \to \infty$, $i/N \to 0$ the asymptotic equality

$$M - \eta_{N-i} \sim (M - \theta_N)(\zeta_0 + \ldots + \zeta_i)^{1/\alpha} \tag{7.1.8}$$

is valid where $\zeta_0, \ldots, \zeta_i$ are as in (7.1.6).

Note that according to classical result,the sum $\zeta_0 + \ldots + \zeta_i$ in (7.1.8) follows gamma distribution with the density

$$x^i e^{-x} / \Gamma(i+1), \qquad x > 0. \tag{7.1.9}$$

Let us prove now two basic auxiliary statements.

Lemma 7.1.2. Let assumption a) hold, $\alpha>1$, $N\to\infty$, $i^2/N\to 0$ (for instance, i may be constant). Then

$$M - E\eta_{N-i} \sim (M - \theta_N) b_i \tag{7.1.10}$$

where

$$b_i = \Gamma(i + 1 + 1/\alpha)/\Gamma(i + 1).$$

Proof. It is well known (cf. e.g. Galambos (1978)) that the density of the order statistic η_{N-i} is

$$p_{N-i}(x) = NC^i_{N-1} F^{N-i-1}(x)\varphi(x)(1 - F(x))^i$$

where $\varphi(x)$ is the density of the c.d.f. $F(x)$. Then it follows that

$$E\eta_{N-i} = NC^i_{N-1} \int_{-\infty}^{\infty} xF^{N-i-1}(x)\varphi(x)(1 - F(x))^i dx.$$

Substituting the variable $y=F(x)$, we obtain

$$E\eta_{N-i} = NC^i_{N-1} \int_0^1 F^{-1}(y) y^{N-i-1}(1 - y)^i dy.$$

Using (7.1.7) we have for $N\to\infty$

$$E\eta_{N-i} \sim MI_1 - (M - \theta_N) I_2$$

where

$$I_1 = NC^i_{N-1} \int_0^1 x^{N-i-1}(1 - x)^i dx = 1,$$

$$I_2 = NC^i_{N-1} \int_0^1 (- N \log x)^{1/\alpha}\, x^{N-i-1}(1 - x)^i dx =$$

$$= \frac{N!N^{1/\alpha}}{i!(N - i - 1)!} \int_0^1 (\log (1/x))^{1/\alpha} x^{N-i-1}(1 - x)^i dx$$

Changing the variable $y=\log(1/x)$, we have

$$I_2 = \frac{N!\,N^{1/\alpha}}{i!\,(N-i-1)!}\int_0^\infty y^{1/\alpha} e^{-Ny}\left(e^y - 1\right)^i dy.$$

Setting x=Ny and using the asymptotic equality

$$\exp\{x/N\} - 1 \sim x/N, \qquad N \to \infty, \tag{7.1.11}$$

we obtain

$$I_2 = \frac{(N-1)!}{i!(N-i-1)!}\int_0^\infty x^{1/\alpha} e^{-x}\left(e^{x/N} - 1\right)^i dx \sim$$

$$\sim \frac{(N-1)!}{i!\,(N-i-1)!\,N^i}\int_0^\infty x^{i+1/\alpha} e^{-x}\, dx = h_{i,N} b_i$$

where

$$h_{i,N} = \frac{N-1}{N}\,\frac{N-2}{N}\cdots\frac{N-i}{N} \to 1 \tag{7.1.12}$$

for $N\to\infty$, $i^2/N\to 0$. Substituting the derived expressions for I_1 and I_2 we obtain (7.1.10): the lemma is proved.

Note that results on the speed of convergence to the extreme value distribution are contained in §2.10 of Galambos (1978), Smith (1982), Falk (1983): the results mentioned show that in the case $\alpha \le 1$ one has

$$M - E\eta_{N-i} = o\left(N^{-1}\right), \qquad N \to \infty$$

instead of (7.1.10). This case is of minor interest and will not be considered.

Lemma 7.1.3. Let assumption a) hold. Then for $\alpha \ge 1$, $N\to\infty$, $i^2/N\to 0$, we have

$$E(\eta_{N-i} - M)(\eta_{N-j} - M) \sim (M - \theta_N)^2 \lambda_{ij} \tag{7.1.13}$$

for $i \ge j$ where

$$\lambda_{ij} = \frac{\Gamma(i+1+2/\alpha)\Gamma(j+1+1/\alpha)}{\Gamma(i+1+1/\alpha)\Gamma(j+1)}. \tag{7.1.14}$$

Proof. The joint probability density of η_{N-i} and η_{N-j} for $i \ge j$ is

$$p_{N-i,N-j}(x,y) = AF^{N-i-1}(x)\varphi(x)(F(y)-F(x))^{i-j-1}\varphi(y)(1-F(y))^{j}$$

where $x \le y$ and for brevity we use the notation

$$A = N!/(j!(N-i-1)!(i-j-1)!).$$

Changing the variables similarly to those in the proof of Lemma 7.1.2, using the asymptotic expressions (7.1.7), (7.1.11) and (7.1.12), introducing the notations

$$B = (M-\theta_N)^2/(j!(i-j-1)!), \qquad D = A(M-\theta_N)^2 N^{2/\alpha},$$

and integrating by parts at the end of the proof we obtain the chain of relations

$$E(\eta_{N-i}-M)(\eta_{N-j}-M) =$$

$$= \int_{-\infty}^{\infty} dy \int_{-\infty}^{y} (x-M)(y-M)p_{N-i,N-j}(x,y)dx =$$

$$= A\int_0^1 dv\int_0^v \left(F^{-1}(u)-M\right)\left(F^{-1}(v)-M\right) u^{N-i-1}(v-u)^{i-j-1}(1-v)^j du \sim$$

$$\sim D\int_0^1 dv\int_0^v \left(\log\tfrac{1}{u}\right)^{1/\alpha}\left(\log\tfrac{1}{v}\right)^{1/\alpha} u^{N-i-1}(v-u)^{i-j-1}(1-v)^j du =$$

$$= D\int_0^{\infty} dy\int_y^{\infty} x^{1/\alpha}y^{1/\alpha}\exp\{-x(N-i)\}(e^{-y}-e^{-x})^{i-j-1}(1-e^{-y})^j e^{-y}dx =$$

$$= D\int_0^{\infty} dv\int_v^{\infty} \left(\tfrac{u}{N}\right)^{1/\alpha}\left(\tfrac{v}{N}\right)^{1/\alpha} e^{-u}(\exp\{(u-v)/N\}-1)^i \times$$

$$\times\left(\frac{1-e^{-v/N}}{e^{-v/N}-e^{-u/N}}\right)^j \frac{1}{1-\exp\{-(u-v)/N\}}\frac{du}{N^2} \sim$$

$$\sim h_{i,N}B\int_0^{\infty} dv\int_v^{\infty} v^{j+1/\alpha}\, u^{1/\alpha}\, e^{-u}(u-v)^{i-j-1}du \sim$$

$$\sim B\int_0^{\infty} v^{j+1/\alpha}dv \int_0^{\infty}(z+v)^{1/\alpha} e^{-(z+v)}z^{i-j-1}dz =$$

$$= B\int_0^{\infty}\int_0^{\infty} v^{j+2/\alpha} e^{-v}(1+z/v)^{1/\alpha} e^{-z} z^{i-j-1} dvdz =$$

$$= B\int_0^{\infty}\int_0^{\infty} (1+y)^{1/\alpha} y^{i-j-1} v^{2/\alpha+i} \exp\{-v(y+1)\}dvdy =$$

$$= B\int_0^{\infty} (1+y)^{1/\alpha} y^{i-j-1} dy \int_0^{\infty} \frac{x^{2/\alpha+i}}{(y+1)^{2/\alpha+i}} e^{-x} \frac{dx}{y+1} =$$

$$= B\Gamma(i+1+2/\alpha)\int_0^{\infty} (y+1)^{-i-1-1/\alpha} y^{i-j-1} dy = (M-\theta_N)^2 \lambda_{ij}.$$

The lemma is proved.

The asymptotic equalities (7.1.10) and (7.1.13) were formulated by Cook (1979), without proof and applying the inexact assumption $i/N \to 0$ instead of $i^2/N \to 0$ for $N \to \infty$. The inadequacy of the condition $i/N \to 0$ ($N \to \infty$) follows by the fact that in this case (7.1.12) does not hold. Indeed, if $N \to \infty$, $i/N \to 0$, but $i^2/N \to 0$ not necessarily holds, we have

$$h_{i,N} = \prod_{j=1}^{i}(1-j/N) = \exp\left\{\sum_{j=1}^{i}\log(1-j/N)\right\} \sim$$

$$\sim \exp\left\{-\sum_{j=1}^{i} j/N\right\} \sim \exp\{-i^2/(2N)\}$$

instead of (7.1.12). So if i^2/N does not approach zero while $N \to \infty$ then

$$\lim_{N\to\infty} h_{i,N} \neq 1$$

and the asymptotic equalities (7.1.10) and (7.1.13) are not valid. Instead of them the asymptotic representations of the following corollary hold.

Corollary 7.1.1. Let assumption (a) hold, $\alpha>1$, $N \to \infty$, $i \geq j$, $i/N \to 0$. Then

$$M - E\eta_{N-i} \sim (M-\theta_N)\exp\{-i^2/(2N)\}b_i,$$

$$E(\eta_{N-i}-M)(\eta_{N-j}-M) \sim (M-\theta_N)^2 \exp\{-i^2/(2N)\}\lambda_{ij}.$$

In particular, if the limit value

$$I = \lim_{N\to\infty} \left(i^2/N\right)$$

exists and is finite then

$$M - E\eta_{N-i} \sim \left(M - \theta_N\right)\exp\{-I/2\}b_i,$$

$$E(\eta_{N-i} - M)(\eta_{N-j} - M) \sim \left(M - \theta_N\right)^2 \exp\{-I/2\}\lambda_{ij}.$$

We shall omit further the multiplicator $\exp\{-i^2/(2N)\}$ or $\exp\{-I/2\}$ supposing i^2/N for $N\to\infty$. Modifications of the statements given below for the more general case, when $i^2/N\to I<\infty$ while $N\to\infty$, are obvious.

7.1.3 *Estimation of M*

We suppose again that condition a) holds and the value of the tail index $\alpha>1$ is known. Under these suppositions we consider below various estimates of the maximum (essential supremum) M=vrai sup y of a random variable y. The estimates use the k+1 upper order statistics (7.1.2) corresponding to an independent sample of y.

The most well-known estimates of M are linear, having the form

$$M_{N,k} = \sum_{i=0}^{k} a_i \eta_{N-i}. \tag{7.1.15}$$

Lemma 7.1.2 states that if a) holds, $\alpha>1$, $N\to\infty$ and $k^2/N\to 0$, then

$$EM_{N,k} = \sum_{i=0}^{k} a_i E\eta_{N-i} = M\sum_{i=0}^{k} a_i - \left(M - \theta_N\right)a'b + o\left(M - \theta_N\right) \tag{7.1.16}$$

where $a=(a_0,\ldots,a_k)'$, $b=(b_0,\ldots,b_k)'$ and

$$b_i = \Gamma(i + 1 + 1/\alpha)/\Gamma(i + 1).$$

Since the c.d.f. F(t) is continuous, thus $M\neq\theta_N$ and $M-\theta_N\to 0$ for $N\to\infty$. Using now (7.1.16), the finiteness of the variances of η_{N-i} for i=1,...,k and the Chebyshev inequality we obtain the following statement.

Proposition 7.1.1. Let assumption a) hold, $N\to\infty$, $k^2/N\to 0$. Then the estimate (7.1.15) is consistent if and only if the equality

$$a'\lambda = \sum_{i=0}^{k} a_i = 1 \tag{7.1.17}$$

holds, where $\lambda=(1,1,...,1)'$.

Note that the statement is true for arbitrary α (not only for the case $\alpha>1$).

Lemmas 7.1.2 and 7.1.3 are immediately followed by the next statement.

Proposition 7.1.2. Let assumption a) hold, $\alpha>1$, $N\to\infty$, $k^2/N\to 0$. Then for consistent linear estimates $M_{N,k}$ of the form (7.1.15), the asymptotic expressions

$$M - EM_{N,k} \sim (M-\theta_N)a'b, \tag{7.1.18}$$

$$E(M_{N,k}-M)^2 \sim (M-\theta_N)^2 a'\Lambda a \tag{7.1.19}$$

are valid, where Λ is the symmetrical matrix of order $(k+1)\times(k+1)$ with elements λ_{ij} defined for $i\geq j$ by (7.1.14).

We refer to (7.1.17) as the consistency condition and to

$$a'b = 0 \tag{7.1.20}$$

as the unbiasedness requirement. Certainly, if (7.1.20) holds, then the estimate (7.1.15) still remains biased, but for $\alpha>1$ its bias has the order $O(N^{-1})$, as $N\to\infty$.

Choose the right hand side of (7.1.19), as the optimality criterion for consistent linear estimates (7.1.15) in the case $\alpha>1$. The optimal consistent estimate $M_{N,k}{}^*$ and the optimal consistent unbiased estimate $M_{N,k}{}^+$ (the word *consistent* will be dropped) are determined by the vectors

$$a^* = \arg\min_{a:a'\lambda=1} a'\Lambda a$$

and

$$a^+ = \arg\min_{\substack{a:a'\lambda=1\\ a'b=0}} a'\Lambda a$$

correpondingly. The explicit forms of a^* and a^+ are

$$a^* = \Lambda^{-1}\lambda/(\lambda'\Lambda^{-1}\lambda), \tag{7.1.21}$$

$$a^{+} = \frac{\Lambda^{-1}\lambda - (b'\Lambda^{-1}\lambda)\Lambda^{-1}b/(b'\Lambda^{-1}b)}{\lambda'\Lambda^{-1}\lambda - (b'\Lambda^{-1}\lambda)^{2}/(b'\Lambda^{-1}b)} \tag{7.1.22}$$

From (7.1.21) there follows also

$$(a^*)'\Lambda a^* = 1/(\lambda'\Lambda^{-1}\lambda). \tag{7.1.23}$$

These expressions are easily derived viz. introducing Lagrange multipliers. (7.1.21) and (7.1.22) are due to Cook (1980) and Hall (1982), respectively.

If k is not small enough, then the vectors (7.1.21) and (7.1.22) are hard to calculate, since the determinant of the matrix $\Lambda=\Lambda_k$ is almost zero. Namely, the following statement holds.

Proposition 7.1.3. For $\alpha>1$, the weak equivalence

$$\det \Lambda_k \asymp \alpha^{-2k}(k!)^{-2(1-1/\alpha)}, \qquad k \to \infty \tag{7.1.24}$$

holds.

The proposition above will be proved in Section 7.3.2. Fortunately, simple expressions for the components of the vectors (7.1.21) and (7.1.22) can be derived (see Section 7.3.1). Using these, the components of $a^*=(a_o^*,\ldots,a_k^*)'$ and $a^+=(a_o^+,\ldots,a_k^+)'$ can be easily calculated for any $\alpha>0$ and $k=1,2,\ldots$. The following tables present them for $\alpha=2,5,10$ and $k=2,4,6$.

Table 3. Components of a*.

k	α	a_o^*	a_1^*	a_2^*	a_3^*	a_4^*	a_5^*	a_6^*	$(a^*)'\Lambda a^*$
2	2	1.636	0.273	-0.909					0.545
4	2	1.314	0.219	0.146	0.109	-0.788			0.438
6	2	1.157	0.193	0.129	0.096	0.077	0.064	-0.716	0.386
2	5	2.598	1.237	-2.835					0.384
4	5	1.811	0.863	0.719	0.634	-3.027			0.268
6	5	1.439	0.685	0.571	0.504	0.458	0.424	-3.082	0.213
2	10	4.246	2.895	-6.140					0.354
4	10	2.766	1.886	1.714	1.607	-6.972			0.231
6	10	2.092	1.427	1.297	1.216	1.158	1.113	-7.303	0.175

Table 4. Components of a^+.

k	α	a_0+	a_1+	a_2+	a_3+	a_4+	a_5+	a_6+	$(a+)'\Lambda a+$
2	2	2.000	0.333	-1.333					0.667
4	2	1.440	0.240	0.160	0.120	-0.960			0.480
6	2	1.224	0.204	0.136	0.102	0.082	0.068	-0.816	0.408
2	5	3.500	1.667	-4.167					0.518
4	5	2.117	1.008	0.840	0.741	-3.706			0.313
6	5	1.598	0.761	0.634	0.560	0.509	0.471	-3.533	0.236
2	10	6.00	4.091	-9.091					0.501
4	10	3.332	2.272	2.065	1.936	-8.606			0.278
6	10	2.377	1.621	1.473	1.381	1.315	1.265	-8.432	0.198

It is also proved in Section 7.3 that the optimal linear estimates are asymptotically Gaussian and efficient. In particular, the following result holds (as being included in Theorem 7.3.1).

Let the (somewhat stricter than a)) condition

$$F(t) = 1 - c_0(M-t)^{\alpha} + o\left((M-t)^{\alpha}\right), \qquad t \to M, \tag{7.1.25}$$

hold where $2 \le \alpha < \infty$ and c_0 is a positive number, $N \to \infty$, $k \to \infty$, $k^2/N \to 0$. Then the asymptotic normality relation

$$\left(M - M^*_{N,k}\right)/\sigma_{N,k} \to N(0,1) \tag{7.1.26}$$

is valid where

$$\sigma^2_{N,k} = \begin{cases} (1-2/\alpha)\, c_0^{-2/\alpha} N^{-2/\alpha} k^{-(1-2/\alpha)} & \text{if } \alpha > 2 \\ (c_0 N \log k)^{-1} & \text{if } \alpha = 2 \end{cases}$$

is the asymptotic mean square error of $M_{N,k}{}^*$ (i.e. $E(M_{N,k}{}^*-M) \sim \sigma_{N,k}{}^2$ for $N \to \infty$, $k \to \infty$, $k^2/N \to 0$).

An analogous result is valid for the estimates $M_{N,k}{}^+$ and for related others. Thus Csörgö and Mason (1989) show it for linear estimates determined by the vectors a with components

$$a_i = \begin{cases} v_i & \text{for } \alpha > 2,\ i = 0,\dots,k-1 \\ v_k + 2 - \alpha & \text{for } \alpha > 2,\ i = k \\ 2/\log(k+1) & \text{for } \alpha = 2, i = 0 \\ (\log(1+1/i))/\log(k+1) & \text{for } \alpha = 2, 1 \le i \le k-1 \\ (\log(1+1/k) - 2)/\log(k+1) & \text{for } \alpha = 2,\ i = k \end{cases} \tag{7.1.27}$$

where

$$v_j = (\alpha - 1)(k+1)^{2/\alpha - 1}\left((j+1)^{1-2/\alpha} - j^{1-2/\alpha}\right).$$

Hall (1982) does the same for the maximum likelihood estimates that are determined by (4.2.22) and (4.2.23).

For practical use, the very simple estimate

$$M^0_{N,k} = (1 + C_k)\eta_N - C_k\eta_{N-k}$$

may be also recommended where $C_k = b_o/(b_k - b_o)$ is found from the unbiasedness condition (7.1.20). For large values of α (this case is of great practical significance)

$$\Gamma(k + 1 + 1/\alpha) - \Gamma(k+1) \sim \alpha^{-1}\Gamma'(k+1), \qquad \alpha \to \infty,$$

and so

$$C_k \sim \frac{\Gamma(1) + \alpha^{-1}\Gamma'(1)}{1 + \alpha^{-1}\psi(k+1) - \Gamma(1) - \alpha^{-1}\Gamma(1)} = \frac{\alpha + \psi(1)}{\psi(k+1) - \psi(1)}, \qquad \alpha \to \infty,$$

where $\psi(.) = \Gamma'(.)/\Gamma(.)$ is the psi-function and $-\psi(1) \approx 0.5772$ is the Euler constant.

7.1.4 Confidence intervals for M

As it was indicated in Section 4.2.3, to construct confidence intervals for M one can use the asymptotic normality of the mentioned estimates or the Chebyshev inequality that together with (7.1.19), (7.1.25) give (4.2.24). The only point to be investigated here is the estimate (4.2.20) of c_o.

Proposition 7.1.4. Let (7.1.25) hold, $N \to \infty$, $k \to \infty$, $k^2/N \to 0$. Then the estimate

$$\hat{c}_0 = k(\eta_N - \eta_{N-k})^{-\alpha}/N \tag{7.1.28}$$

for the constant c_o is consistent, asymptotically unbiased, and

$$\lim_{k\to\infty} kE(\hat{c}_0 - c_0)^2 = c_0^2. \tag{7.1.29}$$

Proof. Using (7.1.25) and Lemma 7.1.1 we have

$$M - \eta_{N-i} \sim (c_0 N)^{-1/\alpha}\left(\sum_{j=0}^{i} \zeta_j\right)^{1/\alpha}$$

for $N\to\infty$ and each $i\le k$, where $\zeta_0, \zeta_1, \ldots$ are independent and exponentially distributed with the density e^{-x}, $x>0$. Therefore

$$E\hat{c}_0 \sim c_0 kE\left(\left((\zeta_0+\ldots+\zeta_k)^{1/\alpha} - \zeta_0^{1/\alpha}\right)^{-\alpha}\right) \sim$$

$$\sim c_0 kE(\zeta_0+\ldots+\zeta_k)^{-1} = c_0, \qquad k\to\infty.$$

Analogously

$$kE(\hat{c}_0 - c_0)^2 \sim k^3 c_0^2 E(\zeta_0+\ldots+\zeta_k)^{-2} - kc_0^2 \sim$$

$$\sim k^3 c_0^2 \frac{1}{k!}\int_0^\infty t^{k-2}e^{-t}dt - kc_0^2 = \left(k^2/(k-1) - k\right)c_0^2 \sim c_0^2, \qquad k\to\infty.$$

The consistency of (7.1.28) immediately follows from (7.1.29) and the Chebyshev inequality. The proposition is proved.

An alternative way of estimating c_0 is due to Hall (1982) and consisting in setting

$$\tilde{c}_0 = (k+1)/\left(N(\hat{M} - \eta_{N-k})^{\alpha}\right) \tag{7.1.30}$$

where $\hat{M}$ is an estimate for M. If $\hat{M}$ is the maximum likelihood estimate (MLE) for M, then (7.1.30) is the MLE for c_0.

The above approach to construct confidence intervals for M can be used only, if N is so large that k also can be chosen large enough. For moderate values of N, this approach is not suitable and another one, due to Cook (1979), Watt (1980) and Weissman (1981) can be recommended. It is based on the following statement which we prove since the above references do not contain the proof.

Lemma 7.1.4. Let assumption a) hold, $N \to \infty$, k be fixed. Then the sequence of random variables

$$D_{N,k} = (M - \eta_N)/(\eta_N - \eta_{N-k}) \quad (7.1.31)$$

converge in distribution to the random variable χ_k with c.d.f.

$$F_k(u) = 1 - (1 - (u/(1+u))^{\alpha})^k, \qquad u \geq 0. \quad (7.1.32)$$

Proof. Set $w=(1+1/u)^{\alpha}-1$ and note that $w \geq 0$.
Using (7.1.8) and (7.1.9) we obtain

$$\Pr\{D_{N,k} < u\} \sim \Pr\left\{\zeta_0^{1/\alpha} / \left[\left(\sum_{i=0}^{k} \zeta_i\right)^{1/\alpha} - \zeta_0^{1/\alpha}\right] < u\right\} =$$

$$= \Pr\{(\zeta_1 + \ldots + \zeta_k)/\zeta_0 > w - 1\} =$$

$$= \frac{1}{\Gamma(k)} \int_0^{\infty} dx \int_{wx}^{\infty} \exp\{-x - t\} t^{k-1} dt =$$

$$= \frac{1}{\Gamma(k)} \int_0^{\infty} dx \int_{w}^{\infty} \exp\{-xy\} x^{k-1} y^{k-1} e^{-x} x dy =$$

$$= k \int_w^{\infty} y^{k-1} (y+1)^{-k-1} dy = 1 - (w/(w+1))^k.$$

The lemma is proved.

The δ-quantile $r_{k,\delta}$ of the c.d.f. (7.1.32) as determined from $F_k(r_{k,\delta})=\delta$, is easily derived:

$$r_{k,\delta} = 1/\left(\left(1 - (1-\delta)^{1/k}\right)^{-1/\alpha} - 1\right).$$

This way the confidence level of the interval

$$\left[\eta_N + r_{k,\delta_1}(\eta_N - \eta_{N-k}), \eta_N + r_{k,\delta_2}(\eta_N - \eta_{N-k})\right] \quad (7.1.33)$$

for M asymptotically equals δ_2-δ_1 where $0 \le \delta_1 < \delta_2 \le 1$ (for $N \to \infty$, $k/N \to 0$. In many applications (including global random search theory), the one-sided confidence intervals

$$[\eta_N, \eta_N + r_{k,1-\gamma}(\eta_N - \eta_{N-k})] \tag{7.1.34}$$

for M (which can be obtained from (7.1.33) by setting $\delta_1=0$, $\delta_2=1-\gamma$) are most naturally used. Let us investigate their average length in order to conclude on the necessary number of the order statistics η_{N-i}.

Proposition 7.1.5. Let condition a) hold, $N \to \infty$, $k^2/N \to 0$, and $\gamma \in (0,1)$ be a fixed number. Then the average length of the confidence interval (7.1.34) for M asymptotically equals $(M-\theta_N)\varphi(k,\gamma)$ where

$$\varphi(k,\gamma) = r_{k,1-\gamma}\left[\frac{\Gamma(k+1+1/\alpha)}{\Gamma(k+1)} - \Gamma(1+1/\alpha)\right] \to (-\log\gamma)^{1/\alpha} \tag{7.1.35}$$

for $k \to \infty$.

Proof. The average interval length equals

$$Er_{k,1-\gamma}(\eta_N - \eta_{N-k}) = r_{k,1-\gamma}(E\eta_N - E\eta_{N-k}) \sim$$

$$\sim (M - \theta_N)\varphi(k,\gamma), \qquad k \to \infty.$$

Using the Stirling representation, we have

$$\Gamma(k+1+1/\alpha)/\Gamma(k+1) \sim k^{1/\alpha}, \qquad k \to \infty.$$

Consequently, for $k \to \infty$ one obtains

$$\varphi(k,\gamma) \sim k^{1/\alpha} / \left(\left(1-\gamma^{1/k}\right)^{-1/\alpha} - 1\right) \sim \left(k\left(1-\gamma^{1/k}\right)\right)^{1/\alpha} \sim$$

$$\sim \left(\frac{\left(1-\gamma^{1/k}\right)'}{(1/k)'}\right)^{1/\alpha} = \left(-\gamma^{1/k}\log\gamma\right)^{1/\alpha} \sim (-\log\gamma)^{1/\alpha}.$$

The proposition is proved.

Numerical results seem to indicate that $\varphi(k,\gamma)$ differs from its limit value $(-\log\gamma)^{1/\alpha}$ insignificantly, even for rather moderate values of k (e.g. for $k \in [5,7]$) and for $k \approx 10$ almost reaches it. This way, even for very large N, the choice k=10 is already *in good match* with the asymptotic requirements $k \to \infty$, $k/N \to 0$ $(N \to \infty)$. The numerical results also demonstrates that the convergence rate in (7.1.35) increases if γ decreases.

Note that analogous conclusion about the selection of k, via numerical analysis of the two-sided confidence intervals (7.1.33), were drawn by Weissman (1981) who did not deduce asymptotic expressions similar to (7.1.35).

7.1.5 *Testing statistical hypotheses about M*

Consider the problem of testing the statistical hypothesis H_0: $M \geq K$ versus the alternative H_1: $M < K$, where K is a fixed value $(K > \eta_N)$.

To accomplish a test, one can construct a one-sided confidence interval for M of a fixed confidence level $1-\gamma$, following Section 7.1.4, and reject H_0 if K would not fall into the interval. We shall investigate below the test determined by the rejection region

$$W = \left\{Y: (K - \eta_N)/(\eta_N - \eta_{N-k}) \geq r_{k,1-\gamma}\right\} \tag{7.1.36}$$

which corresponds to (7.1.34). According to this test, H_0 is rejected if $(K-\eta_N)/(\eta_N-\eta_{N-k}) \geq r_{k,1-\gamma}$, and accepted otherwise. This test was proposed and investigated by Cook (1979) for the particular case k=1, note, however, that Cook's results need some correction. Below we approximate the power function $\beta_N(M,\gamma)=\Pr\{Y \in W\}$ of the test. Set

$$\mathcal{T}(u,v) = \Gamma(u+1,v)/\Gamma(u+1)$$

where $\Gamma(.)$ is the gamma function,

$$\Gamma(u+1,v) = \int_v^\infty t^u e^{-t} dt$$

and introduce the abbreviations

$$\kappa = (K - M)/(M - \theta_N), \qquad z = r_{k,1-\gamma}, \qquad \lambda = (\kappa/z)^\alpha,$$

$$\delta = \alpha\gamma\left(\gamma^{-1/k} - 1\right), \qquad (a)_+ = \max\{0, a\} \qquad \text{for any } a \in \mathbb{R}.$$

Theorem 7.1.2. Let condition a) hold, $\alpha>0$, $N \to \infty$, $k/N \to 0$. Then

1) $\beta_N(M,\gamma) \sim a_N(M,\gamma)$ for $N \to \infty$, where

$$a_N(M,\gamma) = \frac{1}{k!}\int_0^\infty y^k e^{-y}\left(1 - \left(\left(z - \kappa y^{-1/\alpha}\right)/(1+z)\right)_+^\alpha\right)_+^k dy,$$

2) $\beta_N(K,\gamma)\to\gamma$ for $N\to\infty$,
3) $a_N(M,\gamma)$ is a decreasing function of M,
4) the asymptotic equality

$$\beta_N(M,\gamma)\sim 1-(1-\gamma)\mathrm{T}(k,\lambda)+\lambda^{1/\alpha}\delta\Gamma(k+1-1/\alpha,\lambda)/\Gamma(k)$$

is valid for $N\to\infty$ and M<K,

5) for $N\to\infty$ and M>K we have

$$\beta_N(M,\gamma)\sim\gamma\mathrm{T}\left(k,(-\kappa)^{\alpha}\right)+\kappa\delta\Gamma\left(k+1-1/\alpha,(-\kappa)^{\alpha}\right)/(z\Gamma(k)).$$

Proof. Introduce the notations

$$D_0=\left\{(x_0,\dots,x_k):x_i\geq 0,\frac{\kappa+x_0^{1/\alpha}}{(x_0+\dots+x_k)^{1/\alpha}-x_0^{1/\alpha}}\geq z\right\},$$

$$D_1=\left\{(y_0,\dots,y_k):y_i\geq 0,\sum_{i=2}^{k}y_i\leq y_1-y_0,\frac{\kappa+y_0^{1/\alpha}}{y_1^{1/\alpha}-y_0^{1/\alpha}}\geq z\right\},$$

$$D_2=\left\{(y_0,y_1):0\leq y_0\leq y_1,\ y_0\geq\left(\left(zy_1^{1/\alpha}-\kappa\right)/(1+z)\right)^{\alpha}\right\}.$$

Using (7.1.8) one obtains for $N\to\infty$ the chain of (approximate) relations

$$\beta_N(M,\gamma)=\Pr\left\{(K-\eta_N)/(\eta_N-\eta_{N-k})\geq r_{k,1-\gamma}\right\}\sim$$

$$\sim\Pr\left\{\frac{\kappa+\zeta_0^{1/\alpha}}{(\zeta_0+\dots+\zeta_k)^{1/\alpha}-\zeta_0^{1/\alpha}}\geq r_{k,1-\gamma}\right\}=$$

$$= \int_{D_0} \exp\left\{-\sum_{i=0}^{k} x_i\right\} dx_0 \ldots dx_k =$$

$$= \int_{D_1} \exp\{-y_1\} dy_0 \ldots dy_k =$$

$$= \frac{1}{\Gamma(k)} \int_{D_2} (y_1 - y_0)^{k-1} \exp\{-y_1\} dy_0 dy_1 = a_N(M,\gamma).$$

This is the first statement of the theorem.

The second assertion follows from the first one and the third is evident. Let us consider now the fourth statement.

Note that in the case $\lambda \to 0$ for $N \to \infty$, $\kappa > 0$ for $M < K$ and

$$\delta = \alpha(z/(1+z))^{\alpha}\left(1 - (z/(1+z))^{\alpha}\right)^{k-1}.$$

Thus for $N \to \infty$ we obtain

$$a_N(M,\gamma) = \frac{1}{k!}\left[\int_0^{\lambda} y^k e^{-y} dy + \int_{\lambda}^{\infty} y^k e^{-y}\left(1 - \left(\frac{z - \kappa y^{-1/\alpha}}{1+z}\right)^{\alpha}\right)^k dy\right] =$$

$$= 1 - \frac{1}{k!}\sum_{j=1}^{k} C_k^j (-1)^{j-1} (z/(1+z))^{\alpha j} \int_{\lambda}^{\infty} y^k e^{-y}\left(1 - (\lambda/y)^{1/\alpha}\right)^{\alpha j} dy \sim$$

$$\sim 1 - \frac{1}{k!}\sum_{j=1}^{k} C_k^j (-1)^{j-1} (z/(1+z))^{\alpha j} \int_{\lambda}^{\infty} y^k e^{-y}\left(1 - \alpha j \lambda^{1/\alpha} y^{-1/\alpha}\right) dy =$$

$$= 1 - (1-\gamma) T(k,\lambda) + \lambda^{1/\alpha} \delta \Gamma(k+1+1/\alpha, \lambda)/\Gamma(k).$$

Applying the first assertion, we arrive at the desired result.

To prove the fifth statement, note that $\kappa < 0$ for $M > K$ and then proceed analogously: on obtains

$$a_N(M,\gamma) = \frac{1}{k!} \int_{(-\kappa)^\alpha}^{\infty} y^k e^{-y} \left(1 - \left(\frac{z - \kappa y^{-1/\alpha}}{1+z}\right)^{\alpha}\right)^k dy =$$

$$= \frac{1}{k!}\left[\int_{(-\kappa)^\alpha}^{\infty} y^k e^{-y} dy - \sum_{j=1}^{k} (-1)^{j-1} C_k^j (z/(1+z))^{\alpha j} \times \right.$$

$$\left. \times \int_{(-\kappa)^\alpha}^{\infty} \left(1 - \kappa y^{-1/\alpha}/z\right)^{\alpha j} y^k e^{-y} dy \right] \sim$$

$$\sim T\left(k,(-\kappa)^\alpha\right) - \frac{1}{k!} \sum_{j=1}^{k} C_k^j (-1)^{j-1} (z/(1+z))^{\alpha j} \times$$

$$\times \int_{(-\kappa)^\alpha}^{\infty} y^k e^{-y}\left(1 - \kappa\alpha j y^{-1/\alpha}/z\right) dy =$$

$$= \gamma T\left(k,(-\kappa)^\alpha\right) + \kappa\delta\Gamma\left(k + 1 - 1/\alpha, (-\kappa)^\alpha\right)/(z\Gamma(k)).$$

The theorem is proved.

The second and third assertions of the theorem show that the probability of the first type error of the test asymptotically does not exceed γ.

Let us investigate the asymptotic behavior of the second type error probability for $k \to \infty$ which equals $1-\beta_N(M,\gamma)$ where $M<K$.

Lemma 7.1.5. If $k \to \infty$, then

$$\left(r_{k,1-\gamma}\right)^{\alpha} \sim (-\log \gamma)/k \qquad (7.1.37)$$

for each $\gamma \in (0,1)$.

The proof is a straightforward application of the l'Hospital rule.

Lemma 7.1.6. If $c>1$, then

$$\lim_{k \to \infty} T(k, ck) = 0.$$

Proof. Represent $\mathcal{T}(k,ck)$ as

$$\mathcal{T}(k,ck) = \frac{1}{k!}\int_{ck}^{\infty} y^k e^{-y} dy = \frac{(ck)^{k+1}}{k!} I,$$

where

$$I = \int_{1}^{\infty} \exp\{k(\log t - ct)dt\}.$$

We shall apply the saddle-point approximation to the integral I. The function logt-ct attains its maximal value (-c) at the interval $[1,\infty)$ at t=1. This way,

$$I \sim \exp\{-kc\}/(k(c-1)) \qquad \text{for } k \to \infty.$$

Applying the Stirling approximation, one obtains for $k \to \infty$

$$\mathcal{T}(k,ck) \sim c^{k+1} \exp\{-k(c-1)\}/(\sqrt{2\pi k}(c-1)) \to 0.$$

The lemma is proved.

The fourth statement of Theorem 7.1.2 is followed by the asymptotic inequality for the probability of the second type error

$$1 - \beta_N(M,\gamma) < (1-\gamma)\mathcal{T}(k,\lambda), \qquad N \to \infty. \tag{7.1.38}$$

(7.1.37) gives

$$\mathcal{T}(k,\lambda) \sim \mathcal{T}(k, k(\kappa^{\alpha}/(-\log\gamma))) \qquad \text{for } k \to \infty. \tag{7.1.39}$$

Since $\kappa=(K-M)/(M-\theta_N)\to\infty$ for $N\to\infty$, thus for sufficiently large N we have $c=\kappa^{\alpha}/(-\log\gamma)>1$. Lemma 7.1.6 together with (7.1.38) and (7.1.39) shows that the second kind error probability approaches zero, if $k\to\infty$ while $N\to\infty$. At the same time, numerical results demonstrate that the choice $k\approx10$ is already suitable in most practical cases.

Let us note finally that the power function of the test with the rejection region

$$\{Y:(K-\eta_N)/(\eta_N-\eta_{N-k}) < r_{k,j}\}$$

applied for testing the hypothesis H_0: M<K versus H_1: $M \geq K$ is representable as $1-\beta_N(M,1-\gamma)$, and so can be approximated using the above formulas.

7.2 Statistical inference when the tail index is unknown

We shall suppose in the sequel that condition a) of Section 7.1 holds, but the value of the tail index α is unknown. This case is typical for many classes of optimization problems (e.g. when a discrete problem is approximated by a continuous one) as well as in some other applications. As above, we confine ourselves to statistical inference that use only the first k+1 elements of the extreme order statistics (7.1.2), corresponding to the independent sample Y. Unlike in Section 7.1, a satisfactory precision of the statistical inferences can be guaranteed only if k is large enough, i.e. $k=k(N)\to\infty$ for $N\to\infty$ (while maintaining $k/N\to 0$).

Subsection 7.2.1 presents statistical inference procedures for estimating M; the main point in their construction, the estimation of α, is considered in Subsection 7.2.2. Subsection 7.2.3 deals with the construction of confidence intervals and statistical hypothesis testing for α that can be useful in some global optimization problems (investigating whether the objective function attains its maximal value inside a subset of the feasible region, recall Sections 4.2 and 4.3.)

7.2.1 *Statistical inference for M*

An ordinary way of drawing statistical inference concerning M, when the tail index α is unknown, consists of the substitution of some estimator $\hat{\alpha}$ of α for α into the formulas determining the statistical inference for the case of known α. Obviously, the accuracy of such statistical inference is the main problem arising here. The most advanced results in this field were obtained for the case, when α and M are estimated by the maximum likelihood technique: below we shall state some of them.

First let us follow Hall (1982) to construct maximal likelihood estimators for M and formulate their properties.

Suppose that instead of the asymptotic equality (7.1.25), the relation

$$F(t) = 1 - c_0(M-t)^{\alpha}$$

takes place for each t in some interval $[M-\delta, M]$, where $c_0>0$, $\alpha\geq 2$, M are unknown parameters, and the order statistics $\eta_N,\ldots,\eta_{N-k}$ fall into the interval $[M-\delta, M]$. Under these suppositions the likelihood function is

$$L(\eta_N,\ldots,\eta_{N-k}; M, c_0, \alpha) = \frac{N!}{(N-k-1)!}(c_0\alpha)^{k+1} \times$$

$$\times\left[1-c_0(M-\eta_{N-k})^{\alpha}\right]^{N-k-1}\prod_{j=0}^{k}(M-\eta_{N-j})^{\alpha-1}.$$

Maximizing this function with respect to M, c_0 and α, one obtains the maximum likelihood estimators

$$\hat{M}, \hat{c}_0 \text{ and } \hat{\alpha}$$

expressed as follows: $\hat{M}$ is the minimal solution of the equation

$$(k+1)\left[1/\sum_{j=0}^{k-1}\log\left(1+\beta_j(\hat{M})\right)-1/\sum_{j=0}^{k-1}\beta_j(\hat{M})\right]=1$$

provided $\hat{M} \geq \eta_N$ (if the solution does not exist, then η_N is taken as $\hat{M}$),

$$\hat{\alpha}=(k+1)/\sum_{j=0}^{k-1}\log\left(1+\beta_j(\hat{M})\right), \tag{7.2.1}$$

and

$$\hat{c}_0=(k+1)/\left(N(M-\eta_{N-k})^{\alpha}\right);$$

in the above formulas the abbreviation

$$\beta_j(\hat{M})=(\eta_{N-j}-\eta_{N-k})/\left(\hat{M}-\eta_{N-j}\right)$$

is used.

Hall (1982) proved the asymptotic normality of the obtained estimate $\hat{M}$, with mean M and variance $(\alpha-1)^2\sigma_{N,k}^2$, where

$$\sigma^2_{N,k}=\begin{cases}(1-2/\alpha)(c_0N)^{-2/\alpha}k^{-(1-2/\alpha)} & \text{for} \quad \alpha>2\\ (c_0N\log k)^{-1} & \text{for} \quad \alpha=2,\end{cases} \tag{7.2.2}$$

and the asymptotic normality of $\hat{\alpha}$ with mean α and variance

$$\alpha^2(\alpha-1)^2/k \tag{7.2.3}$$

provided that the stronger assumption on F:

$$1-F(t)=c_0(M-t)^{\alpha}\left(1+O((M-t))^{\ell}\right) \quad \text{for} \quad t\to M \tag{7.2.4}$$

holds, where $\ell>0$, $\alpha\geq 2$, and

$$k = k(N) \to \infty, \qquad k(N) = o\left(N^{-\beta}\right) \quad \text{for } N \to \infty \tag{7.2.5}$$

$\beta=\gamma/(\gamma+1/2)$, $\gamma=\min\{1, \ell/\alpha\}$.

Smith (1987) used a different approach to construct the maximum likelihood estimators. To describe it, let us introduce first the so-called generalized Pareto distribution by

$$G(t;\sigma,\nu) = \begin{cases} 1-(1-\nu t/\sigma)^{1/\nu} & \text{for } \nu \neq 0 \\ 1-\exp\{-t/\sigma\} & \text{for } \nu = 0 \end{cases} \tag{7.2.6}$$

where $\sigma>0$, $0<t<\infty$ for $\nu\leq 0$ and $0<t<\sigma/\nu$ for $\nu>0$. Now let y be a random variable with c.d.f. F(t), upper bound $M\leq\infty$ and let $h<M$. Then

$$F_h(t) = (F(h+t) - F(h))/(1-F(h)), \qquad 0 < t < M-h \tag{7.2.7}$$

is the conditional c.d.f. of y-h given y>h. Pickands (1975) showed that (7.2.6) is a good approximation of (7.2.7), in the sense of the relation

$$\lim_{h\to M} \sup_{0<t<M-h} \left| F_h(t) - G(t;\sigma(h),\nu) \right| = 0 \tag{7.2.8}$$

for some fixed ν and function $\sigma(h)$, if and only if F is in the domain of attraction of one of the three limit probability laws (namely,

$$\Psi_\alpha(z) = \exp\left(-(-z)^\alpha\right) \quad \text{for } z<0, \qquad \Phi_\alpha(z) = \exp\left(-z^{-\alpha}\right) \quad \text{for } z>0,$$

and $\Lambda(z)=\exp(-e^{-z})$). In case of c.d.f. Ψ_α, the constant ν in (7.2.6) equals $1/\alpha$.

Now the approach of Smith (1987) is as follows. Let N be sufficiently large, $y_1,\dots,y_N$ be independent realizations of the random variable y with c.d.f. F(t), h=h(N) be a high threshold value, k be the number of exceedances of h, and $x_1,\dots,x_k$ denote the corresponding excesses. That is, $x_i=y_j-h$ where j=j(i) is the index of the i-th exceedance. Under fixed N, the excesses $x_1,\dots,x_k$ are independent and have the c.d.f. (7.2.7). Relying upon (7.2.8), the generalized Pareto c.d.f. $G(t;\nu,\sigma)$ is substituted for (7.2.7) in the construction of the likelihood function. This way, its maximization yields the maximal likelihood estimators

$\hat{\sigma}_N$ and $\hat{\nu}$ for σ and ν,

respectively. The corresponding estimator for M is

$h + \hat{\sigma}_N/\hat{\nu}_N$.

Smith (1987) extensively studied the asymptotic properties (including the asymptotic normality and efficiency) of these estimators for M and $\nu=1/\alpha$) under fairly general conditions on F. Smith's results cover the case $0< \alpha <2$, together with the cases of $\alpha \geq 2$ and the other two limit laws for the extremes.

To construct confidence intervals for M and to test statistical hypothesis about M in the case of unknown α, one can use the above mentioned results of Hall and Smith, concerning the asymptotic normality of the maximal likelihood estimates of M. Recall again that, generally, this approach is applicable in the case when N is very large. The alternative techniques of de Haan (1981) and Weissman (1982) seem more suitable, if k can not be chosen to be very large (this holds e.g. for moderate values of N, say $N \approx 100 \div 200$). De Haan proved that for $N \to \infty$, $k \to \infty$, $k/N \to 0$, the test staistics (4.2.34) converge in distribution to the standard exponential random variable with density e^{-t}, $t>0$. (Similar test statistics were considered by Weissman.)

7.2.2 Estimation of α

The estimation of the tail index α is a major task in drawing statistical inference about M. It is important also in some other tail estimation problems and is often stated not only in connection with the extreme value distribution Ψ_α but including all three extreme value distributions (for references, see Smith (1987)).

A number of estimators of α are known, cf. Csörgö et al (1985), Smith (1987), the above mentioned maximum likelihood estimators as well as the formulas

$$\tilde{\alpha} = (\log 2)/\log\left((\eta_{N-2m} - \eta_{N-4m})/(\eta_{N-m} - \eta_{N-2m})\right) \tag{7.2.9}$$

and

$$\hat{\alpha} = (\log k/N)/\log\left((\eta_N - \eta_{N-k})/(\eta_N - \eta_{N-m})\right) \tag{7.2.10}$$

where $N \to \infty$, $m \to \infty$, $k \to \infty$, $m<k$, $k/N \to 0$. The estimator (7.2.9) was proposed by Pickands (1975) and thoroughly investigated e.g. by Dekkers and de Haan (1987). (7.2.10) was proposed by Weiss (1971) who formulated also some of its asymptotical properties. Below we derive some more general results concerning (7.2.10) and modify it, to reduce its bias.

Theorem 7.2.1. Let condition a) of Section 7.1 hold, $\alpha>1$, $N \to \infty$, $k \to \infty$, $k/N \to 0$, $m/k \to \tau$ where $0<\tau<1$. Then the estimator (7.2.10) for α is consistent, asymptotically unbiased, and there holds the relation

$$E(\hat{\alpha} - \alpha)^2 = \alpha^2\left(v_\tau/k + u_\tau/k^2\right) + o\left(k^{-3}\right), \qquad k \to \infty \tag{7.2.11}$$

where

$$v_\tau = (1-\tau)/(\tau \log^2 \tau),$$
$$u_\tau = \left(9 + 20(1-\tau^2)\log\tau + \tfrac{5}{3}(2+\tau+2\tau^2)\log^2\tau\right)/(\tau^2 \log^4 \tau).$$

Proof. According to (7.1.8), $M-\eta_i \sim (M-\theta_N)\mu_i^{1/\alpha}$ for $i \le k$, $N \to \infty$ where the random variable μ_i has the density $x^i e^{-x}/\Gamma(i+1)$. Using this approximation we have for $N \to \infty$, $k \to \infty$, $k/N \to 0$, $m/k \to \tau$

$$E\hat{\alpha} \sim (-\log\tau) E \log^{-1} \frac{\mu_k^{1/\alpha} - \mu_0^{1/\alpha}}{\mu_m^{1/\alpha} - \mu_0^{1/\alpha}} =$$

$$= (-\log\tau) E \log^{-1} \frac{(\mu_k/k)^{1/\alpha} - (\mu_0/k)^{1/\alpha}}{(\mu_m/k)^{1/\alpha} - (\mu_0/k)^{1/\alpha}} \sim$$

$$\sim \alpha(-\log\tau) \int_0^\infty\int_0^\infty \left(\log^{-1}\left(1+\tfrac{t}{s}\right)\right) \frac{\exp\{-t-s\} s^m}{m!\,(k-m-1)!} t^{k-m-1} dt\, ds =$$

$$= \alpha(-\log\tau) k!\, \Im/(m!\,(k-m-1)!)$$

where for $m/k \to \tau$

$$\Im = \int_0^\infty t^{-1} e^{-kt}(e^t - 1)^{k-m-1} dt \sim \Im_1 = \int_0^\infty g(t) \exp\{ks(t)\}\, dt,$$

$$g(t) = t^{-1}, \qquad s(t) = (1-\tau)\log(e^t - 1) - t.$$

By the Stirling approximation, for $k \to \infty$, $m \to \infty$

$$\frac{k!}{m!\,(k-m-1)!} \sim (k/2\pi)^{1/2} \tau^{-m-1/2} (1-\tau)^{-k+m+1/2}.$$

By the saddle-point approximation

$$\Im_1 \sim \left(2\pi/(k|s''(t_0)|)\right)^{1/2} g(t_0) \exp\{ks(t_0)\}$$

where $t_0 = \log(1/\tau)$ is the maximizer of g,

$$g(t_0) = (1-\tau)\log(1-\tau) + \tau\log\tau, \qquad s''(t_0) = -\tau/(1-\tau).$$

Consequently, we have

$$\mathfrak{I}_1 \sim (2\pi/k)^{1/2}\tau^{m+1/2}(1-\tau)^{k-m-1/2}/(-\log\tau):$$

this gives the asymptotic unbiasedness of (7.2.10). Analogously,

$$E(\hat{\alpha}-\alpha)^2 \sim \frac{\alpha^2 k!}{m!(k-m-1)!}\int_0^\infty \left(1 - t^{-1}\log\frac{1}{\tau}\right)^2 e^{-kt}(e^t-1)^{k-m-1} dt \sim$$

$$\sim \alpha^2(k-m)C_k^m \mathfrak{I}_2$$

where

$$\mathfrak{I}_2 = \int_0^\infty h(t)\exp\{ks(t)\}dt, \qquad h(t) = \left(1 + t^{-1}\log\tau\right)^2.$$

The saddle-point approximation gives for $\mathfrak{I}_2$:

$$\mathfrak{I}_2 \sim k^{-1/2}\exp\{ks(t_0)\}\sum_{j=0}^\infty a_j k^{-j}$$

where

$$a_j = \frac{\Gamma(j+1/2)}{\Gamma(2j+1)}\left(\frac{d}{dt}\right)^{2j}\left[h(t)\left(\frac{s(t_0)-s(t)}{(t_0-t)^2}\right)^{-j-1/2}\right]\Bigg|_{t=t_0}.$$

Since $h(t_0)=0$, thus $a_0=0$. The expressions for a_1 and a_2 are

$$a_1 = \sqrt{2\pi}\,(1/\tau - 1)^{3/2}/\log^2\tau,$$

$$a_2 = \sqrt{\frac{\pi}{2}}\,\frac{(1-\tau)^{5/2}}{2\tau^{5/2}\log^2\tau}\left(\frac{36}{\log^2\tau} + \frac{20(1+\tau)}{(1-\tau)\log\tau} + \frac{5(2+\tau+2\tau^2)}{3(1-\tau)^2}\right).$$

These expressions lead to (7.2.11) which in its turn implies the consistency of (7.2.10). The theorem is proved.

Note that the function $v_\tau=(1-\tau)/(\tau\log^2\tau)$ attains its minimal value (≈1.544) at $\tau_0\approx0.2032$: therefore to approach the minimal asymptotic variance of the estimator (7.2.10), one has to choose $m\sim k/5$ for $k\to\infty$.

Comparing (7.2.11) for $\tau=0.2$ and (7.2.3), one can deduce that for $\alpha\geq2.25$ the estimator (7.2.10) of the tail index α has smaller asymptotic variance, than the maximum likelihood estimator (7.2.1).

Naturally, the estimator (7.2.10) would be better if the exact optimum value M were substituted for η_N. Since M is unknown, η_N replaces it in (7.2.10): this enters a bias into the estimate, for any fixed k and m. To reduce the bias, let us use the estimate

$$M^0_{N,r} = (1+a_r)\eta_N - a_r\eta_{N-r}$$

presented at the end of Section 7.1.2, where r is either k or m and

$$a_r = (\hat{\alpha} + \psi(1))/(\psi(r+1) - \psi(1)). \tag{7.2.12}$$

Substituting $M_{N,k}{}^0$ and $M_{N,m}{}^0$ for η_N into the corresponding arguments of (7.2.10), we obtain the modified estimator

$$(\log k/m)\Big/\left(\log\frac{\psi(m+1)-\psi(1)}{\psi(k+1)-\psi(1)} + \log\frac{\psi(k+1)+\hat{\alpha}}{\psi(m+1)+\hat{\alpha}} + \frac{\log(k/m)}{\hat{\alpha}}\right) \tag{7.2.13}$$

where $\hat{\alpha}$ is the Weiss estimator (7.2.10). The numerical investigations indicate that for moderate values of k, the estimator (7.2.13) is more accurate than (7.2.10) and the accuracy difference increases, if α increases.

7.2.3 *Construction of confidence intervals and statistical hypothesis tests for α*

For suitably large N and k, in constructing confidence intervals and testing statistical hypothesis for α one can apply the results of Section 7.2.1 on the asymptotic normality of maximum likelihood estimators. We shall consider another approach based on the asymptotical properties of the estimator (7.2.10) which seems to be applicable also for moderately large values of k.

Proposition 7.2.1. Let the conditions of Theorem 7.2.1 be fulfilled. Then the sequence of random variables

$$\chi_{N,k} = \exp\{-(\alpha \log \tfrac{1}{\tau})/\hat{\alpha}\}, \tag{7.2.14}$$

where $\hat{\alpha}$ is defined by (7.2.10), converges in distribution to the random variable with the c.d.f.

$$F_k(t) = \begin{cases} 0 & \text{for } t \le 0 \\ t^{m+1} \sum\limits_{i=1}^{k-m-1} C^i_{m+i}(1-t)^i & \text{for } 0 < t < 1 \\ 1 & \text{for } t \ge 1 \end{cases} \tag{7.2.15}$$

Proof. We obtain from (7.2.10):

$$\chi_{N,k} \sim [(\eta_N - \eta_{N-m})/(\eta_N - \eta_{N-k})]^{\alpha} = \nu_{N,k} \qquad \text{for } N \to \infty,\ k \to \infty.$$

Using this and (7.1.8) we get

$$\Pr\{\chi_{N,k} < t\} \sim \Pr\{\nu_{N,k} < t\} \sim \Pr\left\{\frac{\mu_k^{1/\alpha} - \mu_0^{1/\alpha}}{\mu_m^{1/\alpha} - \mu_0^{1/\alpha}} < t^{1/\alpha}\right\} \sim$$

$$\sim \Pr\{\mu_k/\mu_m < t\} = \frac{1}{m!\,(k-m-1)!} \iint\limits_{\substack{y \ge 0, z \ge 0 \\ 1 + y/z \ge 1/t}} \exp\{-z-y\} z^m y^{k-m-1} dz\,dy =$$

$$= \frac{k!}{m!\,(k-m-1)!} \int_{1/t}^{\infty} x^{-k-1}(x-1)^{k-m-1} dx$$

where as earlier $\mu_i = \zeta_0 + \ldots + \zeta_i$. Multiple integration by parts gives (7.2.15): the proposition is proved.

The statement implies that the asymptotic level of the confidence interval ($0 \le \gamma \le \delta \le 1$)

$$[\hat{\alpha}(\log t_{k,\delta})/\log \tau,\ \hat{\alpha}(\log t_{k,\gamma})/\log \tau] \tag{7.2.16}$$

for α equals δ-γ, where $t_{k,\delta}$ denotes the δ-quantile of the c.d.f. (7.2.15): for illustration, some quantile values are given in Table 5.

Table 5. Some quantiles $t_{k,\delta}$ of F_k.

δ	k = 5	k = 10	k = 20	k = 50
0.025	0.052745	0.066739	0.086572	0.115266
0.05	0.076440	0.087264	0.104081	0.128558
0.1	0.112235	0.115825	0.126925	0.144980
0.9	0.583890	0.449603	0.360662	0.291297
0.95	0.657409	0.506902	0.401029	0.315597
0.975	0.716418	0.556096	0.436614	0.337182

In a standard manner the confidence interval (7.2.16) may serve as the base for constructing the statistical hypothesis tests concerning α. The rejection region of the test for the hypothesis H_0: $\alpha \geq \alpha_0$ against the alternative H_1: $\alpha < \alpha_0$ is

$$W = \left\{Y: \hat{\alpha} < (\alpha_0 \log \tau)/\log t_{k,1-\gamma}\right\}.$$

The power function for this procedure can be written as follows

$$\Pr\left\{\hat{\alpha} < (\alpha_0 \log \tau)/\log t_{k,1-\gamma}\right\} = \Pr\left\{\frac{\alpha \log \tau}{\hat{\alpha}} < \frac{\alpha \log t_{k,1-\gamma}}{\alpha_0}\right\} =$$

$$= \Pr\left\{\exp\left((\alpha \log \tau)/\hat{\alpha}\right) < (t_{k,1-\gamma})^{\alpha/\alpha_0}\right\} \sim F_k\left(t_{k,1-\gamma}^{\alpha/\alpha_0}\right)$$

for $k \to \infty$.

Analogously, the rejection region of the test for the hypothesis H_0: $\alpha=\alpha_0$ versus H_1: $\alpha=\alpha_1>\alpha_0$ can be

$$W = \left\{Y: \hat{\alpha} > (\alpha_0 \log \tau)/\log t_{k,\gamma}\right\}$$

For this test

$$\Pr\left\{Y \in W \middle| \alpha = \alpha_0\right\} \sim \gamma, \qquad \Pr\left\{Y \in W \middle| \alpha = \alpha_1\right\} \sim 1 - F_k\left(t_{k,\gamma}^{\alpha_1/\alpha_0}\right)$$

for $k \to \infty$.

7.3 Asymptotic properties of optimal linear estimates

We suppose again that the condition a) of Section 7.1 holds and that the value of the tail index α is known. We shall prove the asymptotic normality and efficiency of the linear estimates, and derive expressions for the components of the vectors (7.1.21) and (7.1.22).

7.3.1 Results and consequences

The main purpose of this section is to state two important theorems.

Theorem 7.3.1. Let the condition a) of Section 7.1 hold, $N \to \infty$, $k \to \infty$, $k^2/N \to 0$, $\alpha > 1$, and α be known. Then for optimal linear estimates $M_{N,k}$, determined by the vectors (7.1.21) and (7.1.22), the asymptotic equality

$$E\left(M_{N,k} - M\right)^2 \sim \sigma^2_{N,k} \tag{7.3.1}$$

holds, where

$$\sigma^2_{N,k} = \begin{cases} \left(M - \theta_N\right)^2 (1 - \alpha/2)/\Gamma(1 + 2/\alpha) & \text{for } 1 < \alpha < 2 \\ \left(M - \theta_N\right)^2 (1 - 2/\alpha) k^{-(1-2/\alpha)} & \text{for } \alpha > 2 \\ \left(M - \theta_N\right)^2 / \log k & \text{for } \alpha = 2 \end{cases} \tag{7.3.2}$$

and for $\alpha \geq 2$ the convergence in distribution holds

$$\left(M - M_{N,k}\right)/\sigma_{N,k} \to N(0,1), \tag{7.3.3}$$

i.e. the sequences $(M-M_{N,k})/\sigma_{N,k}$ are asymptotically Gaussian with zero mean and unit variance.

The theorem will be proved in Section 7.3.3. (The proof was done in collaboration with M. V. Kondratovich.)

Theorem 7.3.2. The components $a_i{}^*$ and $a_i{}^+$ of the vectors (7.1.21) and (7.1.22) are representable as

$$a_i^* = u_i/A \qquad \text{for } i = 0, 1, \ldots, k,$$

$$a_i^+ = u_i/(A - B) \qquad \text{for } i = 0, 1, \ldots, k - 1,$$

$$a_k^+ = \left(u_k - B\right)/(A - B)$$

where $\alpha>0$,

$$u_0 = (\alpha + 1)/\Gamma(1 + 2/\alpha),$$
$$u_i = (\alpha - 1)\Gamma(i + 1)/\Gamma(i + 1 + 2/\alpha) \quad \text{for} \quad i = 1,\dots,k-1,$$
$$u_k = -(\alpha k + 1)\Gamma(k + 1/\Gamma(k + 1 + 2/\alpha),$$

$$A = \begin{cases} (\alpha\Gamma(k+2)/\Gamma(k+1+2/\alpha) - 2/\Gamma(1+2/\alpha))/(\alpha-2) & \text{for } \alpha>0, \alpha\neq 2 \\ \sum_{i=0}^{k} 1/(i+1) & \text{for } \alpha = 2, \end{cases}$$

and also $A = u_0 + \dots + u_k$,

$$B = \Gamma(k+1)/\Gamma(k+1+2/\alpha).$$

The theorem will be proved at the end of Section 7.3.2.

In the case, when the condition (7.1.25) is fulfilled, being somewhat stricter than a), the expression (7.2.2) resulting from (7.3.2), furtherly, (4.2.38) hold instead of (7.3.2). (7.2.2) also occurs in the works of Hall (1982), Smith (1987) and of Csörgö and Mason (1989), in connection with the maximum likelihood estimators and the linear estimators determined by the vector with coefficients (7.1.27). Thus, under suitable conditions, the asymptotic mean square errors of the above estimates coincide. Combining this with the result of Hall (1982) on the asymptotic efficiency of the maximal likelihood estimators in the class of asymptotically unbiased and asymptotically Gaussian estimators under a fixed increase rate of $k=k(N)$, we can conclude that the optimal linear estimators $M_{N,k}{}^*$ and $M_{N,k}{}^+$ are asymptotically efficient under Hall's conditions. Note that these conditions, namely, (7.2.4) and (7.2.5), are generally stricter than ours and that similar results on the asymptotic normality of the maximum likelihood estimates for M are obtained by Hall (1982) and Smith (1987) for the case when α is unknown. In essence, the results for unknown values of $\alpha \geq 2$ coincide with the above results for the case when $\alpha \geq 2$ is known, after the substitution $(\alpha-1)^2\sigma_{N,k}{}^2$ for $\sigma_{N,k}{}^2$.

7.3.2 Auxiliary statements and proofs of Theorem 7.3.2 and Proposition 7.1.3.

In this section all matrices have the order $(k+1)\times(k+1)$, all vectors belong to $\mathbb{R}^{k+1}$, $\lambda=(1,\dots,1)'$, $|\,.\,|$ denotes the determinant, and the abbreviation

$$\Gamma_{i,j} = \Gamma(i + 1 + j/\alpha) \tag{7.3.4}$$

is used.

Lemma 7.3.1. Let $z, d_o,\ldots,d_k$ be vectors in $\mathbb{R}^{k+1}$ and $D=\| d_o,\ldots,d_k \|$ be a nondegenerate matrix. Then

$$\lambda' D^{-1} z = |z, d_1 - d_0, d_2 - d_1, \ldots, d_k - d_{k-1}| / |D|. \tag{7.3.5}$$

Proof. Set $x=D^{-1}z$. Then $Dx=z$ and Cramer's representation gives

$$x = \| |D_i| / |D| \|_{i=0}^{k}$$

where

$$D_i = \| d_0, \ldots, d_{i-1}, z, d_{i+1}, \ldots, d_k \|.$$

This way, we have

$$\lambda' D^{-1} z = \lambda' x = \sum_{i=0}^{k} |D_i| / |D|.$$

Define the matrix

$$D_z = \| d_0 + z, \ldots, d_k + z \|$$

and transform its determinant:

$$|D_z| = |z, d_1 + z, \ldots, d_k + z| + |d_0, d_1 + z, \ldots, d_k + z| =$$

$$= |z, d_1, \ldots, d_k| + |d_0, z, d_2 + z, \ldots, d_k + z| +$$

$$+ |d_0, d_1, d_2 + z, \ldots, d_k + z| = \ldots = |D| + \sum_{i=0}^{k} |D_i|.$$

Hence

$$\lambda' D^{-1} z = \sum_{i=0}^{k} |D_i| / |D| = |D_z| / |D| - 1. \tag{7.3.6}$$

Tranform the determinant $|Dz|$ in another way, subtracting the previous column from each one, beginning with the last column:

$$|D_z| = |d_0 + z, d_1 - d_0, \ldots, d_k + d_{k-1}| =$$

$$= |d_0, d_1 - d_0, \ldots, d_k - d_{k-1}| + |z, d_1 - d_0, \ldots, d_k - d_{k-1}| =$$

$$= |D| + |z, d_1 - d_0, \ldots, d_k - d_{k-1}|.$$

This, together with (7.3.6), gives the desired relation: the lemma is proved.

Lemma 7.3.2. Let the vectors $x=(x_0,\ldots,x_k)'$ and $y=(y_0,\ldots,y_k)'$ consist of positive numbers and the elements d_{ij} of the symmetrical matrix

$$D_k = \|d_{ij}\|_{i,j=0}^{k}$$

be defined by $d_{ij}=x_i y_j$ for $i \geq j$. Then

$$|D_0| = x_0 y_0, \; |D_k| = \mu_k |D_{k-1}|,$$

where

$$\mu_k = x_k (x_{k-1} y_k - x_k y_{k-1}) / x_{k-1}.$$

Proof. Multiplying the last but one row by x_k/x_{k-1} and subtracting it from the last one we obtain

$$|D_k| = \begin{vmatrix} x_0 y_0 & x_1 y_0 & \ldots & x_{k-1} y_0 & x_k y_0 \\ \cdots & \cdots & \cdots & \cdots & \cdots \\ x_{k-1} y_0 & x_{k-1} y_1 & \ldots & x_{k-1} y_{k-1} & x_k y_{k-1} \\ x_k y_0 & x_k y_1 & \ldots & x_k y_{k-1} & x_k y_k \end{vmatrix} =$$

$$= \begin{vmatrix} & & \ldots & x_{k-1} y_0 & x_k y_0 \\ & & & x_{k-1} y_k & x_k y_1 \\ & & & \cdot & \cdot \\ & D_{k-1} & & \cdot & \cdot \\ & & & \cdot & \cdot \\ 0 & 0 & \ldots & 0 & \mu_k \end{vmatrix} = \mu_k |D_{k-1}|.$$

The lemma is proved.

Consider now the determinant

$$\Delta_k = |z, d_1 - d_0, \ldots, d_k - d_{k-1}|$$

where $z=(z_0,\ldots,z_k)' \in \mathbb{R}^{k+1}$ and $d_0,\ldots,d_k$ are the columns of the matrix D_k defined in Lemma 7.3.2.

Lemma 7.3.3. There holds the relation

$$\Delta_k = \mu_1 \cdots \mu_k \left\{ z_0 + \delta \left[-\varphi_1 + \sum_{j=2}^{k} (-1)^j \varphi_j \prod_{i=2}^{j} (\nu_i/\mu_i) \right] \right\}, \tag{7.3.7}$$

where

$$\mu_i = x_i(x_{i-1}y_i - x_i y_{i-1})/x_{i-1},$$
$$\varphi_i = z_i - z_{i-1}x_i/x_{i-1},$$
$$\nu_i = (x_{i-1}y_{i-2} - x_{i-2}y_{i-1})(x_i - x_{i-1})/(x_{i-1} - x_{i-2}),$$
$$\delta = x_0(x_1 y_0 - x_0 y_0)/(x_1(x_0 y_1 - x_1 y_0)).$$

Proof. Multiplying the last but one row by x_k/x_{k-1} and subtracting it from the last one, we obtain

$$\Delta_k = (-1)^k \varphi_k \Delta'_k + \mu_k \Delta_{k-1}$$

where

$$\Delta'_k = \begin{vmatrix} (x_1 - x_0)y_0 & (x_2 - x_1)y_0 & \cdots & (x_k - x_{k-1})y_0 \\ x_1(y_1 - y_0) & (x_2 - x_1)y_1 & \cdots & (x_k - x_{k-1})y_0 \\ \cdots & \cdots & \cdots & \cdots \\ x_{k-1}(y_1 - y_0) & x_{k-1}(y_2 - y_1) & \cdots & (x_k - x_{k-1})y_{k-1} \end{vmatrix}$$

Multiplying the last but one column of Δ_k' by $(x_k-x_{k-1})/(x_{k-1}-x_{k-2})$ and subtracting it from the last one, we have

$$\Delta'_k = [(x_k - x_{k-1})y_{k-1} - (x_k - x_{k-1})x_{k-1}(y_{k-1} - y_{k-2})/(x_{k-1} - x_{k-2})]\Delta'_{k-1} =$$

$$= \nu_k \Delta'_{k-1} = \nu_k \nu_{k-1} \Delta'_{k-2} = \ldots = \nu_k \nu_{k-1} \cdots \nu_2 \Delta'_1.$$

Since $\Delta_1'=\mu_1\delta$, thus $\Delta_k'=\delta\mu_1\nu_2...\nu_k$.

The equality (7.3.7) will be proved inductively. Its validity for k=1 can be checked immediately. Suppose now that it is valid for the determinant Δ_{k-1} and show it for Δ_k:

$$\Delta_k = \mu_k \Delta_{k-1} + (-1)^k \varphi_k \Delta_k' =$$

$$= \mu_1 ... \mu_k \left\{ z_0 + \delta \left[-\varphi_1 + \sum_{j=2}^{k-1} (-1)^j \varphi_j \prod_{i=2}^{j} (\nu_i/\mu_i) \right] \right\} + (-1)^k \varphi_k \delta \mu_1 \nu_2 ... \nu_k =$$

$$= \mu_1 ... \mu_k \left\{ z_0 + \delta \left[-\varphi_1 + \sum_{j=2}^{k-1} (-1)^j \varphi_j \prod_{i=2}^{j} (\nu_i/\mu_i) \right] \right\}.$$

The lemma is proved.

Lemmas 7.3.1 - 7.3.3 will be used for investigating the case, when $k \geq 2$, $\alpha>0$, and vectors x and y consist of the numbers

$$x_i = \Gamma_{i,2}/\Gamma_{i,1}, \qquad y_j = \Gamma_{j,1}/\Gamma_{j,0} \tag{7.3.8}$$

where i,j=0,1,...,k and the symbol $\Gamma_{i,m}$ is determined by (7.3.4). In this case (7.1.14) for $i \geq j$ defines the elements of the matrix D_k which will be denoted by Λ or Λ_k. (The matrix Λ is the same as in Proposition 7.1.2.)

Lemma 7.3.4. There holds for $\alpha>0$

$$\lambda'\Lambda^{-1}\lambda = \begin{cases} \left(\alpha\Gamma(k+2)/\Gamma_{k,2} - 2/\Gamma_{0,2}\right)/(\alpha-2) & \text{for} \quad \alpha \neq 2 \\ \sum_{i=0}^{k} 1/(i+1) = \psi(k) - \psi(1) & \text{for} \quad \alpha = 2 \end{cases} \tag{7.3.9}$$

where ψ is the psi-function.

Proof. According to Lemma 7.3.1, we have $\lambda'\Lambda^{-1}\lambda=\Delta_k/|\Lambda|$ where the first column of the determinant Δ_k is $z=\lambda=(1,...,1)'$. By Lemma 7.3.2, we know that

$$|\Lambda| = \mu_1 ... \mu_k x_0 y_0 = \mu_1 ... \mu_k \Gamma_{0,2}.$$

Taking into account (7.3.8) and that $z=\lambda$, simplify the expression for μ_i, φ_i, ν_i ($i=0,...,k$) and δ of Lemma 7.3.3:

$$\mu_i = \Gamma_{i,2}/\left(\alpha^2(i+1/\alpha)^2\Gamma(i+1)\right),$$
$$\varphi_i = -1/(\alpha(i+1/\alpha)),$$
$$\nu_i = -\Gamma(i+2/\alpha)/\left(\alpha^2(i+1/\alpha)(i-1+1/\alpha)\Gamma(i)\right),$$
$$\delta = \alpha(1+1/\alpha)/(1+2/\alpha).$$

This way,

$$\nu_i/\mu_i = -i(i+1/\alpha)/((i+2/\alpha)(i-1+1/\alpha)),$$

$$(-1)^j\varphi_j\prod_{i=2}^{j}(\nu_i/\mu_i) = \Gamma_{j,0}\Gamma_{1,2}/((\alpha+1)\Gamma_{j,2}).$$

Hence, (7.3.7) yields

$$\lambda'\Lambda^{-1}\lambda = \sum_{i=0}^{k}\Gamma_{i,0}/\Gamma_{i,2}.$$

Now, (7.3.9) can be inductively deduced from this relation: the lemma is proved.

Lemma 7.3.5. For $b=(b_o,...,b_k)'$, where $b_i=\Gamma_{i,1}/\Gamma_{i,o}$, we have

$$\lambda'\Lambda^{-1}b = \Gamma_{k,1}/\Gamma_{k,2}. \tag{7.3.10}$$

Proof. By Lemma 7.3.1, $\lambda'\Lambda^{-1}b=\Delta_k/|\Lambda|$, where the first column of Δ_k equals $z=b$. The expressions for μ_i, ν_i, δ are as in the proof of Lemma 7.3.4 and

$$\varphi_i = \Gamma_{i,1}/\left(\alpha^2(i+1/\alpha)^2\Gamma_{i,0}\right)$$

Hence (7.3.7) gives

$$\lambda'\Lambda^{-1}b = \Gamma_{0,1}/\Gamma_{0,2} - \alpha^{-1}\sum_{i=1}^{k}\Gamma(i+1/\alpha)/\Gamma_{i,2}.$$

By induction, we obtain (7.3.10): the lemma is proved.

Lemma 7.3.6. These holds

$$b'\Lambda^{-1}b = \Gamma^2_{k,1}/(\Gamma_{k,2}\Gamma_{k,0}). \tag{7.3.11}$$

Proof. Let us represent vector b as $b=B\lambda$, where B is the diagonal matrix with the diagonal elements $b_o,\dots,b_k$. We have

$$b'\Lambda^{-1}b = \lambda' B\Lambda^{-1}B\lambda = \lambda'\left(B^{-1}\Lambda B^{-1}\right)^{-1}\lambda.$$

The matrix $D_k=B^{-1}\Lambda B^{-1}$ is symmetric and its elements equal $d_{ij}=x_i'y_j'$ for $i\geq j$, where

$$x'_i = x_i/b_i = \Gamma_{i,2}\Gamma_{i,0}/\Gamma^2_{i,1}, \qquad y'_j = y_j/b_j = 1.$$

Analogously with the proof of Lemma 7.3.4, we obtain

$$b'\Lambda^{-1}b = \alpha^2 \sum_{i=0}^{k} \Gamma^2_{i,1}/\left((i+1/\alpha)^2\Gamma_{i,2}\Gamma_{i,0}\right) = \Gamma^2_{k,1}/\left(\Gamma_{k,2}\Gamma_{k,0}\right)$$

The lemma is proved.

Let us deduce now Proposition 7.1.3 and Theorem 7.3.2 from the above lemmas.

Proof of Proposition 7.1.3. Lemma 7.3.2 gives $|\Lambda_k|=\mu_k|\Lambda_{k-1}|$ where

$$\mu_k = \Gamma_{k,2}/\left(\alpha^2(k+1/\alpha)^2\Gamma_{k,0}\right) \text{ and } \Lambda_{k-1} = \left\|\lambda_{ij}\right\|_{i,j=0}^{k-1}.$$

Due to the Stirling approximation

$$\mu_k \sim \alpha^{-2}(k+1/\alpha)^{-2}k^{2/\alpha} \sim \alpha^{-2}k^{-2(1-1/\alpha)} \qquad \text{for} \quad k\to\infty.$$

This gives (7.1.24): the proposition is proved.

Proof of Theorem 7.3.2. Set

$$u = (u_0,\dots,u_k)' = \Lambda^{-1}\lambda.$$

We have $a_i^*=u_i/A$ for $i=0,1,\dots,k$, where $A=\lambda'\Lambda^{-1}\lambda$ is calculated by (7.3.9). Represent u_i as

$$u_i = \lambda' \Lambda^{-1} e_i$$

where all components of the i-th coordinate vector e_i are zero except the i-th which equals 1. Applying Lemmas 7.3.1 - 7.3.3 with $D_k=\Lambda$ and $z=e_i$, one obtains the expressions for a_i^*.

Turn now to the vector a^+ and represent a_i^+ in the form

$$a_i^+ = \frac{u_i - \beta_i(\lambda'\Lambda^{-1}b)/(b'\Lambda^{-1}b)}{\lambda'\Lambda^{-1}\lambda - (\lambda'\Lambda^{-1}b)^2/(b'\Lambda^{-1}b)} \tag{7.3.12}$$

where β_i are the components of the vector

$$\beta = (\beta_0, \ldots, \beta_k)' = \Lambda^{-1}b.$$

Since the last column of Λ is proportional to b, thus all the components β_i equal zero, except β_k:

$$\beta_i = 0 \text{ for } i = 0, \ldots, k-1, \qquad \beta_k = \lambda'\Lambda^{-1}b. \tag{7.3.13}$$

The expressions for a_i^+ follow from the expressions derived for a_i^* and from (7.3.9) - (7.3.12). The theorem is proved.

7.3.3 Proof of Theorem 7.3.1

Due to (7.1.19), (7.1.23) and (7.3.9), we have

$$E(M^*_{N,k} - M)^2 \sim \delta^2_{N,k} \qquad \text{for } N \to \infty, k \to \infty, k^2/N \to 0 \tag{7.3.14}$$

where

$$\delta^2_{N,k} = \begin{cases} (\alpha-2)(M-\theta_N)^2/(\alpha\Gamma(k+2)/\Gamma_{k,2} - 2/\Gamma_{0,2}) & \text{for } \alpha > 1, \alpha \neq 2 \\ (M-\theta_N)^2/\sum_{i=0}^{k} 1/(i+1) & \text{for } \alpha = 2. \end{cases}$$

By the Stirling approximation we obtain

$$\Gamma(k+2)/\Gamma_{k,2} \sim k^{1-2/\alpha} \qquad \text{for } k \to \infty.$$

Hence, for $k \to \infty$

$$\delta^2_{N,k} \sim \sigma^2_{N,k} = \begin{cases} (M-\theta_N)^2 (1-\alpha/2)/\Gamma_{0,2} & \text{for } 1<\alpha<2 \\ (M-\theta_N)^2 (1-2/\alpha) k^{-1+2/\alpha} & \text{for } \alpha>2 \\ (M-\theta_N)^2 / \log k & \text{for } \alpha=2 \end{cases}$$

that is (7.3.1) and (7.3.2) hold for the estimate $M_{N,k}{}^*$ determined by the vector (7.1.21). Turn now to the estimate $M_{N,k}{}^+$ determined in (7.1.22).

By (7.1.22), $a^+=(\Lambda^{-1}\lambda - p_k\beta)/q_k$, where

$$q_k = \lambda'\Lambda^{-1}\lambda - \left(b'\Lambda^{-1}\lambda\right)^2 / \left(b'\Lambda^{-1}b\right),$$

$$p_k = b'\Lambda^{-1}\lambda / b'\Lambda^{-1}b, \qquad \beta = \Lambda^{-1}b.$$

Using (7.3.9), (7.3.10), (7.3.11) and the Stirling approximation for $k \to \infty$, we obtain

$$\lambda'\Lambda^{-1}b = \Gamma_{k,1}/\Gamma_{k,2} \sim k^{-1/\alpha},$$

$$b'\Lambda^{-1}b = \Gamma^2_{k,1}/\left(\Gamma_{k,2}\Gamma(k+2)\right) \sim 1.$$

With the help of (7.3.13), we get $a^* \sim a^+$ for $k \to \infty$: this yields (7.3.1) and (7.3.2) for the estimate $M_{N,k}{}^+$.

Prove now the asymptotic normality for the optimal linear estimates.

Set

$$D_{N,k} = \left(M - M_{N,k}\right) / \left(M - \theta_N\right).$$

Then, by (7.1.8)

$$D_{N,k} \sim \sum_{i=0}^{k} a_i (\zeta_0 + \ldots + \zeta_i)^{1/\alpha}$$

where either $a_i = a_i{}^*$ or $a_i = a_i{}^+$ are the coefficient of the vectors (7.1.21) and (7.1.22), respectively, further on $\zeta_0, \zeta_1, \ldots$ are independent and have the density e^{-x}, $x>0$.

Set

$$v_i = \zeta_i - 1, \qquad u_i = \sum_{j=0}^{i} v_j \qquad \text{for } i = 0, 1, \ldots$$

Then

$$\sum_{j=0}^{i} \zeta_i = u_i + (i+1), \qquad \left(\sum_{j=0}^{i} \zeta_j\right)^{1/\alpha} = (i+1)^{1/\alpha}\left(1 + \frac{u_i}{i+1}\right)^{1/\alpha}.$$

For each i the random variable $u_i/(i+1)$ has density

$$i^{i+1}(x+1)^i \exp\{-i(x+1)\}/\Gamma(i+1), \qquad x \geq -1,$$

with the maximal value

$$i^{i+1} e^{-i}/\Gamma(i+1) \sim (i/2\pi)^{1/2} \qquad \text{for } i \to \infty$$

at zero. Consequently,

$$\lim_{i \to \infty} u_i/(i+1) = 0 \qquad \text{a.s.}$$

and

$$(1 + u_i/(i+1))^{1/\alpha} = 1 + u_i/(\alpha(i+1)) + O\left(\left(\frac{u_i}{i+1}\right)^2\right) \qquad \text{for } i \to \infty \text{ a.s.}$$

This, together with $a_i \to 0$ for any fixed i and $k \to \infty$, yields

$$D_{N,k} \sim \sum_{i=0}^{k} a_i (i+1)^{1/\alpha} + \sum_{i=0}^{k} a_i \frac{u_i}{\alpha(i+1)^{1-1/\alpha}}$$

for $N \to \infty$, $k \to \infty$, $k/N \to 0$.

The optimal coefficients a_i satisfy the unbiasedness condition (7.1.20), either exactly or asymptotically exactly, i.e.

$$\sum_{i=0}^{k} a_i \Gamma_{i,1}/\Gamma_{i,0} \to 0 \qquad \text{for } k \to \infty.$$

By the Stirling approximation, $\Gamma_{i,1}/\Gamma_{i,o} \sim i^{1/\alpha}$ for $i \to \infty$. Therefore, using once more the relation $a_i \to 0$ for $k \to \infty$, we obtain

$$\sum_{i=0}^{k} a_i (i+1)^{1/\alpha} = \sum_{i=0}^{k} a_i \Gamma_{i,1}/\Gamma_{i,0} + \sum_{i=0}^{k} a_i \left((i+1)^{1/\alpha} - \Gamma_{i,1}/\Gamma_{i,0}\right) \to 0$$

for $k \to \infty$. This gives for $k \to \infty$

$$D_{N,k} \sim \alpha^{-1} \sum_{i=0}^{k} a_i u_i (i+1)^{-1+1/\alpha} = \sum_{i=0}^{k} v_i s_i$$

where

$$s_i = \alpha^{-1} \sum_{j=i}^{k} a_j (j+1)^{-1+1/\alpha}.$$

Using the expression for $a_i = a_i^*$ (for $a_i = a_i^+$ the expressions are asymptotically equivalent) obtained in Theorem 7.3.2, we can derive the asymptotic forms for s_i $(i=0,1,\dots,k)$.

If $\alpha > 2$, then for $k \to \infty$

$$s_0 \sim \alpha^{-2}(\alpha-2)k^{-1+2/\alpha}\left((\alpha+1)/\Gamma_{0,2} + \sum_{j=1}^{k-1}(\alpha-1)j^{-1-1/\alpha} - \alpha k^{-1/\alpha}\right) \sim$$

$$\sim (1-2/\alpha)k^{-1+2/\alpha}\left((1+1/\alpha)/\Gamma_{0,2} + (\alpha-1)\right),$$

$$s_i \sim \alpha^{-2}(\alpha-2)k^{-1+2/\alpha}\left(\sum_{j=i}^{k-1}(\alpha-1)j^{-1-1/\alpha} - \alpha k^{-1/\alpha}\right) \sim$$

$$\sim (1-2/\alpha)k^{-1+2/\alpha}\left((\alpha-1)i^{-1/\alpha} - k^{-1/\alpha}\right), \qquad i = 1,2,\dots$$

If $\alpha = 2$ and $k \to \infty$, then

$$s_0 \sim 1/\log k,$$

$$s_i \sim \left(i^{-1/2} - k^{-1/2}\right)/\log k \qquad \text{for } i = 1,2,\dots$$

Let us show that Lyapunov's condition

$$L_{k,\delta} = B_k^{-2-\delta} \sum_{i=0}^{k} E|v_i s_i|^{2+\delta} \to 0 \qquad (k \to \infty)$$

for $\delta = 2$ holds, where

$$B_k^2 = \sum_{i=0}^{k} s_i^2 \operatorname{var}(v_i) = \sum_{i=0}^{k} s_i^2.$$

We have

$$Ev_i^4 = \int_{-1}^{\infty} x^4 \exp\{-(x+1)\}dx = 9,$$

$$\sum_{i=0}^{k} E|v_i s_i|^4 = \sum_{i=0}^{k} s_i^4 E v_i^4 = 9 \sum_{i=0}^{k} s_i^4.$$

For $\alpha>2$ and $k\to\infty$, we have

$$B_k^2 \sim (1-2/\alpha)k^{-1+2/\alpha}.$$

For $\alpha=2$ and $k\to\infty$ there holds

$$B_k^2 \sim 1/\log k.$$

Turn now to the asymptotic representation of

$$\sum_{i=0}^{k} s_i^4.$$

For $\alpha>2$, $i>0$ and $k\to\infty$ there holds

$$s_i^4 \sim \frac{(1-2/\alpha)^4}{k^{4-8/\alpha}}\left(\frac{(\alpha-1)^4}{i^{4/\alpha}} + \frac{6(\alpha-1)^2\alpha^2}{i^{2/\alpha}k^{2/\alpha}} + \frac{\alpha^4}{k^{4/\alpha}} - \frac{4\alpha(\alpha-1)^3}{i^{3/\alpha}k^{1/\alpha}} - \frac{4(\alpha-1)\alpha^3}{i^{1/\alpha}k^{3/\alpha}}\right).$$

If $\alpha\neq2$, $\alpha\neq4$, $\alpha>2$ then for $k\to\infty$

$$\sum_{i=0}^{k} s_i^4 \sim k^{-3+4/\alpha} 3(\alpha-2)^3(3\alpha^2+\alpha+2)/[\alpha^3(\alpha-3)(\alpha-4)].$$

For $\alpha=3$, $k\to\infty$

$$\sum_{i=0}^{k} s_i^4 \sim (6+4/\Gamma(1+2/3))^4 k^{-4/3}.$$

For $\alpha=4$, $k\to\infty$

$$\sum_{i=0}^{k} s_i^4 \sim \left(\tfrac{3}{2}\right)^4 k^{-2} \log k.$$

For $\alpha=2$, $k\to\infty$

$$\sum_{i=0}^{k} s_i^4 \sim 8/\log^3 k.$$

Consequently, for $k\to\infty$

$$L_{k,2} \asymp \begin{cases} 1/\log k & \text{for } \alpha=2 \\ k^{-2/3} & \text{for } \alpha=3 \\ (\log k)/k & \text{for } \alpha=4 \\ k^{-1} & \text{for } \alpha>2, \alpha\neq 3, \alpha\neq 4 \end{cases}$$

Thus, in all cases $L_{k,2}\to 0$ for $k\to\infty$.
This yields the desired result: the theorem is proved.

Let us remark that there is another method of proving the asymptotic normality (7.3.3), i.e. the second part of Theorem 7.3.1. This method consists in referring to the asymptotic normality result of Csörgö and Mason (1989) and mentioning that the optimal linear estimators $M_{N,k}^*$ and $M_{N,k}^+$ asymptotically, if $k\to\infty$, coincide with Csörgö-Mason estimators which are determined by the coefficient vectors a with components (7.1.27).

Indeed, the asymptotic coincidence of the estimators $M_{N,k}^*$ and $M_{N,k}^+$ was established in the proof of Theorem 7.3.1. Further, applying (7.1.27) and Theorem 7.3.2, that contain explicit expressions for the components a_i and a_i^* of the Csörgö-Mason and optimal linear estimators, respectively, we obtain for $k\to\infty$, $\alpha>2$:

$$a_0 \sim (\alpha-1)k^{2/\alpha-1}, \qquad a_0^* \sim \frac{(\alpha+1)(\alpha-2)}{\alpha\Gamma(1+2/\alpha)} k^{2/\alpha-1},$$

$$a_k \to 2-\alpha, \qquad a_k^* \to 2-\alpha,$$

$$a_i \sim a_i^* \sim (\alpha - 1)(1 - 2/\alpha) i^{-2/\alpha} k^{2/\alpha - 1} \qquad \text{for } i < k, i \to \infty,$$

$$a_i / a_i^* \sim 1 + \frac{2}{\alpha}[1 - i \log(1 + 1/i) + \Psi(i + 1) - \log(i + 1)]$$

$$\text{for } i \geq 0, i = \text{const}, \alpha \to \infty,$$

where Ψ is the psi-function and $0\log(1+1/0)=1$. Analogously, for $\alpha=1$, $k\to\infty$ we have

$$a_0 \sim 2/\log k, \qquad a_0^* \sim 3/\log k,$$

$$a_k \sim a_k^* \sim -2/\log k,$$

$$a_i \sim a_i^* \sim 1/(i \log k) \qquad \text{for } i < k, i \to \infty,$$

$$a_i / a_i^* \sim (i + 1) \log(1 + 1/i) \qquad \text{for } i > 0, i = \text{const}.$$

For $1<\alpha<2$ the Csörgö-Mason estimator is not defined and the asymptotic (while $k\to\infty$) expressions for a_i^* are as follows:

$$a_0^* \to (\alpha + 1)(1 - \alpha/2), \qquad a_k^* \sim -\alpha k^{1-2/\alpha}(1 - \alpha/2)\Gamma(1 + 2/\alpha),$$

$$a_i^* \sim (\alpha - 1)(1 - \alpha/2) i^{-2/\alpha} \Gamma(1 + 2/\alpha) \qquad \text{for } i < k, i \to \infty.$$

Bearing in mind the asymptotic expressions presented above let us introduce a new, very simple, linear estimator determining its coefficients by the formulas: for $1<\alpha<2$

$$a_i = (\alpha - 1)(1 - \alpha/2) i^{-2/\alpha} \Gamma(1 + 2/\alpha), \qquad 0 < i < k,$$

$$a_0 = (\alpha + 1)(1 - \alpha/2), \qquad a_k = 1 - (a_0 + \ldots + a_{k-1});$$

for $\alpha>2$

$$a_i = (\alpha - 1)(1 - 2/\alpha) i^{-2/\alpha} k^{2/\alpha - 1}, \qquad 0 < i < k,$$

$$a_k = 2 - \alpha, \qquad a_0 = 1 - (a_1 + \ldots + a_k).$$

For $\alpha=2$ the expressions for a_i^* are so simple that it is hard to simplify them. Thus, we set $a_i=a_i^*$ for $\alpha=2$, $0\leq i\leq k$.

The discussion concerning the Csörgö-Mason and optimal linear estimators leads to the conjecture that the above defined linear estimator has the asymptotic properties (7.3.1) and (7.3.3), i.e. it is also asymptotically normal and efficient.

CHAPTER 8 SEVERAL PROBLEMS CONNECTED WITH GLOBAL RANDOM SEARCH

As it was pointed out in Section 2.3 and 4.1, the theory of global random search is connected with many important branches of mathematical statistics and computational mathematics. Two of them were already studied in Section 5.3 and Chapter 7, three others will be treated in this chapter (note that their connection with global random search was highlighted in Section 4.1).

8.1 Optimal design in extremal experiments

We shall consider a particular class of local optimization problems related to a regression function, as a class of extremal experiment problems: its features will be discussed in Subsection 8.1.1.

The standard *rule* in extremal experiment algorithms involves the least square estimation of the gradient of an objective function f at the current point and moving in the direction of that estimate. The main purpose of this section is to show that one can construct much simpler algorithms, sequentially replacing the actual optimality property by another.

8.1.1 Extremal experiment design

The search for the local extremum of a regression function is often treated as an extremal experiment problem; naturally the search for the global extremum belongs to the field of global optimization.

The cost (or time-consumption) of the objective function evaluation greatly influences the selection of the extremal experiment algorithm. If the cost is high and the number N of function evaluations can not be large, then a passive grid algorithm may be expedient. If the cost is low and N can be rather high, then adaptive algorithms of the stochastic approximation type are usually applied. All of them are representable by the following general recurrent relation

$$x_{k+1} = x_k - \gamma_k s_k, \tag{8.1.1}$$

where k=1,2,... is the iteration number, x_1 is a given initial point, $x_1, x_2, \dots$ is the sequence of points in $\mathcal{X} \subset \mathbb{R}^n$ generated to approach a local minimizer of f, $\gamma_1, \gamma_2, \dots$ are nonnegative numbers called the step lengths, $s_1, s_2, \dots$ are random vectors in $\mathbb{R}^n$ called the search directions. To construct each subsequent point in (8.1.1), one uses the results of evaluations of the regression function f (i.e. the random values $y(x)=f(x)+\xi(x)$, where $E\xi(x)=0$) at the preceding points of the sequence and possibly also at some auxiliary points. As usual, we shall suppose that different evaluations of f produce independent random values.

The majority of works, devoted to the local optimization of a regression function, studies the asymptotic characteristics of the algorithm-type (8.1.1). The extremal experiment design deals with the one-step characteristics of the algorithms, rather than with their asymptotic properties.

The extremal experiment algorithms (the simplex method of Nelder and Mead (1965) and steepest descent are, probably, the most popular of them) have been developed and applied for optimization of real objects (even in absence of computers) and thus have a number of properties that distinguish them from the stochastic approximation type adaptive algorithms. The pecularities of most extremal experiment algorithms are due to the inclusion of the following elements into each of their step: (i) statistical inference (usually: linear regression analysis for constructing and investigating the local first or second degree polynomial model of f), (ii) specific experimental design for selecting auxiliary points to evaluate f (the design criteria are chosen among the following: symmetricity, orthogonality, saturation, rotatability, simplicity of construction, optimality in some appropriate sense, etc.), (iii) selection of the search direction in accordance with the regression models constructed (the least square estimate of the gradient of f at x_k is customarily used as the search direction), (iv) selection of the step length at random via evaluating f at several auxiliary points along the chosen direction.

As for the step length selection rules, they are thoroughly discussed in many works, see for instance Ermakov et al. (1983). The procedure of Myers and Khuri (1979) and its modification studied in Zhigljavsky (1985) seem to be the most promising and recommendable. Below we shall deal only with the search direction construction problem.

8.1.2 Optimal selection of the search direction

Let f be a regression function, $y(x)=f(x)+\xi(x)$ be the (partially) random result of evaluating f at a point $x\in\mathcal{X}\subset\mathbb{R}^n$, $E\,\xi(x)=0$ for each $x\in\mathcal{X}$, $E\,\xi^2(x)=\sigma^2=\text{const}$, the results corresponding to various evaluations of f be mutually independent, x_k be a fixed point obtained at the k-th iteration, and $z_1,\dots,z_N$ be the points at which f is evaluated at the k-th iteration, in order to determine the search direction $s=s_k$. Without loss of generality, assume that $x_k=0$. We shall introduce the following two suppositions that are standard in the construction of extremal experiment algorithms. First, the points $z_1,\dots,z_N$ are selected in a sufficiently small neighbourhood U of $x_k=0$, in which f is approximately linear, i.e.

$$f(z_j) \approx \theta_0 + \sum_{i=1}^{n} \theta_i z_{ij} \tag{8.1.2}$$

for j=1,...,N where $z_{1j},\dots,z_{nj}$ are the coordinates of the point z_j. (Note that interesting results concerning the choice problem of x_{k+1} and applying the second order polynomial model of f are derived by Mandal (1981, 1989).) Second, the k-th step design, i.e. the point selection $\{z_1,\dots,z_N\}$ is symmetrical, that is the equality

$$\sum_{j=1}^{N} z_{ij} = 0 \tag{8.1.3}$$

holds for i=1,...,n.

The (approximate) gradient of f at $x_k=0$ is a vector

$$\nabla f(x_k) = \theta = (\theta_1,\ldots,\theta_n)'$$

of unknown parameters. If N>n and rankZ=n, then θ can be estimated by the least square estimator

$$\hat{\theta} = (ZZ')^{-1}ZY. \tag{8.1.4}$$

Here

$$Z = \|z_{ij}\|_{i,j=1}^{n,N}$$

is the design matrix and $Y=(y_1,\ldots,y_N)'$ is the vector of evaluation results, in accordance with (8.1.2) consisting of the elements

$$y_j = \theta_0 + \sum_{i=1}^{n}\theta_i z_{ij} + \xi_j, \qquad j = 1,\ldots,N \tag{8.1.5}$$

where $\xi_1,\ldots,\xi_N$ are mutually independent, $E\,\xi_j=0$, $E\,\xi_j^2=\sigma^2$.

The selection of (8.1.4) as the search direction is typical in extremal experiment algorithms (e.g. for steepest descent). The popularity of the choice of an unbiased estimate of $\theta=\nabla f(x_k)$ as the search direction $s=s_k$, is motivated by the fact that the average decrease rate of f in the direction -s is locally maximal. But this is not the unique sensible optimality criterion for choosing s: let us consider another one below.

Every search direction s is constructed using the values of some random variables; thus the function f may locally increase along the selected direction rather than decrease. Since increasing function values in the direction s are undesirable, therefore the probability of decrease of f is a characteristics of obvious importance and can be selected as the optimality criterion in choosing s.

The fact that a function f decreases in some direction -s, can be written in the form

$$\frac{\partial f}{\partial(-s)}(x_k) = [\nabla f(x_k)]'(-s) = \theta'(-s) < 0.$$

Thus, the probability of decrease of f in the direction -s equals

$$\Pr\{\theta' s > 0\}. \tag{8.1.6}$$

We shall maximize (8.1.6) with respect to s, in the class of linear statistics of the form

$$s = s(A) = AY, \qquad A \in \mathcal{A}, \tag{8.1.7}$$

where $\mathcal{A}$ is the set of matrices of the order $n \times N$ and Y consists of the evaluation results (8.1.5), thus satisfying the conditions

$$EY = Z'\theta, \qquad \text{cov}Y = \sigma^2 I_N. \tag{8.1.8}$$

Set

$$t(A) = t(A,\theta) = \theta' s(A), \qquad \eta(A) = \eta(A,\theta) = (t(A) - E\,t(A))(\text{var}\,t(A))^{-1/2}.$$

With these notations, the probability (8.1.6) can be rewritten as

$$\Pr\{t(A) > 0\} = \Pr\left\{\eta(A) > -(\text{var}\,t(A))^{-1/2} E\,t(A)\right\} \tag{8.1.9}$$

Under fixed $A \in \mathcal{A}$ and $\theta \in \mathbb{R}^n$, the random variable $\eta(A,\theta)$ is a linear combination of the random variables $\xi_1, \ldots, \xi_N$, having zero mean and unit variance. If ξ_j are Gaussian, then $\eta(A,\theta)$ follows Gaussian distribution as well and the probability (8.1.9) is completely determined by the magnitude

$$\kappa(A) = \kappa(A,\theta) = (\text{var}\,t(A))^{-1/2} E\,t(A) \tag{8.1.10}$$

which is to be maximized with respect to $A \in \mathcal{A}$. If ξ_j are not necessarily Gaussian, but N is large, then by virtue of the central limit theorem $\eta(A,\theta)$ is approximately Gaussian, i.e. the probability (8.1.9) can be characterized again by $\kappa(A)$. In the general case the distribution of the random variable $\eta(A,\theta)$ depends on A and θ and thus the quantity $\kappa(A)$ determines the probability (8.1.9) within some accuracy, rather than exactly. As it is usual to say in similar situations, the value

$$A^* = \arg\max_{A \in \mathcal{A}} \kappa(A) \tag{8.1.11}$$

gives a quasi-optimal solution of the initial problem of maximizing (8.1.6).

It is usual when solving extremal problems with an objective function depending on parameters (θ in the present case) that the optimizer also depends on these parameters. The important specialty of the investigated extremal problem is that an optimal matrix A^* exists, the same for all $\theta \neq 0$.

Theorem 8.1.1. For each $\theta \neq 0$, the equality

$$\max_{A \in \mathcal{A}} \kappa(A) = \sigma^{-1}(\theta' ZZ' \theta)^{1/2} \tag{8.1.12}$$

holds an the maximum is attained at the matrix $A^* = Z$.

Proof. We have

$$Et(A) = \theta' AZ' \theta,$$

$$\mathrm{var}(t(A)) = \mathrm{var}(\theta' AY) = \theta' A(\mathrm{cov} Y) A' \theta = \sigma^2 \theta' AA' \theta.$$

Denote by $\mathcal{L}(A)$ the functional

$$\mathcal{L}(A) = \sigma^2 \kappa^2(A) = (\theta' AZ' \theta)^2 / (\theta' AA' \theta)$$

and by E an arbitrary matrix in $\mathcal{A}$. Bearing in mind the identity

$$\theta' EA' \theta = \theta' AE' \theta,$$

let us compute the derivetive of $\mathcal{L}$ at the point (i.e. matrix) A in the direction E:

$$\frac{\partial \mathcal{L}(A)}{\partial E} = \lim_{\alpha \to 0} [\mathcal{L}(A + \alpha E) - \mathcal{L}(A)] / \alpha =$$

$$= 2 \frac{\theta' AZ' \theta}{(\theta' AA' \theta)^2} \theta' E(Z' \theta \theta' AA' \theta - A' \theta \theta' AZ' \theta)$$

Let us characterize now each matrix A for which the derivative equals zero, for all $E \in \mathcal{A}$. The equality $\partial \mathcal{L}(A)/\partial E = 0$ for all $E \in \mathcal{A}$ is equivalent to that either

$$\theta' AZ' \theta = 0 \tag{8.1.13}$$

or

$$Z' \theta \theta' AA' \theta = A' \theta \theta' AZ' \theta \tag{8.1.14}$$

takes place. If (8.1.13) holds, then $\mathcal{L}(A) = 0$ and thus $\mathcal{L}$ is being minimized: therefore we shall assume that (8.1.13) does not hold, i.e. $\theta' AZ' \theta \neq 0$. The equality (8.1.14) is equivalent to $A' \theta = cZ' \theta$, where

$$c = A'\theta\theta'A/Z'\theta\theta'A$$

is a positive number. Substitute $cZ'\theta$ for $A'\theta$ in the expression for $\mathcal{l}(A)$: this way, one obtains that if for some $A \in \mathcal{A}$, $\theta \neq 0$ (8.1.14) is valid, then $\mathcal{l}(A) = \theta ZZ'\theta$ the maximal value of $\mathcal{l}$ under a fixed θ. This value is obviously reached for $A=Z$. The theorem is proved.

The theorem implies that the optimal search direction, in the above sense, is given by

$$s^* = ZY \tag{8.1.15}$$

The advantages of s^* compared to $s=\hat{\theta}$, i.e. the least square estimate (8.1.4) of θ, are two-fold: it is much simpler to compute and can be used also for $N \leq n$; moreover, the search direction (8.1.15) is of slight sensibility to violations of the validity of the linear regression model (8.1.5). In fact, (8.1.15) is nothing but a cubature formula for estimating the vector

$$\int_U z f(z) dz$$

which converges to $\theta = \nabla f(x_k)$ when asymptotically decreasing the size of U.

8.1.3 Experimental design applying the search direction (8.1.15)

Theorem 8.1.2 yields that if the search direction $s=s_k$ is chosen at each step of the algorithm (8.1.1) by (8.1.15), then the experimental design (i.e. the selection rule for points $z_1,...,z_N$ in U) is to be selected so that the values $\theta'ZZ'\theta$ are as large as possible, for various θ. Let us formulate the design problem in its standard form for the regression design theory (see e.g. Ermakov and Zhigljavsky (1987)).

Let ε be an arbitrary approximate design, i.e. a probability measure on U,

$$M(\varepsilon) = \int_U zz'\varepsilon(dz) \tag{8.1.16}$$

be the information matrix of the design ε. For the discrete design

$$\varepsilon_N = \begin{Bmatrix} z_1, \ldots, z_N \\ 1/N, \ldots, 1/N \end{Bmatrix}$$

corresponding to evaluations of f at the points $z_1,...,z_N$, the matrix (8.1.16) equals

$$M(\varepsilon_N) = N^{-1}ZZ' = N^{-1}\sum_{j=1}^{N} z_j z_j'.$$

Considering the design problem in the set of all approximate designs, we have the class of criteria

$$\Phi_\theta(\varepsilon) = \theta' M(\varepsilon)\theta$$

depending on the unknown parameters θ. Since the true value of θ is unknown, the true optimality criterion Φ_θ is unknown, too. In a typical way, let us define the Bayesian criterion

$$\Phi_B(\varepsilon) = \int_\Omega \theta' M(\varepsilon)\theta\, \nu(d\theta) = \operatorname{tr} M(\varepsilon) \int_\Omega \theta\theta' \nu(d\theta) \tag{8.1.17}$$

and the maximum criterion

$$\Phi_M(\varepsilon) = \min_{\theta\in\Omega} \theta' M(\varepsilon)\theta \tag{8.1.18}$$

where $\Omega \subset \mathcal{R}^n$ is the set containing the unknown parameters θ with probability one and $\nu(d\theta)$ is a probability measure on Ω reflecting our prior knowledge about θ.

Since the norm $\|\theta\|$ of the gradient $\theta=\nabla f(x_k)$ is of no interest, we may assume that $\|\theta\|=1$, i.e.

$$\Omega \subset S = \{\theta \in \mathcal{R}^n : \|\theta\| = 1\}.$$

If prior information about θ is essentially not available, then it is natural to take S as Ω and the uniform probability measure on Ω as $\nu(d\theta)$. Now, if Ω=S, then by virtue of the maximal matrix eigenvalue properties, (8.1.18) is nothing but the maximal eigenvalue of $M(\varepsilon)$: this way, Φ_M is the well-known E-optimality criterion in the regression design theory. Thus, the optimal design problem for the maximum criterion (8.1.18) is reduced to that of classical regression design theory.

Let us turn now to the Bayesian criterion (8.1.17). Denoting

$$L = \int_\Omega \theta\theta' \nu(d\theta)$$

one can easily derive (see Zhigljavsky (1985)) from the equivalence theorems of regression design theory that the set of optimal designs with respect to the Bayesian criterion (8.1.17) i.e.

$$\left\{\varepsilon^* = \arg\max_\varepsilon \operatorname{tr} LM(\varepsilon)\right\}$$

coincides with the set of all probability measures concentrated on the set

$$\left\{ \arg \max_{z \in U} z'Lz \right\}.$$

Thus, the indicated design problem either is very easy or can be reduced to a standard problem.

8.2 *Optimal simultaneous estimation of several integrals by the Monte Carlo method*

A number of Monte Carlo and experimental design problems can be reduced to optimization of a convex functional of a matrix with the probability density in the denominator of each element. Existence and uniqueness of optimal densities for a large class of functionals are studied. Necessary and sufficient conditions of optimality are given. Algorithms for constructing optimal densities are suggested, the structure of these densities is investigated. The exposition of the section follows Zhigljavsky (1988) and, partly, Mikhailov and Zhigljavsky (1988).

8.2.1 Problem statement

A large number of Monte Carlo and experimental design problems consist in choosing a probability distribution P, given on an arbitrary measurable set $(\mathcal{X},\mathcal{B})$, with the density $p=dP/d\nu$ with respect to a σ-finite measure ν on $(\mathcal{X},\mathcal{B})$, sampling this distribution and estimating some linear functionals (integrals). The efficiency of such algorithms depends to a great extent on the choice of the density p and is defined as a functional of the estimator covariance matrix, with p in the denominator of its elements. These algorithms are considered in Zhigljavsky (1988) and some of them are presented below.

The optimality problem arising in the mentioned algorithms has the general form

$$p^* = \arg\min_{p\in\mathcal{P}} \Phi(D(p)). \tag{8.2.1}$$

Here the matrix

$$D(p) = \int \frac{g(x)g'(x)}{p(x)}\nu(dx) - A \tag{8.2.2}$$

has order $m\times m$ and in most cases is proportional to the covariance matrix of the estimator, $g(x)=(g_1(x),...,g_m(x))'$ is a vector of linearly independent piecewise continuous functions from $L_2(\mathcal{X},\nu)$, $\mathcal{P}$ is the set of densities p such that $\|D(p)\|<\infty$, A is a matrix in the set $\mathcal{N}$ of positive semi-definite matrices of order $m\times m$ for which $D(p)-A\in\mathcal{N}$ for all $p\in\mathcal{P}$ (in some cases A is the zero matrix), $\Phi: \mathcal{N}\rightarrow\mathcal{R}$ is a continuous convex functional.

The most well-known problem leading to (8.2.1) is the Monte Carlo simultaneous estimation of several integrals and is formulated as follows.

Let m integrals

$$\mathcal{I}_k = \int g_k(x)\nu(dx), \qquad k = 1,\dots,m, \tag{8.2.3}$$

be estimated. The ordinary Monte Carlo estimates of the integrals (8.2.3) are constructed in the following way. First a probability distribution $P(dx)=p(x)\nu(dx)$ on the measurable space $(\mathcal{X},\mathcal{B})$ is chosen: here the density $p=dP/d\nu$ is positive modulo ν on the set

$$\mathcal{X}_* = \{x \in \mathcal{X} : g_1^2(x) + \ldots + g_m^2(x) > 0\}.$$

Then N independent elements $x_1,\ldots,x_N$ are generated from the distribution P. Finally, the integrals (8.2.3) are estimated by the formulas

$$\hat{\iota}_k = \frac{1}{N}\sum_{j=1}^{N} g_k(x_j)/p(x_j), \qquad k = 1,\ldots,m. \tag{8.2.4}$$

Set $\quad \iota = (\iota_1,\ldots,\iota_m)', \quad \hat{\iota} = (\hat{\iota}_1,\ldots,\hat{\iota}_m)'$

and prove the following auxiliary assertion.

Proposition 8.2.1. If the density $p = dP/d\nu$ is positive modulo ν on the set $\mathcal{X}_*$, then the estimators (8.2.4) are unbiased

(i.e. $\quad E\hat{\iota} = \iota$)

and their covariance matrix equals

$$\operatorname{cov}\hat{\iota} = E(\hat{\iota} - \iota)(\hat{\iota} - \iota)' = \frac{1}{N}D(p)$$

where the matrix D(p) is defined by formula (8.2.2) with $A = \iota\iota'$.

Proof. The unbiasedness of the estimators (8.2.4) is an evident result, well-known in the Monte Carlo theory. We have for variances and covariances the relations

$$E(\hat{\iota}_j - \iota_j)(\hat{\iota}_k - \iota_k) = E\hat{\iota}_j\hat{\iota}_k - \iota_j\iota_k,$$

$$E\hat{\iota}_j\hat{\iota}_k = \frac{1}{N^2}\sum_{\substack{i,\ell=1\\ i\neq\ell}}^{N} E\frac{g_j(x_i)g_k(x_\ell)}{p(x_i)p(x_\ell)} + \frac{1}{N^2}\sum_{i=1}^{N} E\frac{g_j(x_i)g_k(x_i)}{p^2(x_i)} =$$

$$= \left(1 - \frac{1}{N}\right)\iota_j\iota_k + \frac{1}{N}\int \frac{g_j(x)g_k(x)}{p(x)}\nu(dx).$$

The formulas above imply the desired form of

$\operatorname{cov}\hat{\iota}$:

the proof is complete.

It follows from the proposition that the matrix (8.2.2) with $A=\mathfrak{l}\mathfrak{l}'$ represents the quality of the Monte Carlo estimators (8.2.4) depending on the density p. The problem of optimal density selection was earlier investigated in the above framework for the two optimality criteria depending only on diagonal elements of the matrix D(p). Evans (1963) solved the problem for the criterion

$$\Phi(B) = \sum_{i=1}^{m} a_i b_{ii}, \tag{8.2.5}$$

where b_{ii} are the diagonal elements of a matrix $B \in \mathcal{N}$ and $a_1,\ldots,a_m$ are fixed nonnegative numbers: this problem is rather simple. Mikhailov (1984, 1987) solved the extremal problem (8.2.1) for the MV-optimality criterion

$$\Phi(B) = \max_{1 \le i \le m} b_{ii}. \tag{8.2.6}$$

Section 8.2.6 describes Mikhailov's results.

Zhigljavsky (1985) studied the problem for general optimality criteria, but he has not exhaustively investigated the existence of optimal densities. Besides, here we study the minimax type optimality criteria more thoroughly than earlier.

Another important problem related to global optimization consists in estimating the Fourier coefficients of a regression function by its observations at random points and is described as follows.

Let f be a regression function on $\mathcal{X}$ with uncorrelated observations at N points x_j $j=1,\ldots,N$:

$$y(x_j) = f(x_j) + \xi(x_j), \qquad E\xi(x) = 0, \qquad E\xi^2(x) = \sigma^2(x)$$

and let the Fourier coefficients

$$\mathfrak{l}_k = \int f(x) f_k(x) \nu(dx), \qquad k = 1,\ldots,m,$$

of the function f with respect to a set of functions $\{f_1,\ldots,f_m\}$ be estimated.

Suppose now that the points $x_1,\ldots,x_N$ are randomly and independently chosen, having the same distribution $P(dx)=p(x)\nu(dx)$, where the density $p=dP/d\nu$ is positive on the set

$$\{x \in \mathcal{X}: f^2(x) + \sigma^2(x) > 0\}.$$

If one uses the Monte Carlo estimators

$$\hat{\mathfrak{l}}_k = \frac{1}{N}\sum_{j=1}^{N} y(x_j) f_k / p(x_j) \tag{8.2.7}$$

for ϑ_k then one can prove (analogously to Proposition 8.2.1) that these estimators are unbiased; further on, that their covariance matrix is equal to

$$\operatorname{cov}\hat{\vartheta} = N^{-1}D(p),$$

where D(p) is determined by (8.2.2), $A=\vartheta\vartheta'$ and

$$g_k(x) = \left(f^2(x) + \sigma^2(x)\right)^{1/2} f_k(x), \qquad k = 1,\ldots,m.$$

Therefore the optimal density choice problem is a particular case of the problem (8.2.1).

Let $r(x)=f(x)-(\vartheta_1 f_1(x)+\ldots+\vartheta_m f_m(x))$. If the least square estimates are used instead of (8.2.7), then their covariance matrix is represented as follows

$$\operatorname{cov}\hat{\vartheta} = \frac{1}{N}D(p) + o\left(N^{-1}\right), \qquad N \to \infty$$

where A=0,

$$g_k(x) = \left(f^2(x) + \sigma^2(x)\right)^{1/2} f_k(x), \qquad k = 1,\ldots,m,$$

(see later Theorem 8.3.4). This way, we have arrived to the problem (8.2.1) again.

The density p is interpreted as the experimental design; the problem of its optimization is similar to the approximate optimal design problem in classical regression design theory. The main difference between these problems lies in the following. In the classical theory experimental design stands in the numerator of the matrix elements and a convex functional of this matrix is minimized, while the design in the extremal problem (8.2.1) stands in the denominator of the elements. From the theoretical point of view, the problem (8.2.1) is a little more complicated than the regression optimal design problem. The main additional complexity is in the existence of the optimal design.

The task of selecting the optimality criterion Φ is analogous to the corresponding one in classical regression design theory. The main difference between them lies in imposing stronger conditions on the optimality criterion in the problem (8.2.1), than the convexity and monotonicity required in the regression design theory. Subsection 8.2.2 will describe these conditions.

Subsection 8.2.3 covers the existence and uniqueness problems: their solution is based on general convex analysis.

Subsection 8.2.4 presents the necessary and sufficient conditions for densities to be optimal. The basis for the results of the subsection is the equivalence theory for optimal regression design developed by J. Kiefer, J, Wolfowitz, V. Fedorov and others: its statements are also analogous to the equivalence theorems.

Subsection 8.2.5 describes algorithms for constructing optimal densities and the structure of optimal densities. Nondifferentiable MV- and E-optimality criteria are treated in Subsection 8.2.6.

Subsection 8.2.7 highlights the difference and similarities between classical regression design theory and the results presented here.

8.2.2 Assumptions

Below we shall suppose that the functional $\Phi: \mathcal{N} \to \mathcal{R}$ is nonnegative, continuous, convex and increasing. The increase of Φ is defined as follows: if $B, C \in \mathcal{N}$ and $B > C$ (or $B \geq C$), then $\Phi(B) > \Phi(C)$ or, respectively, $\Phi(B) \geq \Phi(C)$.

Let $\mathcal{M}$ be the closure of the set $\mathcal{N}\text{-}A = \{C-A, C \in \mathcal{N}\}$ containing some matrices with infinite elements. Extend the functional Φ from $\mathcal{N}$ onto $\mathcal{M}$ preserving its continuity and convexity. Suppose that this extension $\Phi: \mathcal{M} \to \mathcal{R} \cup \{+\infty\}$ has the property

a) $\Phi(B) < \infty$ for $B \in \mathcal{M}$, if and only if all elements of B are finite.

We also suppose that the following simple condition is satisfied:

b) there exists a density p in $\mathcal{P}$ for which the matrix D(p) consists of finite elements only.

The above suppositions are required to hold everywhere. Many widely used criteria satisfy them, for instance, the linear criterion

$$\Phi(B) = \operatorname{tr} LB, \qquad L \in \mathcal{N}, \tag{8.2.8}$$

its special case (8.2.5), the MV-criterion (8.2.6), the E-criterion

$$\Phi(B) = \lambda_{max}(B) \tag{8.2.9}$$

where $\lambda_{max}(B)$ is the maximal eigenvalue of the matrix B, and the so-called Φ_r-criterion

$$\Phi(B) = \left(\operatorname{tr} B^r/m\right)^{1/r} \tag{8.2.10}$$

for $1 \leq r < \infty$. If $r = 1$ then (8.2.10) and (8.2.8) with $L = I_m$ coincide. If $r \to \infty$ then (8.2.10) converges to (8.2.9). On the contrary if $-\infty < r < 1$, then according to Pukelsheim (1987) the criterion (8.2.10) is increasing and concave which case is unsuitable here. The same is true for the well known D-criterion

$$\Phi(B) = \left(m^{-1} \det B\right)^{1/m} \tag{8.2.11}$$

which can be regarded as the limit of (8.2.10) under $r \to 0$.

Sometimes we need the condition of strict convexity of the functional $\Phi: \mathcal{N} \to \mathcal{R}$ and also its differentiability. The last condition means the existence of the matrix

$$\overset{o}{\Phi}(B) = \left(\frac{d\Phi}{dB}\right)(B)$$

consisting of the partial derivatives of $\Phi(B)$ with respect to the elements b_{ij} of the matrix $B \in \mathcal{N}$.

All the above suppositions are related to the functional Φ. Besides them, we need two further assumptions concerning the functions $g_1,...,g_m$. It will be supposed throughout that they are piecewise continuous linearly independent functions from $L_2(\mathcal{X},\nu)$. Sometimes we shall also use the following condition of their ν-regularity:

c) functions $g_1,...,g_m$ are linearly independent on any measurable subset Z of the set $\mathcal{X}$ with $\nu(Z)>0$.

In a slightly different form the ν-regularity condition c) was used by Ermakov (1975), when investigating random quadrature formulas.

Note also that we do not require more concerning the set $\mathcal{X}$, than its measurability.

8.2.3 Existence and uniqueness of optimal densities

Set $p_+(x)=\max\{0,p(x)\}$ for all p of $L_1(\mathcal{X},\nu)$ and

$$Q = \{p \in L_1(\mathcal{X},\nu) : \Phi(D(p_+)) < \infty\}.$$

It is evident that the set **Q** is convex. Define the functional

$$\varphi : L_1(\mathcal{X},\nu) \to Q \cup \{+\infty\}$$

by $\varphi(p)=\Phi(D(p_+))$: we shall prove first that φ is convex.

Proposition 8.2.2. For all p,q form **Q** and any $0<t<1$ we have

$$D(t,p,q) = \int u(x)g(x)g'(x)\nu(dx) \tag{8.2.12}$$

where the matrix D(t,p,q) is defined by formula

$$D(t,p,q) = t\,D(p_+) + (1-t)D(q_+) - D\big((t\,p + (1-t)q)_+\big), \tag{8.2.13}$$

$$u(x) = \frac{t(1-t)(p_+(x) - q_+(x))^2}{p_+(x)q_+(x)(t\,p_+(x) + (1-t)q_+(x))} \ge 0,$$

and the matrix D(t,p,q) is positive semi-definite.

The proof is based on elementary algebraic transformations, see Lemma 8.1.2 in Zhigljavsky (1985).

Proposition 8.2.3. If condition c) holds, then the matrix (8.2.13) is positive definite for any t from (0,1) and all p,q from $\mathbf{Q}$, $p \neq q$ (modulo ν).

Proof. By virtue of Proposition 8.2.3, the matrix (8.2.13) can be written as

$$D(t,p,q) = \int_Z u(x)g(x)g'(x)\nu(dx) \tag{8.2.14}$$

where $Z=\{x \in \mathcal{X}: u(x)>0\}$. Since $p \neq q$ (modulo ν) and $t \in (0,1)$, then $\nu(Z)>0$. Using the supposition c), we obtain that functions $g_1,\ldots,g_m$ are linearly independent on Z. Evidently, the functions

$$\sqrt{u(x)}\, g_i(x), \qquad i = 1,\ldots,m,$$

have the same property. The positive definiteness of matrix (8.2.14) follows now directly from the definition of positive definiteness: the proof is completed.

Proposition 8.2.4. If the functional Φ is convex, increasing and the conditions a), b) are fulfilled, then the functional φ is convex on $L_1(\mathcal{X},\nu)$. Besides, if either the functional Φ is strictly convex on $\mathcal{N}$ or the condition c) holds, then φ is a strictly convex functional on $\mathcal{P}$.

Proof. Let $p,q \in L_1(\mathcal{X},\nu)$, $p \neq q$ (modulo v), and $t \in (0,1)$. The convexity of φ implies that the inequality

$$t\varphi(p) + (1-t)\varphi(q) \geq \varphi(tp + (1-t)q) \tag{8.2.15}$$

holds. If either $p \notin \mathbf{Q}$ or $q \notin \mathbf{Q}$, then the inequality (8.2.15) is certainly fulfilled; therefore we may suppose that p and q belong to $\mathbf{Q}$.
By virtue of Proposition 8.2.2, we have

$$tD(p_+) + (1-t)D(q_+) \geq D\big((tp + (1-t)q)_+\big). \tag{8.2.16}$$

This and the increase of Φ yields the inequality

$$\Phi\big(tD(p_+) + (1-t)D(q_+)\big) \geq \varphi(tp + (1-t)q). \tag{8.2.17}$$

The convexity of Φ implies that

$$t\varphi(p) + (1-t)\varphi(q) \geq \Phi\big(tD(p_+) + (1-t)D(q_+)\big). \tag{8.2.18}$$

Coupling the inequalities (8.2.17) and (8.2.18) we obtain (8.2.15), i.e. the convexity of φ.

The strict convexity of φ on $\mathcal{P}$ is equivalent to the strict inequality in (8.2.15) for p and q from $\mathcal{P}$. This inequality follows from the strict inequalities in either (8.2.17) or (8.2.18) or both. The strict inequality in (8.2.18) follows from the strict convexity of Φ. On the other hand, if the condition c) holds then, by virtue of Proposition 8.2.3, the strict inequality in (8.2.16) takes place. This way, by virtue of the increase of Φ the strict inequality in (8.2.17) is valid. The proof is completed.

Now we are able to investigate the existence and uniqueness of the optimal density.

Theorem 8.2.1. If the functional Φ is continuous, convex, increasing and conditions a), b) are fulfilled, then the optimal density p^* in $\mathcal{P}$ exists.

Proof. The set

$$S = \left\{p \in L_1(\mathcal{X},\nu): p \geq 0, \int p(x)\nu(dx) \leq 1\right\}$$

contains $\mathcal{P}$, is bounded in norm and (due to the Fatou theorem) closed by measure.

Let us show that the functional φ is lower semi-continuous in measure: it means that the inequality

$$\liminf_{i\to\infty} \varphi(p_i) \geq \varphi(p) \tag{8.2.19}$$

holds for any sequence $p_1, p_2, \ldots$ of elements of $L_1(\mathcal{X},\nu)$ which converges in measure ν to an element p of $L_1(\mathcal{X},\nu)$.

Using the Fatou theorem, we have for any vector $b \in \mathbb{R}^m$:

$$b'\left(\liminf_{i\to\infty} D\left((p_i)_+\right)\right)b = \liminf_{i\to\infty} \int \frac{b'g(x)g'(x)b}{(p_i(x))_+}\nu(dx) - b'Ab \geq$$

$$\geq \int \frac{b'g(x)g'(x)b}{p_+(x)}\nu(dx) - b'Ab = b'D(p_+)b.$$

This way, the inequality

$$\liminf_{i\to\infty} D\left((p_i)_+\right) \geq D(p_i)$$

holds. From this, we obtain the inequality (8.2.19) and the continuity and monotonicity of the functional Φ.

Theorem 6 from §5 Chapter 10 of Kantorovich and Akilov (1977) states a generalized Weierstrass-type theorem, viz. that a convex lower semicontinuous in measure functional on $L_1(\mathcal{X},\nu)$ attains its minimum value in any subset of $L_1(\mathcal{X},\nu)$ closed by measure and bounded in norm. Using Proposition 8.2.4 and the above results we get that the functional

φ attains its minimum value in S at a certain point $p^* \in S$. By a) and b) we have $\Phi(D(p^*))<\infty$ and $\|D(p^*)\|<\infty$. Let us show that $\int p^*(x)\nu(dx)=1$. It follows by the incidence $p^* \in \mathcal{P}$ and the statement of Theorem, too. Assuming the contrary, let

$$r = \int p^*(x)\nu(dx) < 1$$

and put $q^*(x)=p^*(x)/r$. We have $q^* \in \mathcal{P}$,

$$D(q^*) = r(D(p^*) + A) - A, \qquad D(p^*) - D(q^*) = (1-r)(D(p^*) + A).$$

Since $\|D(p^*)\|<\infty$ and the functions $g_1,\ldots,g_m$ are linearly independent on $\mathcal{X}$, therefore the functions

$$(p^*(x))^{-1/2} g_i(x), \qquad i = 1,\ldots,m,$$

have the same property. Analogously to Proposition 8.2.3, we can see that the matrix $D(p^*)+A$ is positively defined. From the monotonicity of the functional Φ we get $\Phi(D(p^*))>\Phi(D(q^*))$, but this inequality contradicts to the optimality of p^*. Consequently, $r=1$ and $p^* \in \mathcal{P}$: the theorem is proved.

Let us turn now to the uniqueness of the optimal density.

Proposition 8.2.5. Let the conditions of Theorem 8.2.1 be satisfied and assume that either the functional Φ is strictly convex or the condition c) holds. Then the optimal density $p^* \in \mathcal{P}$ exists and is unique modulo ν.

This statement is a simple corollary of Theorem 8.2.1 and Proposition 8.2.4.

8.2.4 Necessary and sufficient optimality conditions

The following statement is analogous to the classical equivalence theorem of regression design theory.

Theorem 8.2.2. Let the assumptions of Theorem 8.2.1 be fulfilled and the functional Φ be differentiable. Then a necessary and sufficient optimality condition, for a density p^*, is the fulfilment of the equality

$$\psi(x,p^*) = c(p^*) \tag{8.2.20}$$

for ν-almost all x in $\mathcal{X}$. Here

$$c(p) = \operatorname{tr} \overset{\circ}{\Phi}(D(p))(D(p) + A), \qquad \psi(x,p) = g'(x)\, \overset{\circ}{\Phi}(D(p))\, g(x)/p^2(x) \tag{8.2.21}$$

Proof. We shall use the necessary and sufficient condition of optimality, for a convex differentiable (along all admissible directions) functional on a convex set (see e.g. Ermakov (1983), p.55 or Ermakov and Zhigljavsky (1987), p.105). Let us compute the derivative

$$\Pi(p,q) = \frac{\partial}{\partial t}\varphi((1-t)p + tq)\Big|_{t=0+} \tag{8.2.22}$$

and find the density p^* in the set $\mathcal{P}$ such that $\Pi(p,h)\geq 0$, for all densities h on $\mathcal{X}$. Simple calculations (see Zhigljavsky (1985)) yield

$$\Pi(p,q) = c(p) - \int \psi(x,p)q(x)\nu(dx)$$

So $\Pi(p,h)\geq 0$ for all h, if and only if the inequality $\psi(x,p)\leq c(p)$ holds for ν-almost x in $\mathcal{X}$. The statement of the theorem follows from the equality

$$\int \psi(x,p)p(x)\nu(dx) = c(p)$$

which is easily verified and is valid for all densities $p\in P$. The theorem is proved.

Suppose now that the optimality criterion is nondifferentiable and minimax, i.e. it is expressed by

$$\Phi(B) = \max_{v\in V} \Phi_v(B), \tag{8.2.23}$$

where V is a compact set and all functionals Φ_v are convex and differentiable.

Theorem 8.2.3. Let the functional Φ have the form (8.2.23), where all functionals $\Phi_v (v\in V)$ satisfy the assumptions of Theorem 8.2.2 and the function $\Phi_v(B)$ is continuous for any fixed matrix B from $\mathcal{N}$. Then the fulfilment of the inequality

$$\sup_{x\in \mathcal{X}} \psi_v(x,p^*) \leq c_v(p^*)$$

for a certain v from the set

$$V(p^*) = \left\{\arg\max_{v\in V} \Phi_v(D(p^*))\right\}$$

is a necessary and sufficient condition for the optimality of density p^*: here ψ_v and c_v are defined by (8.2.21), with the substitution of Φ_v for Φ.

The proof is analogous to the proof of Theorem 8.2.2 and uses the formula

$$\Pi(p,q) = \max_{v \in V(p)} \Pi_v(p,q)$$

where the derivative Π_v is defined in (8.2.22), with the substitution of Φ_v for Φ, and equals

$$\Pi_v(p,q) = c_v(p) - \int \psi_v(x,p) q(x) v(dx)$$

Note finally that Theorem 8.2.3 is also analogous to the corresponding statement in regression design theory.

8.2.5 *Construction and structure of optimal densities*

Consider first the structure of optimal densities for the linear criterion (8.2.8). In this case

$$\overset{o}{\Phi}(B) = d\,\mathrm{tr}\,LB/dB = L$$

and the optimality condition (8.2.20) is equivalent to

$$p^*(x) = (g'(x)Lg(x))^{1/2} / \int (g'(z)Lg(z))^{1/2} v(dz). \qquad (8.2.24)$$

It should be noted that the expression (8.2.24) can be obtained by simpler tools, using the Cauchy-Schwarz inequality, and that the linearly optimal density is always unique, modulo v. The following example presents a case, when the linearly optimal density does not belong to the set $\mathcal{P}$.

Example. Let the functional Φ have the form (8.2.5), $m=2$, $a_1=1$, $a_2=0$. The condition a) does not hold and the optimal density has the form $p^* = |g_1| / \int |g_1| dv$. If the function g_1 vanishes on a subset Z of $\mathcal{X}$ with measure $v(Z)>0$, but the function g_2 does not vanish on the subset then p^* does not belong to $\mathcal{P}$. At the same time, if g_1 does not vanish on $\mathcal{X}$ and $g_1^2/|g_2| \in L_1(\mathcal{X},v)$, then $p^* \in \mathcal{P}$.

Consider now the structure of the optimal density and algorithms of its construction, for an arbitrary differentiable criterion.

If the functional Φ is differentiable, but can not be represented as (8.2.8), then the optimal density has the form (8.2.24) again, although now the matrix

$$L = \overset{o}{\Phi}(D(p^*))$$

is unknown and depends upon the optimal density p^*. This is a simple corollary of Theorem 8.2.2.

Hence, the problem of optimal density construction may be considered as the problem of constructing the optimal matrix L. It can be solved by general global optimization techniques. If the values $\Phi(B)$ depend on the diagonal elements of matrices B only, then the matirx L is diagonal and the extremal problem is not very complicated.

If the number m is large and the set $\mathbf{X}$ is either discrete or has a small dimension, then the above algorithms may be of smaller efficiency than those described below. They use the features of the problem and are analogous to the construction methods in optimal regression design. They are pseudogradient-based algorithms in $\mathbf{P}$, using the expression for the derivative Π.

The general form of these methods is

$$p_{k+1}(x) = (1-\alpha_k)p_k(x) + \alpha_k h_k(x), \qquad k = 1,2,\ldots \tag{8.2.25}$$

Here $p_1 \in \mathbf{P}$ is an initial density, $\alpha_1, \alpha_2, \ldots$ is a numerical sequence the choice of which may be the same as in classical regression design. If $\alpha_k > 0$, then the density h_k has to satisfy the inequality $\Pi(p_k, h_k) \leq 0$: for example, one such density is proportional to the positivity indicator of the function

$$\psi(x, p_k) - c(p_k) + \varepsilon_k, \tag{8.2.26}$$

where $\varepsilon_k \geq 0$, $\varepsilon_k \to 0$ for $k \to \infty$. If $\alpha_k < 0$, then the inequalities $\Pi(p_k, h_k) \leq 0$ and

$$|\alpha_k| \leq 1/\left(\sup_x (h_k(x)/p_k(x)) - 1\right)$$

are to be satisfied. The former is satisfied, for instance, for the density h_k proportional to the negativity indicator of the function (8.2.26) with $\varepsilon_k \leq 0$; the latter relation is equivalent to the nonnegativity of the density $p_{k+1}(x)$.

8.2.6 *Structure of optimal densities for nondifferentiable criteria*

Unlike in case of differentiable criteria, the necessary and sufficient condition of optimality for nondifferentiable criteria (see Theorem 8.2.3) is nonconstructive and can not be used for the construction of optimal densities, but only for verifying their optimality. Nevertheless, for a large class of nondifferentiable criteria, we are able to show that the structure of optimal densities is the same as above (cf. (8.2.24)).

Theorem 8.2.4. Suppose that there exist a functional Φ and a sequence $\{\Phi_i\}$ of functionals on $\mathbf{N}$ which are convex, increasing, continuous, and for which the conditions a), b) are valid. Let also the functionals Φ_i be differentiable, the condition c) hold and

$$\Phi_{i+1}(B) \leq \Phi_i(B), \qquad i = 1,2,\ldots, \qquad \lim_{i\to\infty} \Phi_i(B) = \Phi(B)$$

for each $B \in \mathbf{N}$. Then the Φ-optimal density p^* exists, it is unique modulo ν, and has the form (8.2.24).

Proof. Denote the Φ_i-optimal density by p_i^*. The existence and uniqueness of the densities p^* and p_i^* follow from Theorem 8.2.1 and Proposition 8.2.5. Theorem 8.2.2 gives

$$p_i^*(x) = \left(g'(x)L_i g(x)\right)^{1/2} / \int \left(g'(z)L_i g(z)\right)^{1/2} \nu(dz), \qquad (8.2.27)$$

where $L_1, L_2, \ldots$ are matrices from $\mathbf{N}$. Without loss of generality, we can assume that $\|L_i\| = 1$ for all $i = 1, 2, \ldots$ Let us choose now a subsequence $\{L_{ij}\}$ from the sequence $\{L_i\}$ converging to a certain matrix L; notice that $L \in \mathbf{N}$ and $\|L\| = 1$.

Define the density p by formula (8.2.24) and note that the pointwise limit of the sequence $\{p_i^*\}$ exists.

Analogously to the proof of Lemma 1.11 in Fedorov (1979), one is able to show that if a limit point of the sequence $\{p_i^*\}$ exists, then this point is the Φ-optimal density p^*. By virtue of the uniqueness of the limit, we have $p^* = p$. Finally, from the uniqueness of the Φ-optimal density p^* we obtain that the limit of the sequence $\{L_i\}$ exists and equal L. The theorem is proved.

It follows from the theorem that, for many nondifferentiable criteria, the problem of optimal density construction is reduced again to the optimal choice of the matirx L in the representation (8.2.24).

Two nondifferentiable criteria of special importance are considered below.

Let Φ have the form of (8.2.9), i.e. it is an E-optimality criterion. Determine the functionals Φ_i by formula

$$\Phi_i(B) = \left(\operatorname{tr} B^i\right)^{1/i}$$

where $B \in \mathbf{N}$. We can apply the classical inequality

$$\left(\sum_{k=1}^{m} a_k^i\right)^{1/i} \geq \left(\sum_{k=1}^{m} a_k^j\right)^{1/j}, \qquad 1 \leq i \leq j \leq \infty \qquad (8.2.28)$$

for any nonnegative numbers $a_1, \ldots, a_m$. Now we obtain the monotonicity of the convergence of $\Phi_i(B)$ to $\Phi(B)$ from the inequality (8.2.28) and the representation

$$\operatorname{tr} B^i = \sum_{k=1}^{m} \lambda_k^i$$

where $\lambda_1,...,\lambda_m$ are the eigenvalues of the matrix B. Hence, the analogue of Theorem 8.2.4 is applicable and the E-optimal density has the form (8.2.24).

Let us turn now to the MV-optimality criterion (8.2.6). First, let us simplify the statement of Theorem 8.2.2 for the MV-criterion. In the representation (8.2.23) we have

$$V = \{1,\ldots,m\}, \qquad \Phi_j(B) = \mathrm{tr}\, E_j B$$

where all elements of the matrix E_j equal zero except the (j,j)-element which equals one. So

$$\overset{o}{\Phi}_j(B) = E_j, c_j(p) = \int \left(g_j^2(z)/p(z) \right) \nu(dz), \qquad \psi_j(x,p) = g_j^2(x)/p^2(x).$$

We shall show now that Theorem 8.2.4 can be applied to the MV-criterion. Determine the sequence $\{\Phi_i\}$ by

$$\Phi_i(B) = \left(\sum_{k=1}^{m} b_{kk}^i \right)^{1/i}.$$

The convexity of these criteria follows from the Minkovsky inequality and their monotonous convergence to Φ follows by (8.2.28). Theorem 8.2.4 says that if the condition c) holds, then the MV-optimal density p* exists, it is unique modulo ν, and has the form (8.2.24), where the matrix L is diagonal. This means that if the condition c) holds, then the MV-optimal density exists and has the form

$$p^* = p_u = \left(u_1 g_1^2 + \ldots + u_m g_m^2 \right)^{1/2} / \int \left(u_1 g_1^2 + \ldots + u_m g_m^2 \right)^{1/2} d\nu \quad (8.2.29)$$

where

$$u = (u_1,\ldots,u_m) \in U = \left\{ v = (v_1,\ldots,v_m) : v_i \geq 0, \sum_{i=1}^{m} v_i = 1 \right\}.$$

The following statement expresses a somewhat stronger result.

Theorem 8.2.5. The MV-optimal density exists and can be represented in the form (8.2.29), where

$$u = \arg\max_{v \in U} G(v), \qquad G(v) = \left[\int \left(\sum_{k=1}^{m} v_k g_k^2(x) \right)^{1/2} \nu(dx) \right]^2 - \sum_{k=1}^{m} v_k a_{kk}^2,$$

(here a_{kk} is the k-th element of the matrix A).

Proof. By virtue of Theorem 8.2.1, the MV-optimal density exists. Consider the value

$$I = \min_{p \in \mathcal{P}} \max_{1 \le i \le m} d_{ii}(p),$$

where $d_{ii}(p)$ are the diagonal elements of the matrix D(p). We have

$$I = \min_{p \in S} \max_{1 \le i \le m} d_{ii}(p) = \min_{p \in S} \max_{v \in U} \sum_{i=1}^{m} v_i d_{ii}(p).$$

The proof of Theorem 8.2.1 indicates that the set S is closed by measure and bounded in norm and the functional φ is convex and lower semicontinuous in the measure. Using now the minimax theorem of Levin (1985), p.293, we have

$$I = \max_{v \in U} I(v),$$

$$I(v) = \min_{p \in S} \sum_{i-1}^{m} v_i d_{ii}(p)$$

By virtue of Section 8.2.5, the minimum I(v) is attained at the density p_v and equals G(v). This fact completes the proof.

Let us remark that under some additional restrictions on the functions $g_1,\ldots,g_m$, Mikhailov (1984) proved a statement similar to Theorem 8.2.5: the latter is due to Mikhailov and Zhigljavsky (1988).

8.2.7 Connection with the regression design theory

First we shall point out some general differences between the above exposition and the classical regression design.

The first difference is the existence of an optimal design. Two conditions a) and b) concerning the optimality criterion, are required to ensure the existence of the optimal density. These conditions do not figure in the regression design theory; however, they are not restrictive.

The second one is that the discreteness of optimal measures in the present case is not desired, even being not admissible.

The third is that if the functions $g_1,\ldots,g_m$ are ν-regular (see condition c)), then according to Proposition 8.2.4 a convex functional Φ on $\mathcal{N}$ induces the strictly convex functional φ on the set of densities $\mathcal{P}$. This allows to guarantee the uniqueness of optimal densities, e.g. for the E- and MV-criteria.

The fourth is that the structure of optimal densities is the same: for a large class of criteria, the optimal densities have the form (8.2.24).

Finally, the fifth difference is that there are no conditions concerning the set $\mathcal{X}$ except its measurability, and the functions $g_1,\ldots,g_m$ are not necessarily continuous.

In some cases the matrix A depends on unknown integrals $\mathcal{I}$ and, hence, the optimal densities depend on them, too. In these cases the extremal problem (8.2.1) is analogous to the nonlinear regression design problem and the optimal density corresponds to the locally optimal design. Sequential, Bayesian or minimax approaches can be used to determine the optimal densities.

The optimality results (Theorems 8.2.2 and 8.2.3) are analogous to the equivalence theorems in classical regression design: apparently, theorems analogous to the dulaity theorems of Pukelsheim (1980) can also be proved.

8.3 *Projection estimation of multivariate regression*

This section deals with some theoretical aspests concerning the asymptotically optimal projection estimation of a multivariate regression. In particular, it is shown that, in a rather general situation, the uniform random choice of points to evaluate the regression function, together with Monte Carlo estimates of its Fourier coefficients provide an asymptotically optimal design and projection estimation procedure. (For a detailed exposition, reference is made to Zhigljavsky (1985).)

8.3.1 *Problem statement*

A general scheme of the regression experiment is as follows. Let the evaluation result at a point $x \in \mathcal{X}$ be a realization of a (partly) random function

$$y(x) = f(x) + \xi(x),$$

where ξ is a zero mean random variable (the error at x), f is a regression function given on $\mathcal{X} \subset \mathbb{R}^n$, belonging to a functional class $\mathcal{F}$. Obtaining for given $x_j \in \mathcal{X}$ $(j=1,\ldots,N)$ the values $y_j=y(x_j)$, one is interested in estimating the regression function $f(x)=Ey(x)$. The problem requires *a priori* information for its correct statement, consisting in indication of the functional class $\mathcal{F}$ and properties of the distributions of the random errors $\xi(x_j)=y(x_j)-f(x_j)$. We shall suppose that the errors are uncorrelated, their variance $\sigma^2(x)=E\xi^2(x)$ is continuous and bounded but may be unknown, and that $\mathcal{F}$ is an infinite dimensional functional class (in this case the estimation problem is nonparametric).

Many nonparametric regression function estimation approaches are known, see e.g. Prakasa Rao (1983): we shall confine ourselves to the projection estimation of a multivariate regression function, as this problem is not thoroughly studied.

To construct a projection estimate of a regression function f belonging to a functional class $\mathcal{F}$, one supposes that an increasing sequence $\{L_m\}$ of m-dimensional spaces is given such that

$$\mathcal{F} \subset \lim_{m\to\infty} L_m \qquad (8.3.1)$$

and the evaluation number N tends to infinity (or is sufficiently large). The projection estimation problem consists of choosing for the spaces L_m dimensions $m=m(N)$, the sequence of passive designs $\{x_1,\ldots,x_N\}$, and a parametric regression estimation method under the supposition $f \in L_m$ in such a way that the obtained estimate

$$\hat{f}_N \in L_m$$

approximates the true regression function $f \in \mathcal{F}$ in a well-defined sense. Sometimes the algorithm of selecting the points $x_1,\ldots,x_N$ is assumed to be given and thus it is not subject to optimization.

A projection estimate of f is representable as

$$\hat{f}_N(x) = \sum_{i=1}^{m} \hat{\theta}_i f_i(x), \tag{8.3.2}$$

where $\{f_1,...,f_m\}$ is the basis of the space L_m,

$$\hat{\theta}_1, \ldots, \hat{\theta}_m$$

are estimates of the unknown parameters of the linear regression

$$\sum_{i=1}^{m} \theta_i f_i(x) \tag{8.3.3}$$

computed using x_j and y_j (j=1,...,N). We shall assume that the estimation method of θ_i is linear with respect to $y_1,...,y_N$.

8.3.2 *Bias and random inaccuracies of nonparametric estimates*

Let the number N of regression function evaluations as well as the estimation algorithm be fixed. We have the following decomposition of the inaccuracy

$$f - \hat{f}$$

corresponding to an estimate of f:

$$f(x) - \hat{f}(x) = \left(f(x) - E\hat{f}(x)\right) + \left(E\hat{f}(x) - \hat{f}(x)\right).$$

Consequently, for an arbitrary metric ρ on $\mathcal{F}$ we have

$$\rho(f - \hat{f}) \le \rho(f - E\hat{f}) + \rho(E\hat{f} - \hat{f}). \tag{8.3.4}$$

A parametric problem of mathematical statistics is often connected with the second summand in the right-hand side of (8.3.4) and this term does not depend explicitly on the unknown function f. The first summand in the right-hand side of (8.3.4) contains the unknown function f and with this term a compromise minimax criterion

$$\sup_{f \in \mathcal{F}} \rho(f - E\hat{f}) \tag{8.3.5}$$

is usually associated. Analogously, instead of the right-hand side of (8.3.4)

$$\sup_{f \in \mathcal{F}} \{\rho(f - E\hat{f}) + \rho(E\hat{f} - \hat{f})\}$$

is often considered.

If a measure λ may be determined on the set $\mathcal{F}$ reflecting additional information about the unknown function $f \in \mathcal{F}$, then - instead of (8.3.5) - the first summand in the right-hand side of (8.3.4) is usually replaced by the Bayesian criterion

$$\int_{\mathcal{F}} \rho(f - E\hat{f}) \lambda(df).$$

As for ρ, the quadratic metric is considered below: for this, as easily seen, instead of the inequality (8.3.4) the equality

$$E\int (f(x) - \hat{f}(x))^2 \nu(dx) = \int (f(x) - E\hat{f}(x))^2 \nu(dx) +$$

$$+ E\int (E\hat{f}(x) - \hat{f}(x))^2 \nu(dx) \tag{8.3.6}$$

takes place, where $\nu(dx)$ is a given probability measure on $(\mathcal{X}, \mathcal{B})$. Consequently, a mean square summed inaccuracy

$$B^2 + V^2 = \sup_{f \in \mathcal{F}} \int (f - \hat{f})^2 d\nu + E\int (E\hat{f} - \hat{f})^2 d\nu \tag{8.3.7}$$

completely characterizes the error of the method. The first term in (8.3.7), i.e. the quantity

$$B^2 = \sup_{f \in \mathcal{F}} \int (f - E\hat{f})^2 d\nu$$

is the square of the so-called bias inaccuracy, while the second term

$$V^2 = E\int (E\hat{f} - \hat{f})^2 d\nu$$

is the mean square of the random inaccuracy. Since for all $x \in \mathcal{X}$ the variance of the estimate

$$\hat{f}(x)$$

is equal to

$$\operatorname{var} \hat{f}(x) = E\left(E\hat{f}(x) - f(x)\right)^2,$$

the square of the random inaccuracy V can be written in the form

$$V^2 = \int \left(\operatorname{var} \hat{f}(x)\right) \nu(dx). \qquad (8.3.8)$$

Consider now some properties of the random inaccuracy of projection estimates.

Let the number N and the space L_m be fixed, $F(x)=(f_1(x),...,f_m(x))'$ be the vector of the ν-orthonormal base functions of the space L_m, the evaluations of the regression function be preformed at points $x_1,...,x_N$,

$$\theta_i = \int f(x) f_i(x) \nu(dx)$$

be the Fourier coefficients of f with respect to the functions f_i $(i=1,...,m)$, $\theta=(\theta_1,...,\theta_m)'$,

$$\hat{f}(x) = \hat{\theta}' F(x)$$

be the regression function estimate where

$$\hat{\theta}$$

is the vector of linear estimates of θ. Assume that $\sigma^2(x)= \sigma^2$=const. for all $x \in \mathcal{X}$.

By virtue of (8.3.2) and (8.3.8), V^2 is representable as

$$V^2 = \int F'(x)\left(\operatorname{cov}\hat{\theta}\right) F(x) \nu(dx) = \operatorname{tr} \operatorname{cov}\hat{\theta}.$$

According to the classical Gauss-Markov theorem applied in regression analysis, the best linear unbiased estimates are the least square estimates for which, in particular, the quantity

$$\operatorname{tr} \operatorname{cov} \hat{\theta}$$

is minimal. The main object of this subsection is the lower estimation of V: therefore we shall consider only the least square estimates for which

$$\operatorname{cov}\hat{\theta} = (\sigma^2/N)\left[M(\varepsilon_N)\right]^{-1}, \qquad (8.3.9)$$

where

$$M(\varepsilon_N) = N^{-1}\sum_{j=1}^{N} F(x_j)F'(x_j)$$

is the normalized information matrix of the experimental design that can be written as follows

$$\varepsilon_N = \begin{Bmatrix} x_1, \dots, x_N \\ 1/N, \dots, 1/N \end{Bmatrix} \tag{8.3.10}$$

Proposition 8.3.1. Let the functions $f_1, \dots, f_m$ be ν-orthonormal and uniformly lower bounded by a constant K, $\sigma^2(x)=\sigma^2<\infty$, an estimate

$$\hat{f}$$

of the regression function f be the projection (8.3.2), where

$$\hat{\theta}_i$$

are linear statistics, and let the matrix $M(\varepsilon_N)$ be nondegenerate. Then the inequality

$$\text{tr cov } \hat{\theta} \geq \sigma^2 m/(K^2 N) \tag{8.3.11}$$

holds.

The proof is based on the following statement.

Lemma 8.3.1. If A is a positive definite matrix of order m×m, then the inequality

$$\text{tr } A^{-1} \geq m^2/\text{tr } A \tag{8.3.12}$$

holds.

Proof. Let B be a positive definite matrix of order m×m and $\lambda_1, \dots, \lambda_m$ be its eigenvalues. By virtue of

$$\text{tr } B = \sum_{i=1}^{m} \lambda_i, \qquad \det B = \prod_{i=1}^{m} \lambda_i$$

the inequality between geometric and arithmetic means

$$\left(\prod_{i=1}^{m} \lambda_i\right)^{1/m} \leq m^{-1} \sum_{i=1}^{m} \lambda_i$$

can be rewritten like

$$\operatorname{tr} B \geq m(\det B)^{1/m}. \tag{8.3.13}$$

Applying the latter inequality to the matrices $B=A^{-1}$ and $B=A$, we obtain that

$$\operatorname{tr} A^{-1} \geq m\left(\det A^{-1}\right)^{1/m} = m(\det A)^{-1/m} \geq$$

$$\geq m((\operatorname{tr} A)/m)^{-1} = m^2/\operatorname{tr} A.$$

The lemma is proved.

Proof of Proposition 8.3.1. Let $\varepsilon(dx)$ be any approximate design, i.e. any probability measure on $(\mathcal{X},\mathcal{B})$ and

$$M(\varepsilon) = \int F(x)F'(x)\varepsilon(dx).$$

In particular, $\varepsilon(dx)$ may be of the form (8.3.10).

For all approximate designs ε such that $\det M(\varepsilon) \neq 0$ the application of inequality (8.3.12) to the matrix $A=M(\varepsilon)$ yields

$$\operatorname{tr}(M(\varepsilon))^{-1} \geq m^2/\operatorname{tr} M(\varepsilon) = m^2/\sum_{i=1}^{m} \int f_i(x)\varepsilon(dx) \geq m/K^2$$

Considering also (8.3.9), we obtain (8.3.11): the proof is completed.

Inequality (8.3.11) implies that there exists a constant $c>0$, independent of m and N, such that $V^2 \geq cm/N$. Therefore the following statement (that will play a substantial role below) holds.

Theorem 8.3.1. Let $\sigma^2(x)=\sigma^2=\text{const}$, $N\to\infty$, $m=m(N)\to\infty$, $m/N\to 0$. Then the convergence order of the mean square of the inaccuracy (8.3.7) can not be less than

$$\mathcal{E}^2(\mathcal{F},L_m) + m/N \tag{8.3.14}$$

for any projection estimate of the form of (8.3.2), any method of linear parametric estimation and any passive design $\{x_1,\ldots,x_N\}$ sequence where

$$\mathcal{E}(\mathcal{F},L_m)=\left[\sup_{f\in\mathcal{F}}\ \inf_{f_{(m)}\in L_m}\int\left(f-f_{(m)}\right)^2 d\nu\right]^{1/2}$$

is the minimal value of the bias inaccuracy B.

8.3.3 *Examples of asymptotically optimal projection procedures with deterministic designs*

Omitting the proofs which are rather tedious, we formulate here two examples of asymptotically optimal projection procedures involving deterministic designs (i.e. rules for the choice of points $x_1,\ldots,x_N$).

Let Z be the set of nonnegative integers, $K=Z^n$ be the multi-index set, $k=(k_1,\ldots k_n)\in K$ be a multi-index, $\{f_k(x),\ k\in K\}$ be a complete ν-orthonormal set of the functions $f_k(x)=\exp\{-2\pi i(k,x)\}$ on $\mathcal{X}=[0,1]^n$, L_m be the set of linear combinations of the functions $f_k(x)$ corresponding to subsets K_q of K:

$$L_m=L(q)=\left\{g_m: g_m(x)=\sum_{k\in K_q} c_k f_k(x)\right\},$$

where

$$K_q=\{k\in K:\|k\|\le q\},\qquad m=\operatorname{card} K_q=\sum_{k\in K_q} 1,$$

$\|.\|$ is a given positive function on K, and (k,x) is the scalar product of $k\in K$ and $x\in\mathcal{X}$.

The functional set H_n^α for integer $\alpha\geq 1$ consists of real functions defined on $\mathcal{X}=[0,1]^n$ continuous partial derivatives

$$\frac{\partial^\ell f(x_1,\ldots,x_n)}{\partial x_1^{\ell_1}\ldots\partial x_n^{\ell_n}}\qquad\left(0\le\ell\le\alpha n,\ 0\le\ell_i\le\alpha,\ \ell=\sum_{i=1}^n\ell_i\right).$$

Define first the functional class $\mathcal{F}=\mathcal{F}(H_n^\alpha)$, as the subset of H_n^α containing all periodical functions of H_n^α with period 1, respectively to each coordinate. For each $f\in\mathcal{F}(H_n^\alpha)$, one has $|\theta_k|\leq L\|k\|^{-\alpha}$, where L is a constant,

$$\theta_k=\int f(x)\exp\{-2\pi i(k,x)\}dx,\qquad k\in K$$

are the Fourier coefficients of f and

$$\|k\| = \prod_{j=1}^{n} \max\{1, k_j\}. \tag{8.3.15}$$

Theorem 8.3.2. Set $q=N^{1/2\alpha}$, select the lattice grids $\Xi_N^{(5)}$ of Section 2.2.1 as the experimental design and estimate the Fourier coefficients θ_k by the least square algorithm. Then the summed mean square inaccuracy (8.3.7) of the projection estimate of a regression function $f \in \mathcal{F}(H_n^{\alpha})$ decreases with the rate

$$N^{-1+1/2\alpha} \log^{n-1} N, \qquad N \to \infty. \tag{8.3.16}$$

This is the optimal decrease rate of B^2+V^2 on $\mathcal{F}(H_n^{\alpha})$ for any linear estimation method, choice $q=q(N)$, and any experimental design sequence: the proof of this statement can be found in Zhigljavsky (1985) and Ermakov and Zhigljavsky (1984).

Turn now to the functional set W_n^{α} consisting of functions f on $\mathcal{X}=[0,1]^n$ which have the derivatives

$$\frac{\partial^{\ell} f(x_1,\ldots,x_n)}{\partial x_1^{\ell_1} \ldots \partial x_n^{\ell_n}} \qquad \left(0 \le \ell \le \alpha n, 0 \le \ell_i \le \alpha n, \ell = \sum_{i=1}^{n} \ell_i\right),$$

define $\mathcal{F} = \mathcal{F}(W_n^{\alpha})$ as the subset of W_n^{α} containing all periodical functions with period 1, respectively to each coordinate.

Zhigljavsky (1985) proved the analogue of Theorem 8.3.2 for the class $\mathcal{F}(W_n^{\alpha})$. Its formulation would coincide with that of Theorem 8.3.2, if we substitute $\mathcal{F}(W_n^{\alpha})$, the cube grids $\Xi_N^{(1)}$,

$$\|k\| = \max\{k_1,\ldots,k_n\}$$

and $N^{-1+n/2\alpha}$ $(N\to\infty)$ for $\mathcal{F}(H_n^{\alpha})$, the lattice grids $\Xi_N^{(5)}$, and the relations (8.3.15) and (8.3.16), respectively.

8.3.4 Projection estimation via evaluations at random points

We shall suppose below that the points $x_1, x_2, \ldots$ at which the regression function f is evaluated are random. In this case the most popular linear parametric estimation methods are the least square and ordinary Monte Carlo. The least square method minimizes the random inaccuracy V, but introduces a bias leading to a slight increase of the bias inaccuracy B. Usually the increase of B is insignificant from the asymptotic point of view and thus the least square method may usually be included into an asymptotically optimal projection estimation procedure. The drawback of the least square estimation method is its numerical complexity. The standard Monte Carlo method is much simpler than the least square approach and produces unbiased estimates of the Fourier coefficients. This way, its use gives the minimal value of the bias inaccuracy B, i.e.

$$B = \mathfrak{E}(\mathfrak{F}, L_m). \tag{8.3.17}$$

On the other hand, the random inaccuracy V is larger than for the least square method. We shall show that the increase of V is often insignificant from the asymptotic point of view and thus the Monte Carlo estimation method may also be included into the asymptotically optimal projection estimation procedure.

Let $x_1,...,x_N$ be independent and identically distributed with a positive (on $\mathfrak{X}$) density $p(x)$ and $\{f_1,...,f_m\}$ be a base of L_m. The standard Monte Carlo estimates of the Fourier coefficients

$$\theta_i = \int f(x) f_i(x) \nu(dx), \qquad i = 1, \dots, m,$$

are of the form

$$\hat{\theta}_i = N^{-1} \sum_{j=1}^{N} y(x_j) f_i(x_j)/p(x_j). \tag{8.3.18}$$

Set

$$\theta = (\theta_1, \dots, \theta_m)', \qquad \hat{\theta} = (\hat{\theta}_1, \dots, \hat{\theta}_m)',$$

$$g_i(x) = \left(f^2(x) + \sigma^2(x)\right)^{1/2} f_i(x), \qquad g(x) = (g_1(x), \dots, g_m(x))'.$$

The following statement is valid.

Lemma 8.3.2. Let $x_1,...,x_N$ be independent and distributed according to a positive probability density p on $\mathfrak{X}$, the evaluation errors $\xi(x_j)=y(x_j)-f(x_j)$ be uncorrelated, their variation $\mathrm{var}(\xi(x))=\sigma^2(x)$ be bounded. Then the estimates (8.3.18) are unbiased and

$$\mathrm{cov}\,\hat{\theta} = E(\hat{\theta} - \theta)(\hat{\theta} - \theta)' = N^{-1}\left[\int \frac{g(x)g'(x)}{p(x)} \nu(dx) - \theta\theta'\right] \tag{8.3.19}$$

The proof can be found in Zhigljavsky (1985).

The unbiasedness of

$\hat{\theta}$ yields (8.3.17) for the bias inaccuracy B. Consider now the random error V:

$$V^2 = V^2(p) = E\int\left(\sum_{i=1}^{m}\theta_i f_i(x) - \sum_{i=1}^{m}\hat{\theta}_i f_i(x)\right)\nu(dx) =$$

$$= \sum_{i,\ell=1}^{m} E\left(\hat{\theta}_i - \theta_i\right)\left(\hat{\theta}_\ell - \theta_\ell\right)\int f_i(x) f_\ell(x)\nu(dx) =$$

$$= N^{-1}\operatorname{tr} D(p)\int F(x)F'(x)\nu(dx),$$

where

$$D(p) = N\operatorname{cov}\hat{\theta}$$

is the normalized covariance matrix representable in the form (8.2.3) with the substitution θ for $\mathbb{1}$. Hence, it follows that the optimal choice criterion for the density p is linear and has the form (8.2.8) with

$$L = \int F(x)F'(x)\nu(dx);$$

and the optimal density has the form (8.2.24). The minimal value of the squared random inaccuracy is attained at this density and equals

$$V^2(p^*) = N^{-1}\left[\int (g'(x)Lg(x))^{1/2}\nu(dx) - \theta' L\theta\right] \tag{8.3.20}$$

Due to Theorem 8.3.1, (8.3.20) is not less than cm/N, where c is a positive number. At the same time, in a rather general case, there exist a density p and a constant $C \geq 1$ such that

$$V^2(p) \leq Cm/N. \tag{8.3.21}$$

The inequality (8.3.21) is valid, for instance, for the cases considered in Section 8.3.3 and in the common case (in optimization theory) when $\nu(\mathcal{X})<\infty$, functions $f(x)$ and $\sigma^2(x)$ are bounded and the base functions f_i (i=1,...,m) are ν-orthonormal and uniformly bounded with respect to i. In the latter case, the uniform (on $\mathcal{X}$) density can be chosen as p. It follows that for orthonormal functions $f_1,...,f_m$ there holds

$$V^2(p) = N^{-1}\int\left(\left(\left(f^2(x) + \sigma^2(x)\right)\sum_{i=1}^{m} f_i^2(x)\right)/p(x)\right)\nu(dx) - \sum_{i=1}^{m}\theta_i^2$$

If (8.3.21) is fulfilled, then the asymptotic relation

$$v^2(p) \asymp m/N \qquad \text{for } m \to \infty, N \to \infty \tag{8.3.22}$$

is valid, i.e. the random error reaches the optimal order of decrease for m, $N \to \infty$. Taking into account that the bias inaccuracy is minimal we get the following statement.

Theorem 8.3.3. Let (8.3.21) be fulfilled for some constant $C \geq 1$ and positive density p on $\mathcal{X}$. Then the independent random choice of points $x_1,...,x_N$ in accordance with the density p, the method of parametric estimation (8.3.18), and a suitable choice of $m=m(N)$ (these leading to equal decrease orders of $\mathcal{E}^2(\mathcal{F},L_m)$ and (8.3.22)) form an asymptotically optimal procedure of projection estimation of a regression function $f \in \mathcal{F}$.

As indicated above, for the functional classes $\mathcal{F}$ and sequences of spaces $\{L_m\}$ considered in Section 8.3.3 the condition (8.3.21) holds, in particular, for the uniform density on $\mathcal{X}$. So the projection estimation procedure of Theorem 8.3.3 with the uniform density is asymptotically optimal. Note also that the above random designs have advantages, compared with the deterministic ones given in Section 8.3.3, viz., they are simpler to construct and they posses the composite property described in Section 2.2.1.

Let us turn now to the construction and study of the least square parametric estimation through regression function evaluations at independent random points.

Let m be fixed, the base functions $f_1,...,f_m$ be ν-orthonormal, the points $x_1,...,x_N$ at which a regression function is being evaluated be random, independent and identically distributed with a density p (with respect to the measure ν) which is ν-almost everywhere positive ($p>0$ (modν), $\int p d\nu=1$) on $\mathcal{X}$, the evaluations of f be uncorrelated and their variance $\sigma^2(x)=E\xi^2(x)$ be uniformly bounded.

Denote by $\theta=(\theta_1,...\theta_m)'$ the vector of Fourier coefficients of f with respect to the base functions $f_1,...,f_m$ and set

$$F(x) = (f_1(x),\ldots,f_m(x))', \qquad r(x) = f(x) - \theta' F(x).$$

The estimation of the unknown parameters (i.e. the Fourier coefficients θ_i) by the least square method is as follows. Supposing that f is a linear combination of the functions $f_1,...,f_m$ let us derive the simultaneous equations

$$\sum_{i=1}^{m} \theta_i f_i(x_j) = y(x_j), \qquad j = 1,\ldots,N,$$

multiply now the i-th equation by $(p(x_j))^{-1/2}$ and obtain $A\theta=Y$, where

$$A = \left\| (p(x_j))^{-1/2} f_i(x_j) \right\|_{i,j=1}^{m,N}, \quad Y = \left\| y(x_j)/\sqrt{p(x_j)} \right\|_{j=1}^{N}.$$

Suppose now that the matrix A'A is nondegenerate. Then the least square estimate of θ will have the form

$$\hat{\theta} = (A'A)^{-1} A'Y. \tag{8.3.23}$$

Since, in general, f can not be represented as a linear combination of $f_1,\ldots,f_m$, the estimate (8.3.23) may be biased. The main purpose of the result given below is the estimation of the inaccuracy of (8.2.23).

According to the central limit theorem, the sequence of matrices

$$N^{-1}A'A = N^{-1}\sum_{j=1}^{N} F(x_j)F'(x_j)/p(x_j)$$

converges for $N\to\infty$ to the unit matrix I_m; the order of convergence rate equals $O(N^{-1/2})$. It follows that in case of existence of the inverse matrix $(A'A)^{-1}$, the asymptotic relation

$$\left(N^{-1}A'A\right)^{-1} = I_m + O\left(N^{-1/2}\right), \quad N\to\infty \tag{8.3.24}$$

is valid.

Consider now the case of a ν-regular system of functions $f_1,\ldots,f_m$ with respect to the measure ν, i.e. such a system that for all $N\geq m$ the measure of the point set $\{x_1,\ldots,x_N\}$ for which the matrix A'A is degenerate equals zero.

Theorem 8.3.4. Let $f, f_i \in L_2(\mathcal{X},\nu)$, $i=1,\ldots,m$, m be fixed, $N\to\infty$, and the collection of ν-orthonormal functions $f_1,\ldots,f_m$ be regular. Then for the estimate (8.3.23) the asymptotic representations

$$E\hat{\theta} = \theta + N^{-1}\int\left[r(x)F(x)\left(\sum_{i=1}^{m} f_i^2(x)\right)\Big/p(x)\right]\nu(dx) + O\left(N^{-2}\right), \tag{8.3.25}$$

$$\operatorname{cov}\hat{\theta} = N^{-1}\int \frac{r^2(x)+\sigma^2(x)}{p(x)} F(x)F'(x)\nu(dx) + O\left(N^{-2}\right) \tag{8.3.26}$$

are valid for $N\to\infty$.

The proof can be found in Korjakin (1983) or in Zhigljavsky (1985).

Let us comment on the above assertion. First, if for the inverse matrix $(A'A)^{-1}$ in (8.3.23) its initial approximation $N^{-1}I_m$ is substituted, then the standard Monte Carlo estimate (8.3.18) is obtained. Second, if p is chosen as

$$p(x) = m^{-1}\sum_{i=1}^{m} f_i^2(x), \tag{8.3.27}$$

then from (8.3.25) we obtain

$$E\hat{\theta} = \theta + O(N^{-2}), \qquad N \to \infty.$$

It is not difficult to show that for the density (8.3.27), the exact unbiasedness

$$E\hat{\theta} = \theta$$

also takes place. Third, analogously with the case when the Fourier coefficients are estimated by standard Monte Carlo method, the minimzation problem of the convex functional of the matrix

$$\int \frac{r^2(x) + \sigma^2(x)}{p(x)} F(x)F'(x)\nu(dx) = N \operatorname{cov}\hat{\theta} + O(N^{-1}), \qquad N \to \infty,$$

with respect to p can be stated. Ignoring the biasedness of the estimates obtained, the results will be similar to those stated in Section 8.2.

Note finally that if the function collection $\{f_1,\ldots,f_m\}$ is ν-regular, N=km, the so-called Ermakov-Zolotukhin points that have the joint density (see Ermakov (1975))

$$p(x_1,\ldots,x_m) = \left[\det\left(\left\| f_1(x_i),\ldots,f_m(x_i)\right\|_{i=1}^{m}\right)\right]^2 / m!$$

for k groups of m points, and the least square estimates of the Fourier coefficients of f with respect to $f_1,\ldots,f_m$ are used, then by virtue of results in Ermakov (1975) one has

$$V^2 + B^2 \asymp \mathcal{E}^2(\mathcal{F}, L_m) + m^2/N \qquad (N \to \infty,\ m \to \infty)$$

for the summed inaccuracy (8.3.7), for details see Zhigljavsky (1985). Comparing this estimate with (8.3.14), we obtain that the Ermakov-Zolotukhin points can not be generally included into an asymptotically optimal projection estimation procedure.

REFERENCES

Aluffi-Pentini F., Parisi V. and Zirilli F. (1985) Global optimization and stochastic differential equations. J. Optimiz. Theory and Applic., 47, No.1, 1-16.

Anderson A. and Walsh G.R. (1986) A graphical method for a class of Branin trajectories. J. Optimiz. Theory and Applic., 49, No.3,367-374.

Anderssen R.S. and Bloomfield P. (1975) Properties of the random search in global optimization. J. Optimiz. Theory and Applic., 16, No.5/6, 383-398.

Anily S. and Federgruen A. (1987) Simulated annealing methods with general acceptance probabilities. J. Appl. Probab., 24, No.3, 657-667.

Archetti F. and Betro B. (1979) A probabilistic algorithm for global optimization. Calcolo, 16, No.3, 335-343.

Archetti F. and Betro B. (1980) Stochastic models and optimization. Bolletino della Unione Matematica Italiana, 17-A, No.5, 225-301.

Archetti F. and Schoen F. (1984) A survey on the global optimization problem: general theory and computational approaches. Ann. Oper. Res., 1, 87-110.

Ariyawansa K.A. and Templeton J.G.C. (1983) On statistical control of optimization. Optimization, 14, No.2, 393-410.

Avriel M. (1976) Nonlinear Programming: Analysis and Methods, Prentice-Hall, Englewood Cliffs, New Jersey e.a.

Baba N. (1981) Convergence of a random oprimization method for constrained optimization problems. J. Optimiz. Theory and Applic., 1981, 33, No.4, 451-461.

Basso P. (1982) Iterative methods for the localization of the global maximum. SIAM J. Numer. Anal., 19, No.4, 781-792.

Bates D. (1983) The derivative of X'X and its uses. Technometrics, 25, No.4, 373-376.

Batishev D.I. (1975) Search Methods of Optimal Construction. Soviet Radio, Moscow, 216 p. (in Russian).

Batishev D.I. and Lyubomirov A.M. (1985) Application of pattern recognition methods to searching for a global minimum of a multivariate function. Problems of Cybernetics, Vol.122, (ed. V.V.Fedorov), 46-60 (in Russian).

Beale E.M.L. and Forrest J.J.H. (1978) Global optimization as an extension of integer programming. In: Towards Global Optimization Vol.2. North Holland, Amsterdam e.a., 131-149.

Bekey G.A. and Ung M.T. (1974) A comparative evaluation of two global search algorithms. IEEE Trans. on Systems, Man and Cybernetics, No.1, 112-116.

Berezovsky A.I. and Ludvichenko V.A. (1984) Sequential searching algorithms of minimal values of differential functions from some classes. Numerical Methods and Optimization. Kiev, 27-32 (in Russian).

Betro B. (1983) A Bayesian nonparametric approach to global optimization. Methods of Oper. Res., 45, No.1, 45-79.

Betro B. (1984) Bayesian testing of nonparametric hypothesis and its application to global optimization. J. Optimiz. Theory and Applic., 42, No.1, 31-50.

Betro B. and Schoen F. (1987) Sequential stopping rules for the multistart algorithm in global optimization. Mathem. Programming, 38, No.2, 271-286.

Betro B. and Vercellis C. (1986) Bayesian nonparametric inference and Monte Carlo optimization. Optimization, 17, No.5, 681-694.

Betro B. and Zielinski R. (1987) A Monte Carlo study of a Bayesian decision rule concerning the number of different values of a discrete random variable. Commun. Statist.: Simulation, 16, No.4, 925-938.

Bhanot G. (1988) The Metropolis algorithm. Rep. Progr. Phys., 429-457.

Billingsley. P. (1968) Convergence of Probability Measures. Wiley, N.Y. e.a.

Blum Z.R. et al.(1959) Central limit theorems for interchangeable processes. Canadian J. Mathem., 10, 222-229.

Boender C.G.E., and Zielinski R. (1985) A sequential Bayesian approach to estimating the dimension of a multinomial distribution. In: Sequential Methods in Statistics. Banach Center Publications, Vol.16, Polish Scientific Publishers, Warsaw, 37-42.

Boender G., Rinnoy Kan A., Stougie L., and Timmer G. (1982) A stochastic method for global optimization. Mathem. Programming, 22, No.1, 125-140.

Boender, C.G.E. (1984) The generalized multinomial distribution: A Bayesien analysis and applications. Ph. D. Thesis, Erasmus University, Rotterdam.

Boender, C.G.E., and Rinnoy Kan A.H.G. (1987) Bayesian stopping rules for multistart global optimization methods. Mathem. Programming, 37, No.1, 59-80.

Bohachevsky I.O., Johnson M.E., and Stein M.L. (1986) Generalized simulated annealing for function optimization. Technometrics, 28, No.3, 209-217.

Branin F.H. (1972) A widely convergent method for finding multiple solutions of simultaneous non-linear equations. IBM J. Res. Develop., 16, 504-522.

Branin F.H., and Hoo S.K. (1972) A method for finding multiple extrema of a function of n variables. In: Numerical Methods for Nonlinear Optimization. Academic Press, London e.a., 231-237.

Bremerman H.A. (1970) Method of unconstrained global optimization. Mathem. Biosciences, 9, No.1, 1-15.

Brent R.P. (1973) Algorithms for Minimization without Derivatives. Prentice-Hall, New Jersey, 195 p.

Brooks S.H. (1958) Discussion of random methods for locating surface maxima. Oper. Res., 6, 244-251.

Brooks S.H. (1959) A comparison of maximum-seeking methods. Operations Research, 7, 430-457.

Bulatov V.P. (1987) Methods of solving of multiextremal problems. In: Methods of Numerical Analysis and Optimization (eds B.A. Beltjukov, and V.P. Bulatov), Nauka, Novosibirsk, 133-157 (in Russian).

Chen J., and Rubin H. (1986) Drawing a random sample from a density selected at random. Comput. Statist. and Data Analys. 4, No.4, 219-227.

Chernousko F.L. (1970) On the optimal search of extremum for unimodal functions. USSR Comput. Mathem. and Mathem. Phys., 10,No.4, 922-933.

Chichinadze V.K. (1967) Random search to determine the extremum of a function of several variables. Engineering Cybernetics, No.5, 115-123.

Chichinadze V.K. (1969) The Ψ-transform for solving linear and nonlinear programming problems. Automatica, 5, No.3, 347-355.

Chuyan O.R. (1986) Optimal one-step maximization of twice differentiable functions. USSR Comput. Mathem. and Mathem. Phys., 26, No.3, 381-397.

Clough D.J. (1969) An asymptotic extreme value sampling theory for the estimation of a global maximum. Canad. Oper. Res. Soc. J. 7, No.1, 105-115.

Cohen J.P. (1986) Large sample theory for fitting an approximate Gumbel model to maxima. Sankhya, A48, 372-392.

Cook P. (1979) Statistical inference for bounds of random variables. Biometrika, 66, No.2, 367-374.

Cook P. (1980) Optimal linear estimation of bounds of random variables. Biometrika, 67, No.1, 257-258.

Corana A., Marchesi M., Martini C., and Ridella S. (1987) Minimizing multimodal functions of continuous variables with the *simulated annealing* algorithm. ACM Trans. on Mathem. Software, 13, No.3, 262-280.

Corles C. (1975) The use of regions of attractions to identify global minima. In: Towards Global Optimization. Vol.1, North Holland, Amsterdam e.a., 55-95.

Crowder H.P., Dembo R.S. and Mulvey J.M. (1978) Reporting computational experiments in mathematical programming. Mathem. Programming, 15, 316-329.

Csendes T. (1985) A simple but hard-to-solve global optimization test problem. Presented at the IIASA Workshop on Global Optimization (held in Sopron, Hungary).

Csörgö S. and Mason D.M. (1989) Simple estimators of the endpoint of a distribution. In: Extreme Value Theory, Oberwolfach, 1987 (eds.Hüsler J. and Reiss R.-D.) Lecture Notes in Statistics, Springer, Berlin e.a.

Csörgö S., Deheuvels P. and Mason D.N. (1985) Kernel estimates of the tail index of a distribution. Ann. Statist., 13, No.3, 1050-1077.

Dannenbring D.G. (1977) Procedures for estimating optimal solution values for large combinatorial problems. - Management Science, 23, 1273-1283.

de Biase L. and Frontini F. (1978) A stochastic method for global optimization: its structure and numerical performance. In: Towards Global Optimization Vol.2. North Holland, Amsterdam e.a., 85-102.

de Haan L. (1970) On Regular Variation and its Application to the Weak Convergence of Sample Extremes. North Holland, Amsterdam e.a., 104 p.

de Haan L. (1981) Estimation of the minimum of a function using order statistics, J. Amer. Statist. Assoc., 76, No.374, 467-475.

Deheuvels P. (1983) Strong bounds for multidimensional spacings. Z.Wahrsch. verw. Gebiete, 64, 411-424.

Dekkers A.L.M. and de Haan L. (1987) On a consistent estimate of the index of an extreme value distribution. Rept. Cent. Math. and Comput. Sci., No. MS - R8710, 1-15.

Demidenko E.Z. (1989) Optimization and Regression. Nauka, Moscow (in Russian).

Demyanov V.F. and Vasil'ev L.V. (1985) Nondifferentiable Optimization. Springer-Verlag, New York e.a., 452 p.

Dennis J.E. and Schnabel R.B. (1983) Numerical Methods for Unconstrained Optimization and Nonlinear Equations. Prentice-Hall, Englewood Cliffs, New Jersey e.a.

Devroye L. (1978) Progressive global random search of continuous functions. Mathem Programming, 15, 330-342.

Devroye L. (1986) Nonuniform Random Variate Generation. Springer, Berlin e.a., 843 p.

Devroye L. and Györfi L. (1985) Nonparametric Density Estimation: the L_1 View. Wiley, New York e.a.

Devroye L.P. (1976) On the convergence of statistical search. IEEE Transactions on System, Man and Cybernetics, 6, 46-56.

Dixon L.C.W. and Szegö G.P., eds (1975) Towards Global Optimization, Vol.1. North Holland, Amsterdam e.a., 472 p.

Dixon L.C.W. and Szegö G.P., eds (1978) Towards Global Optimization, Vol.2. North Holland, Amsterdam e.a., 364 p.

Dorea C.C.Y. (1983) Expected number of steps of a random optimization method. J. Optimiz. Theory and Applic., 39, No.2, 165-171.

Dorea C.C.Y. (1986) Limiting distribution for random optimization methods. SIAM J. Control and Optimization, 24, No.1, 76-82.

Dorea C.C.Y. (1987) Estimation of the extreme value and the extreme points. Ann. Inst. Statist. Mathem., 39, No.1, 37-48.

Dunford N. and Schwartz J.T. (1958) Linear Operators, Part I.: General Theory, Interscience Publishers, New York e.a.

Duran B.S., and Odell P.L. (1977) Cluster Analysis: A Survey. Springer, Berlin e.a.

Ermakov S.M. (1975) Monte Carlo Methods and Related Problems. Nauka, Moscow, 472 p. (in Russian).

Ermakov S.M. (1983) Mathematical Theory of Experimental Design. Nauka, Moscow, 392 p.(in Russian).

Ermakov S.M. and Mitioglova L.V. (1977) On a global search method based on estimation of the covariance matrix. Automatika and Computers, No.5, 38-41 (in Russian).

Ermakov S.M. and Zhigljavsky A.A. (1983) On the global search of a global extremum. Theory of Probab. and Applic., 28,No.1, 129-136.

Ermakov S.M. and Zhigljavsky A.A. (1984) Nonparametric estimation and the asymptotical optimum design of an experiment. Vestnik of Leningrad University, No.7, 20-27.

Ermakov S.M. and Zhigljavsky A.A. (1985) Monte Carlo method for estimating functionals of eigen-measures of linear integral operators. USSR Comput. Mathem. and Mathem Phys., 25, No.5, 666-679.

Ermakov S.M. and Zhigljavsky A.A. (1987) Mathematical Theory of Optimal Experiments. Nauka, Moscow, 320 p.(in Russian).

Ermakov S.M., Zhigljavsky A.A. and Kondratovich M.V. (1988) Reduction of the problem of random estimation of the function extremum value. Dokl. Akad. Nauk SSSR, 302, No.4, 796-798.

Evans O.H. (1963) Applied multiplex sampling. Technometrics, 5, No.3, 341-359.

Evthusenko Yu.G. (1985) Numerical Optimization Techniques. Springer, Berlin e.a.

Evtushenko Yu.G. (1971) Algorithm for finding the global extremum of a function (case of a non-uniform mesh). USSR Comput. Mathem. and Mathem Phys., 11, No.6, 1390-1403.

Falk M. (1983) Rates of uniform convergence of extreme order statistics. Ann. Instit. Statist. Mathem. 38, part A, No.2, 245-265.

Faure H. (1982) Discrepance de suites associées a un systéme de numeration (en dimension s). Acta Arithm., 41, p.337-351.

Fedorov V.V. (1979) Numerical Maximin Methods. Nauka, Moscow, 280 p. (in Russian).

Fedorov V.V., ed. (1985) Problems of Cybernetics, No. 122: Models and Methods of Global Optimization. USSR Academy of Sciences, Moscow (in Russian).

Feller W. (1966) An Introduction to Probability Theory and its Applications, Vol.2, Wiley, New York e.a.

Fletcher R. (1980) Practical Methods of Optimization. Wiley, New York e.a.

Gabrielsen G. (1986) Global maxima of real-valued functions. J. Optimiz. Theory and Applic., 50, No.2, 257-266.

Galambos J. (1978) The Asymptotic Theory of Extreme Order Statistics. Wiley, New York e.a.

Galperin E.A. (1988) Precision, complexity and computational schemes of the cubic algorithm. J. Optimiz. Theory and Applic., 57, No.2, 223-238.

Galperin E.A. and Zheng Q. (1987) Nonlinear observation via global optimization methods: measure theory approach. J. Optimiz. Theory and Applic., 54, No.1, 63-92.

Ganshin G.S. (1976) Calculation of the maximal value of functions. USSR Comput. Mathem. and Mathem. Phys., 16, No.1, 30-39.

Ganshin G.S. (1977) Optimal algorithms of calculation of the highest value of functions. USSR Comput. Mathem. and Mathem. Phys., 17, No.3, 562-571.

Ganshin G.S. (1979) The optimization of algorithms on classes of nets. USSR Comput. Mathem. and Mathem. Phys., 19, No.14, 811-821.

Gaviano M. (1975) Some general results on the convergence of random search algorithms in minimization problems. In: Towards Global Optimization, Vol.1, North Holland, Amsterdam e.a., 149-157.

Ge R.P. (1983) A filled function method for finding a global minimizer. Presented at the Dundee Biennial Conference on Numerical Analisys.

Ge R.P. and Qin Y.F. (1987) A class of filled functions for finding global minimizers of a function of several variables. J. Optimiz. Theory and Applic., 54, No.2, 241-252.

Geman S. and Hwang C.-R. (1986) Diffusions for global optimization. SIAM J. Control and Optimization, 24, No.5, 1031-1043.

Golden B.L. and Alt F.B. (1979) Interval estimation of a global optimum for large combinatorial problems. Naval Res. Logistics Quaterly, 26, No.1, 69-77.

Gomulka J. (1978) Two implementations of Branin's method: numerical experience. In: Towards Global Optimization, Vol.2, North Holland, Amsterdam e.a., 151-163.

Griewank A.O. (1981) Generalized descent for global optimization. J. Optimiz. Theory and Applic., 34, No.1, 11-39.

Gumbel E.J. (1958) Statistics of Extremes. Columbia University Press.

Haines L.M. (1987) The application of the annealing algorithm to the construction of exact optimal design for linear regression models. Technometrics, 29, No.4, 439-448.

Hall P. (1982) On estimating the endpoint of a distribution. Ann. Statist., 10, No.2, 556-568.

Hansen E. (1979) Global optimization using interval analysis: the one-dimensional case. J. Optimiz.Theory and Applic., 29, No.3, 331-334.

Hansen E. (1980) Global optimization using interval analysis: the multidimensional case. Numer. Math., 34, 247-270.

Hansen E. (1984) Global optimization with data perturbations. Comput. Oper. Res., 11, No.2, 97-104.

Hardy J. (1975) An implemented extension of Branin's method. In: Towards Global Optimization. North Holland, Vol.1, Amsterdam e.a., 117-142.

Harris T.E. (1963) The Theory of Branching Processes. Springer, Berlin e.a.

Hartley H.O. and Pfaffenberger P. (1971) Statistical control of optimization. In: Optimizing Methods in Statistics. Academic Press, New York e.a., 281-300.

Hartley H.O. and Ruud P.G. (1969) Computer optimization of second order response surface designs. In: Statistical Computations, Proceedings of a Conference held at the University of Wisconsin, Madison, Wisconsin, April 29-30, 1969, 441-462.

Hock W. and Schittkowski K. (1981) Test Examples for Nonlinear Programming Codes. Lecture Notes in Economics and Mathematical Systems, No.187, Springer, Berlin e.a., 177 p.

Horst R. (1986) A general class of branch-and-bound methods in global optimization with some new approaches for concave minimization. J. Optimiz. Theory and Applic., 51, No.2, 271-291.

Horst R. and Tuy H. (1987) On the convergence of global methods in multiextremal optimization. J. Optimiz. Theory and Applic., 54, No.2, 253-271.

Hua L.K. and Wang Y. (1981) Applications of Number Theory to Numerical Analysis. Springer, Berlin e.a.

Ichida K. and Fujii Y. (1979) An interval arithmetic method for global optimization. Computing, 23, 85-97.

Incerti S., Parisi V. and Zirilli F. (1979) A new method for solving nonlinear simultaneous equations. SIAM J. on Numerical Analysis, 16, 779-789.

Isaac R. (1988) A limit theorem for probabilities related to the random bombardment of a square. Acta Mathematica Hungarica, 51, No.1-2, 85-97.

Ivanov V.V. (1972) On Optimal Algorithms of minimization in the class of functions with the Lipschitz condition, Information Processing, 2, p. 1324-1327.

Ivanov V.V., Girlin S.K. and Ludvichenko V.A. (1985) Problems and results of global optimization for smoothing functions. Problems of Cybernetics, Vol.122.(ed.Fedorov V.V.), 3-13 (in Russian).

Jacobsen S. and Torabi M. (1978) A global minimization algorithm for a class of one-dimensional functions. J. of Mathem. Analysis and Applic., 62, 310-324.

Jacson R. and Mulvey J. (1978) A critical review of comparisons of mathematical programming algorithms and software (1953-1977). J. of Research of the National Bureau of Standards, 83, No.6, 563-584.

Janson S. (1986) Random coverings in several dimensions. Acta Mathematica, 156, 83-118.

Janson S. (1987) Maximal spacings in several dimensions. Ann. Probab., 15, No.1, 274-280.

Kantorovich L.V. and Akilov G.P. (1977) Functional Analysis. Nauka, Moscow, 744 p. (in Russian).

Kashtanov, Yu. N. (1987) The rate of convergence towards its own distribution of an integral operator in the generation method. Vestnik of Leningrad University, No.1, 17-21.

Katkovnik V.Ya. (1976) Linear Estimators and Stochastic Optimization Problems. Nauka, Moscow, 488 p. (in Russian).

Khairullin R.H. (1980) On the estimation of the critical parameter of the branching processes of a special kind. Izvestiya Vuzov, Matematika. No.8, 77-84 (in Russian).

Khasminsky R.Z. (1965) Application of random noise in optimization and recognition problems. Problems of Information Transfer. 1, No.3, 113-117 (in Russian).

Kiefer J. (1953) Sequential minimax search for a maximum. Proc. Amer. Math. Soc., 4, No.3, 502-506.

Kirkpatrick S., Gelatt C.D. and Vecchi M.P. (1983) Optimization by simulated annealing. Science, 220, pp. 671-680.

Kleibohm K. (1967) Bemerkungen zum Problem der Nichtkonvexen Programmierung. Unternehmensforschung, 11, pp. 49-60.

Kolesnikova C.N., Kornikov V.V., Rozhkov N.N. and Khovanov N.V. (1987) Stochastic processes with equiprobable monotone realizations as models of information deficiency. Vestnik of Leningrad University, No.1, 21-26.

Korjakin A.I. (1983) The estimation of a function from randomized observations. USSR Comput. Mathem. and Mathem. Phys., 23, No.1, 21-28.

Korobov N.M. (1963) Number-Theoretical Methods in Approximate Analysis. Fizmatgiz, Moscow, 224 p. (in Russian).

Kushner H.J. (1964) A new method of locating the maximum point of an arbitrary multipeak curve in the presence of noise. Trans.ASME, ser. D, 86, No.1, 97-105.

Kushner H.J. (1987) Asymptotic global behavior of stochastic approximation and diffusion with slowly descreasing noise effects: global minimization via Monte Carlo. SIAM J. Appl. Math., 47, No.1, 169-184.

Lbov G.S. (1972) Training for extremum determination of a function of variables measured on nominal names scale. In: Second Intern. Conf. on Artificial Intelligence. London, 418-423.

Levin V.L. (1985) Convex Analysis in Spaces of Measurable Functions and Applications in Mathematics and Economics. Nauka, Moscow, 316 p. (in Russian).

Levy A.V. and Montalvo A. (1985) The tunneling algorithm for the global minimization of functions. SIAM J. Sci. Statist. Comput., 6, No.1, 15-29.

Lindgren G. and Rootzen H. (1987) Extreme values: Theory and technical applications. Scand. J. Statist., 14, No.4, 241-279.

Loève M. (1963) Probability Theory. D. Van Nostrand, New York e.a.

Low W. (1984) Estimating an endpoint of a distribution with resampling methods. Ann. Statist., 12, No. 4, 1543-1550.

Makarov I.M., Radashevich Yu. B. (1981) Statistical estimation of accuracy in constrained extremal problems. Automation and Remote Control, No.3, 41-48.

Mancini L.J. and McCormick G.P. (1979) Bounding global minima with interval arithmetic. Oper. Res. 27, No.4, 743-754.

Mancini L.J., and McCormick G.P. (1976) Bounding global minima. Mathem. of Oper. Res., 1, No.1, 50-53.

Mandal N.K. (1981) D-optimal designs for estimating the optimum point in a multifactor experiment. Calcutta Statist. Assoc. Bull., 30, No.119-120, 145-169.

Mandal N.K. (1989) D-optimal designs for estimating the optimum point in a quadratic response surface - rectangular region. J. Statistical Planning and Inference, 23, 243-252.

Marti K. (1980) On accelerations of the convergence in random search methods. Oper. Res. Verfahren, 37, 391-406.

Masri S.F., Bekey G.A. and Safford F.B (1980) A global optimization algorithm using adaptive random search. Appl. Mathem. Comput. 7, 353-375.

McCormick G.P. (1972) Attempts to calculate global solutions of problems which may have local minima. In: Numerical Mehods for Nonlinear Optimization. Academic Press, London e.a., 209-221.

McCormick G.P. (1983) Nonlinear Programming. Theory, Algorithms, and Applications. Wiley, N.Y., 444 p.

McMurty G.J. and Fu K.S. (1966) A variable structure automaton used as a multimodal searching technique. IEEE Trans. on Automatic Control, 11, No. 3, 379-387.

Meewella C.C., and Mayne D.Q. (1988) An algorithm for global optimization of Lipschitz continuous functions. J. Optimiz. Theory and Applic., 57, No.2, 307-322.

Metropolis N., Rosenbluth A.W., Rosenbluth M.N. and Teller A.H. (1953) Equations of state calculations by fast computing machines. J. Chem. Phys., 21, 1087-1091.

Mikhailov G.A. (1966) Calculation of critical systems by the Monte Carlo method. USSR Comput. Mathem. and Mathem. Phys., 6,No.1, 71-80.

Mikhailov G.A. (1984) Minimax theory of weighted Monte Carlo methods, USSR Comput. Mathem. and Mathem. Phys., 24, No.9, 1294-1302.

Mikhailov G.A. (1987) Optimizing Weighted Monte Carlo Methods. Nauka, Moscow, 240 p. (in Russian).

Mikhailov G.A., Zhigljavsky A.A. (1988) Uniform optimization of weighted Monte Carlo estimates. Dokl. Akad. Nauk SSSR, 303, No.2, 290-293.

Mikhalevich V.S., Gupal A.M. and Norkin V.I. (1987) Methods of Nonsmooth Optimization. Nauka, Moscow (in Russian).

Mitra D., Romeo F. and Sangiovanni-Vincentelli A. (1986) Convergence and finite-time behavior of simulated annealing. Adv. Appl. Probab. 18, No.3, 747-771.

Mockus J. (1967) Multiextremal Problems in Design. Nauka, Moscow, 214 p. (In Russian).

Mockus J. (1989) Bayesian Approach to Global Optimization. Kluwer Academic Publisher, Dordrecht e.a.

Mockus J., Tiesis V. and Zilinskas A. (1978) The application of Bayesian methods for seeking the etremum. In: Towards Global Optimization, Vol. 2. North Holland, Amsterdam e.a., 117-130.

Moiseev E.V and Nekrutkin V.V. (1987) On the rate of convergence of some random search algorithms. Vestnik of Leningrad University, No.1, 29-34.

Moore R.E. (1966) Interval Analysis. Prentice-Hall, Englewood Cliffs, New Jersey e.a.

Myers R.H. and Khuri A.I. (1979) A new procedure for steepest ascent. Commun. Statist., A8, No.14, 1359-1376.

Nagaraja H.N. (1982) On the non-Markovian structure of discrete order statistics. J. Statist. Planning and Inference, 7, No.1, 29-33.

Nefedov V.N. (1987) Searching the global maximum of a function with several arguments on a set given by inequalities. USSR Comput. Mathem. and Mathem. Physics, 27, No1, 35-51.

Nelder,J.A. and Mead,R. (1965) A simplex method for functional minimization. Computer Journal 7, 308-313.

Neveu J. (1964) Bases Mathematiques du Calcul des Probabilites. Masson et Cie, Paris.

Niederreiter H. (1978) Quasi-Monte-Carlo methods and pseudorandom numbers. Bull. Amer. Mathem. Soc., 84, N 6, 957-1041.

Niederreiter H. (1986) Low-discrepancy point sets. Monatsh. Mathem. 102, N 2, 155-167.

Niederreiter H. and McCurley M. (1979) Optimization of functions by quasi-random search methods. Computing, 22, 119-123.

Niederreiter H. and Peart P. (1982) A comparative study of quasi-Monte Carlo methods for optimization of functions of several variables, Caribbean J. Math., 1, 27-44.

Pardalos,P.M. and Rosen,J.B. (1987) Constrained global optimizations algorithms and applications. Lecture Notes in Computer Science 268, Springer-Verlag, Berlin Heidelberg New York.

Pickands J. III. (1975) Statistical inference using extreme order statistics. Ann. Statist., 3, No.1, 119-131.

Pinkus M. (1968) A closed form solution of certain programming problems. Oper. Res. 16, 690-694.

Pintér J. (1986a) Extended univariate algorithms for dimensional global optimization. Computing, 36, No.1, 91-103.

Pintér J. (1986b) Globally convergent methods for dimensional multiextremal optimization. Optimization, 17, No. 2, 187-202.

Pintér J. (1988) Branch-and-bourd methods for solving global optimization problems with Lipschitzian structure. Optimization 19, No.1, 101-110.

Pintér,J. (1983) A unified approach to globally convergent one-dimensional optimization algorithms. Research Report JAMI 83-5. Inst. Apl. Math. Inf. CNR, Milan.

Pintér,J. (1984) Covergence properties of stochastic optimization procedures. Optimization, 15, No.3, 405-427.

Piyavskii, S.A. (1967) An algorithm for finding the absolute minimum of a function. Theory of Optimal Solutions, No.2. Kiev, IK AN USSR, pp. 13-24. (In Russian).

Piyavsky, S.A. (1972) An algorithm for finding the absolute extremum of a function. USSR Comput. Mathem. and Mathem. Phys., 12, No.4, 888-896.

Prakasa Rao B.L.S. (1983) Nonparametric Functional Estimation. Academic Press. N.Y.e.a.

Price W.L. (1983) Global optimization by controlled random search. J. Optimiz. Theory and Applic., 40, No.2, 333-348.

Price W.L. (1987) Gl,obal optimization algorithms for a CAD workstation. J. Optimiz. Theory and Applic., 55, No.1, 133-146.

Pronzato L., Walter E., Venot A. and Lebrichec J.F. (1984) A general purpose global optimizer: Implementation and applications. Mathem. Comput. Simul., 26, 412-422.

Pshenichny B.N. and Marchenko D.I. (1967) On an approach to evaluation of the global minimum. Theory of Optimal Decisions, Vol. 2. Kiev, 3-12 (in Russian).

Pukelsheim F. (1980) On linear regression design which maximizes the information, J.Statist. Planning and Inference, 4, 339-364.

Pukelsheim F. (1987) Information increasing orderings in experimental design theory. International statist. Review, 55, No.2, 203-219.

Rastrigin L.A. (1968) Statistical Search Methods. Nauka, Moscow, 376 p. (in Russian).

Ratschek H. (1985) Inclusion functions and global optimization. Mathem. Programming, 33, No.3, 300-317.

Ratschek, H. and Rokra, J. (1984) Computer methods for the range of functions. Ellis Harwood-Wiley, New York.

Resnick S.I. (1987) Extreme Values, Regular Variation and Point Processes. Springer, Berlin e.a.

Revus D. (1975) Markov Chains. North Holland, Amsterdam e.a., 334 p.

Rinnooy Kan A.H.G and Timmer G.T. (1987a) Stochastic global optimization methods. I. Clustering methods. Mathem. Programming, 39, No.1, 27-56.

Rinnooy Kan A.H.G. (1987) Probabilistic analysis of algorithms. In: Surveys in Combinatiorial Optimization, North Holland Mathem. Stud., 132. North Holland, Amsterdam e.a., 365-384.

Rinnooy Kan A.H.G. and Timmer G.T. (1985) A stochastic approach to global optimization. In:Numerical Optimization, 1984. SIAM, Philadelphia, Pa., 245-262.

Rinnooy Kan A.H.G. and Timmer G.T. (1987b) Stochastic global optimization methods. II. Multilevel methods. Mathem. Programming, 39, No.1, 5778.

Robson D.S. and Whitlock J.H. (1964) Estimation of a truncation point. Biometrika, 51, No.1, 33-39.

Rosen J.B. (1983) Global minimization of a linearly constrained concave function by partition of the feasible domain. Mathem. Res., 8, 215-230.

Sager T. (1983) Estimating modes and isoplets. Commun. Statist.: Theory and Math., 12, No.5, 529-557.

Saltenis (1989) Analysis of optimization problem structure. Mokslas, Vilnius (in Russian).

Schoen F. (1982) On a sequential search strategy in global optimization problems. Calcolo, 19, No.3, 321-334.

Schwartz L. (1967) Analyse Mathematique, V.1, Hermann, Paris.

Seneta E.(1976) Regularly-varying functions. Lecture Notes in Mathem., 508, Springer, Berlin e.a.

Shen Z. and Zhu Y. (1987) An interval version of Shubert's iterative method for the localization of the global maximum. Computing. 38, 275-280.

Shubert B.O. (1972) A sequential method seeking the global maximum of a function. SIAM J. on Numer. Analysis, 9, No.3, 379-388.

Silverman B.W. (1983) Some properties of a test for multimodality based on kernel density estimates. London Mathem. Soc. Lecture Note Ser., No.79, 248-259.

Smith R.L. (1982) Uniform rates of convergence in extreme value theory. Adv.Appl. Probab., 14, 600-622.

Smith R.L. (1987) Estimating tails of probability distributions. Ann. Statist., 15, No.3, 1174-1207.

Snyman J.A. and Fatti L.P. (1987) A multistart global minimization algorithm with dynamic search trajectories. J. Optimiz. Theory and Applic., 54, No. 1, 121-141.

Sobol I.M. (1969) Multivariate Quadrature Formulas and Haar Functions. Nauka, Moscow, 288 p. (in Russian).

Sobol I.M. (1982) On an estimate of the accuracy of a simple multidimensional search. Dokl. Akad. Nauk SSSR, 266, 569-572 (in Russian).

Sobol I.M. (1987) On functions satisfying a Lipschitz condition in multidimensional problems of numerical mathematics. Dok. Akad. Nauk SSSR, 293, N 6, 1314-1319 (in Russian).

Sobol I.M. and Statnikov R.B. (1981) Optimal Choice of Parameters in Multicriteria Problems. Nauka, Moscow, 110 p. (in Russian).

Solis F and Wets R. (1981) Minimization by random search techniques. Mathem. Oper. Res., 6, No.1, 19-30.

Spircu L. (1978) Cluster analyisis in global optimization. Economic Computation and Economic Cybernetic Studies and Research, 13, 43-50.

Strongin R.G. (1978) Numerical Methods in Multiextremal Optimization. Nauka, Moscow, 240 p. (in Russian).

Sukharev A.G. (1971) On optimal strategies for an extremum search. USSR Comput.Mathem. and Mathem. Phys., 11, N. 4, 910-924.

Sukharev A.G. (1975) Optimal Search of the Extremum. Moscow University Press. 100 p. (in Russian).

Sukharev A.G. (1981) Construction of a one-step-optimal stochastic algorithm seeking the maximum. USSR Comput. Mathem. and Mathem. Phys., 21, No.6, 1385-1401.

Suri R. (1987) Infinitesimal perturbation analysis for general discrete event systems. Journ. of ACM, 34, N 3, 686-717.

Törn A. (1978) A search-clustering approach to global optimization. In: Towards Global Optimization, Vol. 2 North Holland, Amsterdam e.a., 49-62.

Törn, A.A. and Zilinskas, A. (1989) Global Optimization. Lecture Notes in Computer Science 350 Springer-Verlag, Berlin Heidelberg New York.

Trecanni G. (1978) A global descent optimization strategy. In: Towards Global Optimization, Vol. 2. North Holland, Amsterdam e.a., 151-163.

Turchin V.F. (1971) On the calculation of multivariate integrals by the Monte Carlo method, Theory of Probab. and Applic., 16, No.4, 738-741.

Ustyuzhaninov V.G. (1983) Possibilities of random search in solution of discrete extremal problems, Kibernetika, No.2, 64-71, 77.

van Laarhoven P.J.M. and Aarts E.H.L. (1987) Simulated Annealing: Theory and Applications, Kluwer Academic Publishers, Dordrecht e.a., 198 p.

Vasiliev F.P. (1981) Methods for Solving Extremal Problems. Nauka, Moscow, 400 p. (in Russian).

Vaysbord E.M and Yudin D.B. (1968) Multiextremal stochastic approximation. Engineering Cybernetics, 5, No.1, 1-10.

Vilkov A.V., Zhidkov N.P and Shchedrin B.M. (1975) A method for finding the global minimum of a function of one variable. USSSR Comput. Mathem. and Mathem Phys., 15, No.4, 1040-1042.

Walster G.W., Hansen E.R. and Sengupta S (1985) Test results for a global optimization algorithm. In: Numerical Optimization (eds. Boggs P.T. and Byrd R.H.), SIAM, New York 272-287.

Watt V.P. (1980) A note on estimation of bounds of random variables. Biometrika, 67, No.3, 712-714.

Weisman I. (1981) Confidence intervals for the threshold parameter. Commun. Statist., A10, No.6, 549-557.

Weisman I. (1982) Confidence intervals for the threshold parameter II: unknown shape parameter. Commun. Statist.: Theory and Meth., 21, 2451-2474.

Weiss L. (1971) Asymptotic inference about a density function at the end of its range. Naval Res. Logistic Quaterly, 18, No.1, 111-115.

Wood G.R. (1985) Multidimensional bisection and global minimization. Presented at the IIASA Workshop on Global Optimization. (Held in Sopron, Hungary).

Yakowitz S.J. and Fisher L. (1973) On sequential search for the maximum of an unknown function. J. Math. Anal. and Appl., 41, No.1, 234-259.

Yamashita H. (1979) A continuous path method of optimization and its application to global optimization. In: Survey of Mathematical Programming, Vol.1/A, Budapest, 539-546.

Zaliznyak N.F. and Ligun A.A. (1978) On optimum strategy in search of global maximum of function. USSR Comput Mathem. and Mathem. Phys., 18, No.2, 314-321.

Zanakis S.H and Evans J.R. (1981) Heuristic *optimization* : why, when and how to use it. Interfaces, 11, No.5, 84-91.

Zang I. and Avriel M. (1975) On functions whose local minima are global. J. Optimiz. Theory and Applic., 16, No.3/4, 183-190.

Zhidkov N.P. and Schedrin B.M. (1968) A certain method of search for the minimum of a function of several variables. Computing Methods and Programming, Moscow University Press, V.10., 203-210 (in Russian).

Zhigljavsky A.A. (1985) Mathematical Theory of Global Random Search. Leningrad University Press, 296 p. (in Russian).

Zhigljavsky A.A. (1987) Monte Carlo methods for estimating functionals of the supremum type. Doctoral's thesis. Leningrad University. (In Russian.)

Zhigljavsky A.A. and Terentyeva M.V. (1985) Statistical methods in global random search. Testing of statistical hypotheses. Vestnik of Leningrad University, No.15, 89-91.

Zhigljavsky a.A. (1981) Investigation of probabilistic global optimization methods. Candidate's Thesis. Leningrad University (in Russian).

Zhigljavsky A.A.(1988) Optimal designs for estimating several integrals. Optimal Design and Analysis of Experiments (eds Y.Dodge, V.V. Fedorov and H.P. Wynn). North Holland, Amsterdam e.a., 81-95.

Zielinski R. (1980) Global stochastic approximation: a review of results and some open problems. In :Numerical Techniques for Stochastic Systems (eds Archetti F. and Cugiani M.). North Holland, Amsterdam e.a., 379-386.

Zielinski R. (1981) A statistical estimate of the structure of multiextremal problems. Mathem. Programming, 21, 348-356.

Zielinski R., and Neumann P. (1983) Stochastische Verfahren zur Suche nach dem Minimum einer Funktion. Akademie-Verlag, Berlin, 136 p.

Zilinskas A. (1978) Optimization of one-dimensional multimodal functions. Algorithm AS 133. Applied Statistics, 23, 367-375.

Zilinskas A. (1981) Two algorithms for one-dimensional multimodal minimization. Optimization, 12, No.1, 53-63.

Zilinskas A. (1982) Axiomatic approach to statistical models and their use in multimodal optimization theory. Mathem. Programming, 22, No.1, 104-116.

Zilinskas A. (1984) On justification of use of stochastic functions for multimodal optimization models. Ann. Oper. Res. 1, 129-134.

Zilinskas A.G. (1986) Global Oprimization: Axiomatics of Statistical Models, Algorithms, Application. Mokslas, Vilnius, 166 p. (in Russian).

SUBJECT INDEX